Lecture Notes in Mechanical Engineering

Lecture Notes in Mechanical Engineering (LNME) publishes the latest developments in Mechanical Engineering - quickly, informally and with high quality. Original research reported in proceedings and post-proceedings represents the core of LNME. Volumes published in LNME embrace all aspects, subfields and new challenges of mechanical engineering. Topics in the series include:

- Engineering Design
- Machinery and Machine Elements
- Mechanical Structures and Stress Analysis
- Automotive Engineering
- Engine Technology
- Aerospace Technology and Astronautics
- Nanotechnology and Microengineering
- Control, Robotics, Mechatronics
- MEMS
- Theoretical and Applied Mechanics
- Dynamical Systems, Control
- Fluid Mechanics
- Engineering Thermodynamics, Heat and Mass Transfer
- Manufacturing
- Precision Engineering, Instrumentation, Measurement
- Materials Engineering
- Tribology and Surface Technology

To submit a proposal or request further information, please contact the Springer Editor of your location:

China: Dr. Mengchu Huang at mengchu.huang@springer.com
India: Priya Vyas at priya.vyas@springer.com
Rest of Asia, Australia, New Zealand: Swati Meherishi at swati.meherishi@springer.com
All other countries: Dr. Leontina Di Cecco at Leontina.dicecco@springer.com

To submit a proposal for a monograph, please check our Springer Tracts in Mechanical Engineering at http://www.springer.com/series/11693 or contact Leontina.dicecco@springer.com

Indexed by SCOPUS. The books of the series are submitted for indexing to Web of Science.

More information about this series at http://www.springer.com/series/11236

Muhammad Aizzat Zakaria ·
Anwar P. P. Abdul Majeed ·
Mohd Hasnun Arif Hassan
Editors

Advances in Mechatronics, Manufacturing, and Mechanical Engineering

Selected articles from MUCET 2019

Springer

Editors
Muhammad Aizzat Zakaria
Faculty of Manufacturing and Mechatronic
Engineering Technology
Universiti Malaysia Pahang
Pekan, Pahang, Malaysia

Anwar P. P. Abdul Majeed
Faculty of Manufacturing and Mechatronic
Engineering Technology
Universiti Malaysia Pahang
Pekan, Pahang, Malaysia

Mohd Hasnun Arif Hassan
Faculty of Mechanical and Automotive
Engineering Technology
Universiti Malaysia Pahang
Pekan, Pahang, Malaysia

ISSN 2195-4356 ISSN 2195-4364 (electronic)
Lecture Notes in Mechanical Engineering
ISBN 978-981-15-7308-8 ISBN 978-981-15-7309-5 (eBook)
https://doi.org/10.1007/978-981-15-7309-5

This Springer imprint is published by the registered company Springer Nature Singapore Pte Ltd.
The registered company address is: 152 Beach Road, #21-01/04 Gateway East, Singapore 189721, Singapore

Preface

The 11th edition of Malaysian Technical Universities Conference on Engineering and Technology (MUCET2019) was held in Kuantan, Malaysia, from 19th November 2019 to 22nd November 2019. It was jointly organized by the Malaysian Technical Universities Network (MTUN) comprising of four universities namely Universiti Tun Hussein Onn (UTHM), Universiti Teknikal Malaysia Melaka (UTeM), Universiti Malaysia Perlis (UniMAP), and the 11th edition's host, Universiti Malaysia Pahang (UMP). MUCET 2019 aims at serving the researchers and practitioners in related fields with timely dissemination of the recent progress of the innovative research in science, engineering, and technology.

The 11th edition of the conference bears a theme of "Communitising Technology in the context of Industrial Revolution 4.0". The advancement of industrial revolution 4.0 will be driven by smart, interconnected devices that will be affecting local communities. Bringing the communities to adapt to the pervasive environment is indeed a towering undertaking for researchers, innovators, technologists, and scientists alike.

MUCET2019 received more than 100 submissions. All submissions were reviewed in a single-blind manner, and the best 30 papers recommended by the reviewers are published in this volume which focuses on advancement in mechatronics, manufacturing, and mechanical engineering technology. The publication is selected based on several criteria; the content of the paper, the reviewers' feedback, and the relevancy of the topics with may interest the readers.

The editors hope that readers find this volume informative. We thank Springer for undertaking the publication of this volume. We also would like to thank the conference organization staff and the members of the International Program Committees for their hard work.

Muhammad Aizzat Zakaria
Anwar P. P. Abdul Majeed
Mohd Hasnun Arif Hassan

Contents

Surface Roughness Study on Mild Steel Under Multi Cooling Condition

M. A. Sulaiman[1(✉)], A. A. Halimnizam[2], M. S. Asiyah[1], R. Shahmi[1], E. Mohamad[1], and M. R. Salleh[1]

[1] Faculty of Manufacturing Engineering, Universiti Teknikal Malaysia Melaka, Hang Tuah Jaya, 76100 Durian Tunggal, Melaka, Malaysia
mohdamri@utem.edu.my

[2] Kolej Kemahiran Tinggi MARA Sri Gading, Batu 11, Jalan Kluang, 86400 Parit Raja, Batu Pahat, Johor, Malaysia

Abstract. Surface integrity is the surface condition of a workpiece after being modified by a manufacturing process and it can change the material's properties. In surface topography, surface roughness (Ra) was concerned with the geometry of the outmost layer of the workpiece texture and interfaces exposed with the environment affects several functional attributes of parts, such as friction, wear and tear, heat transmission, ability of distributing and holding a lubricant, etc. Therefore, the desired surface finish was usually specified and appropriate processes were required to assess and maintain the quality of a component. The research was to investigate the influence of machining parameters and optimum process parameter to the surface roughness value of mild steel material in conventional turning using CVD (chemical vapor deposition) coated carbide insert in three cutting conditions (dry, wet and oil). Optimization of the cutting parameters is very important in determining the optimum cutting conditions, thus reduce machining cost and time consumed. From the results obtained, better surface roughness value was determined by a combination of cutting speed 150 m/min, and feed rate 0.1 mm/rev with under coolant oil condition.

Keywords: Mild steel · Surface roughness · Turning · Multi coolants · Coated carbide

1 Introduction

1.1 Cutting Fluids

Cutting fluids are most fundamental and important part in the metalworking industries. It widely employed due to their ability to reduce friction, cutting temperature, thus enhance the workpiece surface quality, Dearnley et al. [1]. The most common metalworking fluids use today belong to one of two categories; emulsion type coolant (water base), and cooling oil fluids including synthetics and semi synthetics, Debnath et al. [2]. An important feature of cutting fluids is the easiest to remove the contaminants from it, which leads to a longer fluid's life and lower environmental impact and cost, Gajrani

M. A. Zakaria et al. (Eds.): *Advances in Mechatronics, Manufacturing, and Mechanical Engineering*, LNME, pp. 1–9, 2021.
https://doi.org/10.1007/978-981-15-7309-5_1

and Sankar [3]. With this respect, cooling oil, coolant is better in tool life and product surface finish and allow the removal of contaminants with minimal alterations on their properties. However, the application of emulsion type coolant (water base) will cause environmental problems such as dermatological problems to the operators when physical contact with the cutting fluid, water pollution when disposes to the earth, and extra space for a pump, filtering, recycling and storage required.

The secondary function of coolants is to flush away chips from the tool or workpiece interface to prevent a finished surface from becoming micro multi-layer and also to reduce the occurrence of built-up edge (BUE). Monitoring and maintenance of coolants are required due to contamination and degradation where coolant require disposal once their efficiency is lost, Raj et al. [4].

Meanwhile, mild steel, also known as low carbon steel is now the most common form of steel because its price relatively low while it provides material properties that are acceptable for many applications, Das et al. [5]. It contains only a small percentage of carbon (low carbon steel) and is strong and easily worked but not readily tempered or hardened.

This paper aims to investigate the influence of machining parameters and optimum process parameter to the surface roughness value of mild steel material in conventional turning using CVD (chemical vapor deposition) coated carbide insert in three cutting conditions (dry, wet and oil).

2 Experimental Works

The cylindrical workpiece mild steel will machine with three different environments of coolants. Each machining process uses different inserts, but still use the same type of insert. Same feed rate, depth of cut and spindle speed applied for each testing process. All workpiece machine continuously until reach desired amount to be cut. The cutting tool or insert was set to move automatically to make sure that the cutting process is controllable and consistent. After that, to analyse the optimization of the tool life and surface roughness value for every cutting condition which was analysed by using Design Expert 6.0 software.

2.1 Experimental Equipment and Materials

Experimental equipment which are conventional lathe machine, Toolmaker's microscope and surface roughness tester were used. Meanwhile, mild steel with diameter 30 mm was used as workpiece in this research.

Conventional Lathe Machine

A conventional lathe machine was used in this research. The model of the lathe machine is NC DEN 250.

Workpiece Material

The material chosen for machining test was mild steel with Japan Standard of steel

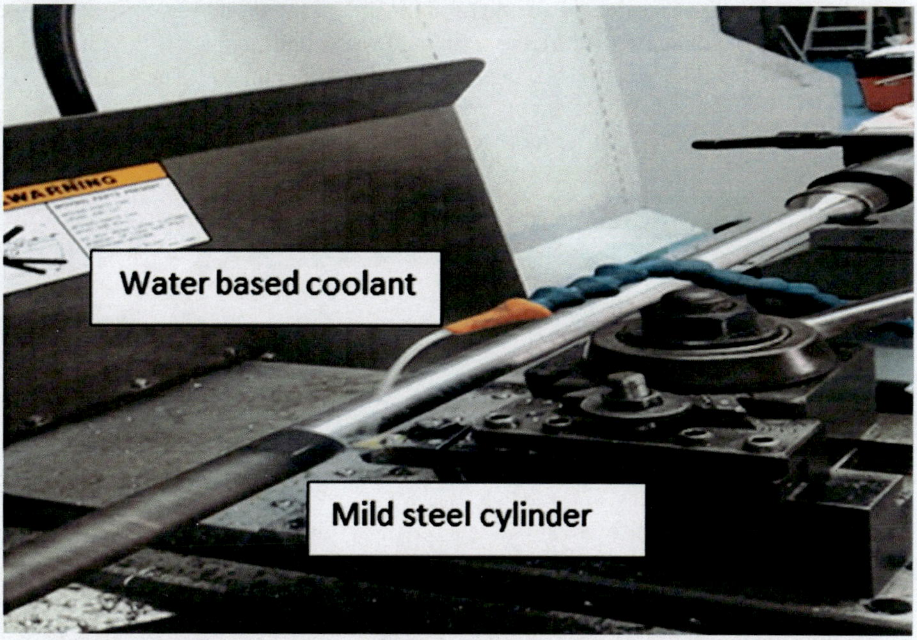

Fig. 1. Experimental setup with water based coolant

grades JIS G3101 SS400. The mild steel in a cylindrical shaped with the 30 mm diameter, 700 mm in length and turning by three cutting conditions. Figure 1 shows the experimental setup with water based coolant (one of the three cutting condition).

Cutting Tool
The cutting tool used in this experiment is CVD coated carbide inserts. This cutting tool made from Yamaloy Tooling Japan with rhombus shape DNMG 150408-NM5. Figure 2 shows the coated carbide tool with nose radius 0.8 mm.

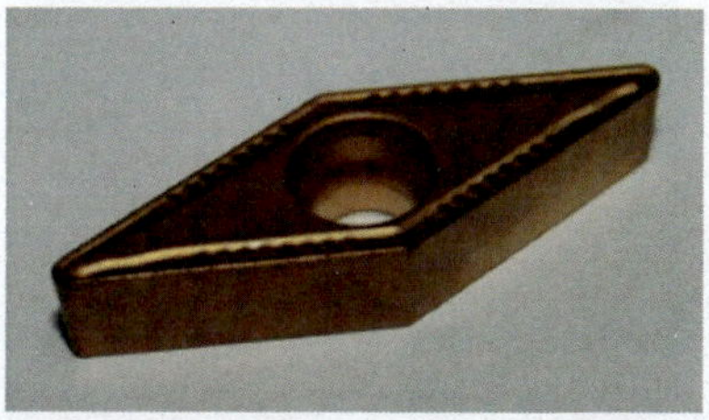

Fig. 2. DNMG 150408-NM5 coated carbide insert tool (Yamaloy Tooling Japan Industry Co. Ltd, 2011)

Surface Roughness Tester
Surface roughness tester also known as portable profilometer is an instrument that used

to measure and assesses value of surface roughness as shown in Fig. 3. Roughness profilometer get in contact with surfaces within a few seconds and show the roughness value in average roughness value (Ra) in μm or roughness depth (Rz). The Mitutoyo SJ-301 will be used to measure the average surface roughness value (Ra) of the sub-surface machined workpiece material after tool wear value 0.2 μm obtained.

Fig. 3. Mitutoyo SJ-301 surface roughness profilometer in measuring surface roughness values on surface machined.

2.2 Experimental Design

In this research, the design of experiments used three factor with three levels, where three factors are varied (cutting speed, feed rate and cutting conditions), meanwhile the factor of depth of cut was kept constant, 1 mm. Table 1 shows the cutting parameters and conditions. The model designed with $k = 3$. The number of runs (z) can be determined by Eq. 1. Three centre points were selected to provide a measure of process stability and inherent variability and to check for curvature.

Table 1. Cutting parameters and conditions.

Symbol	Factor	Unit	−1	0	1
v	Cutting speed	m/min	70	110	150
f	Feed rate	mm/rev	0.1	0.15	0.2
c	Conditions	–	Dry	Water	Oil
d	Depth of cut	mm	1.0		

$$z = 3^k \tag{1}$$

where, k is the number of factors.

The parameters will be set for the design of experiment (design expert software) and the experiment matrix will be generated.

3 Results

3.1 Surface Roughness

In this experiment, three level factorial design was used to generate the design summary with the combination of all the selected parameters. Figure 4 shows the result of the surface roughness in μm. There were 27 runs of experiment conducted and the combination of the parameter was generated by the software. From the Fig. 4, it shows that the highest value of the surface roughness was 22.171 μm with cutting speed of 110 m/min and the feed rate 0.15 mm/rev in dry condition methods. Meanwhile the minimum value of surface roughness was obtained 3.726 μm at cutting speed of 70 m/min and the feed rate 0.1 mm/rev in coolant water condition methods.

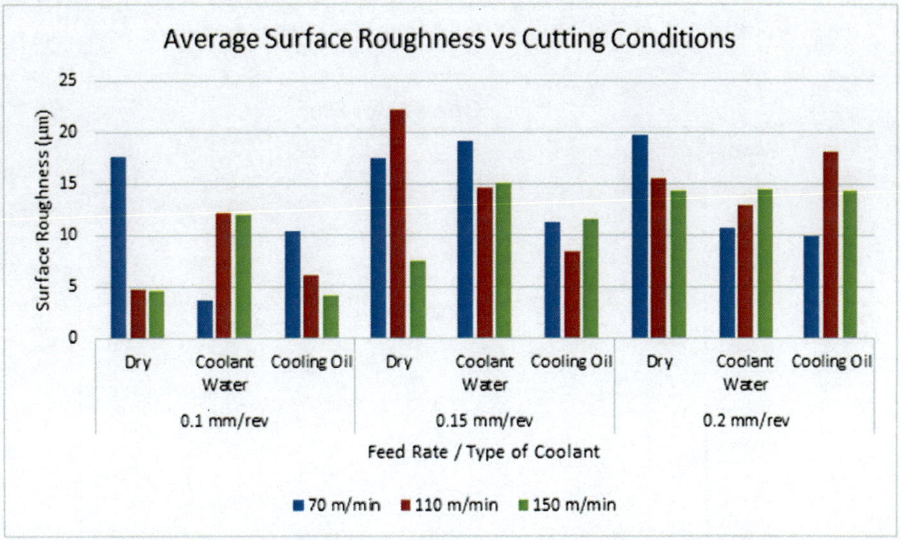

Fig. 4. Average surface roughness value for each cutting condition

Surface Roughness Modelling

Figure 5, 6 and 7 show one factor plot of cutting speed, feed rate and types of coolant for surface roughness. From these three graphs, it shows that the feed rate has a higher slope compared to cutting speed. This means the feed rate were more significant to the surface roughness effect compared to cutting speed factor. Figure 8 shows the 3D and contour plot for the surface roughness model. From all these figures, it can be concluded that the feed rate given more significant effect rather than the cutting speed. The finer surface roughness (lowest values) could be achieved with the reducing in feed rate value but at high speed cutting.

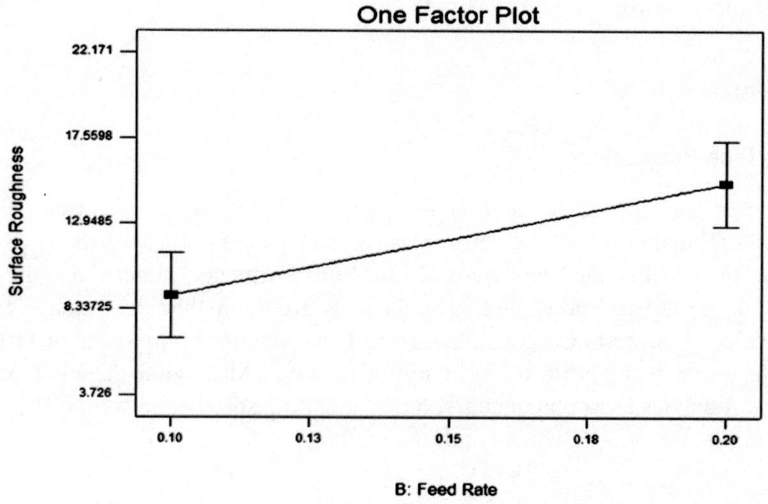

Fig. 5. One factor plot of surface roughness versus feed rate

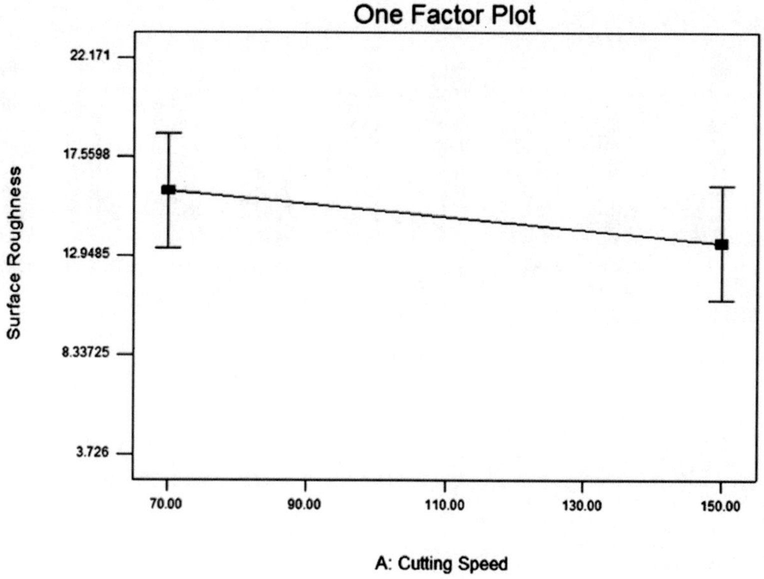

Fig. 6. One factor plot of surface roughness versus cutting speed

Optimization of Parameter

Parameter optimization was conducted to identify the combination of factor levels that

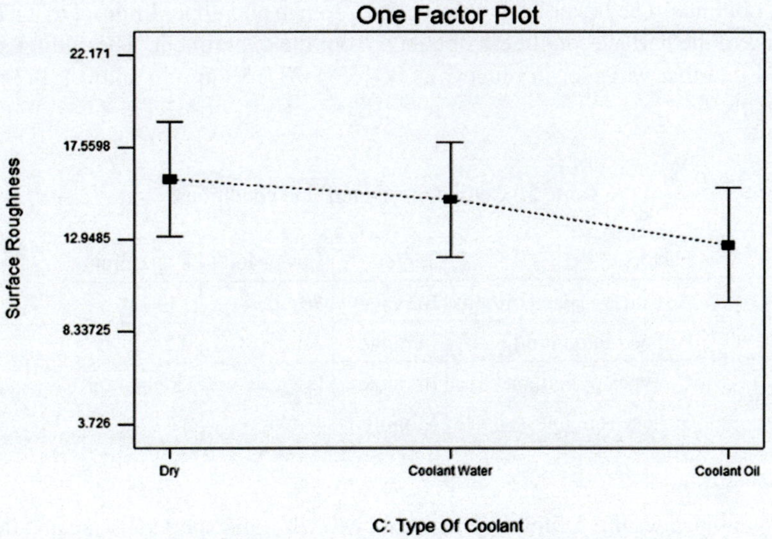

Fig. 7. One factor plot of surface roughness versus types of coolant

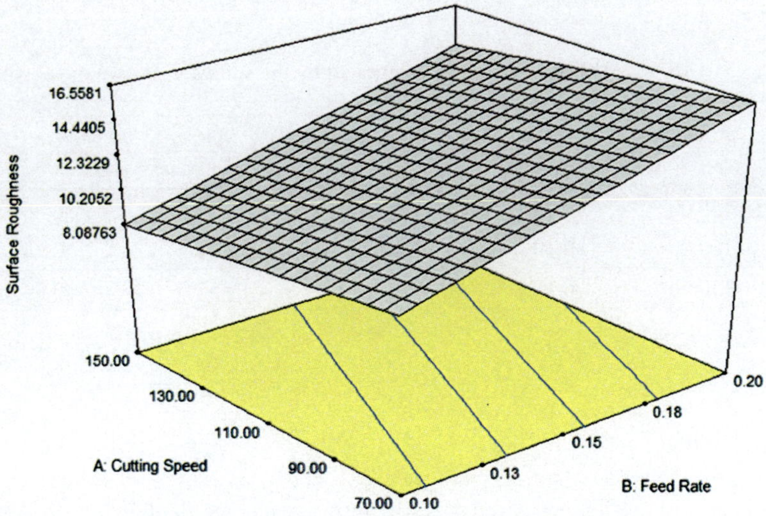

Fig. 8. 3D surface plot for surface roughness model

satisfy the requirements place on each response and factors, Nordin et al. [6]. The desired goal for each of the factor and response are set for the numerical optimization with the possible goal.

The main objectives of the optimization on this experiment are to obtain minimum surface roughness Ra in μm. The numerical optimization for surface roughness model improvement that was conducted are shown in Table 2. The goal of this optimization was set to minimize indicating that the surface roughness will be optimize until the minimum value is obtained. The lower limit was set to 3.726 μm and upper limit set to 22.171 μm regarding to the surface roughness obtained from the experiment. The cutting speed is and feed rate that were set in range was between 70–150 m/min and 0.1–0.2 mm/rev respectively.

Table 2. Cutting parameters and conditions.

Factors	Target	Lower limit	Upper limit
A: Cutting speed, m/min	In range	70	150
B: Feed rate, mm/rev	In range	0.1	0.2
C: Types of coolant	In range	Dry	Coolant oil
Surface roughness, μm	Minimum	3.726	22.171

Table 3 indicates the solutions generated by the design expert software and there are three solutions are suggested. The desirability ranked by highest to lowest of surface roughness values. The solution number one is the best. The high desirability will be selected for the confirmatory trial.

Table 3. Solutions suggested by the software.

Number	A: Cutting speed, m/min	B: Feed rate, mm/rev	Type of coolant	Surface roughness, μm	Desirability
1	150.00	0.10	Coolant oil	6.24	0.86
2	150.00	0.10	Coolant water	8.51	0.74
3	150.00	0.10	Dry	9.52	0.69

4 Conclusion

The tool wear progression increased steadily until the average flank wear, $V_b = 0.2$ mm achieved. High cutting speed and lowest feed rate effected on lowest surface roughness value. Meanwhile, low cutting speed, and high feed rate significantly influenced surface integrity and resulted to a rougher surface roughness.

ANOVA analysis indicated feed rate is more significant factor in affecting a rougher surface roughness compared to cutting speed. High speed machining produced fined surface roughness and suitable for conventional machining with used coolant oil.

Optimization of parameters shows that coolant oil condition in cutting speed with 150 m/min and feed rate of 0.1 mm/rev obtain a better (minimum) surface roughness of 6.24 μm.

Acknowledgements. The authors would like to thank Universiti Teknikal Malaysia Melaka (UTeM) and Institut Kemahiran MARA Johor Bahru for technical support and the facilities of instruments in implementing the experiment.

References

1. Dearnley, P.A., Matthews, A., Leyland, A.: Tribological behavior of thermochemical surface engineered steels. Thermochem. Surf. Eng. Steels, 241–266 (2015)
2. Debnath, S., Reddy, M.M., Yi, Q.S.: Environmental friendly cutting fluids and cooling techniques in machining: a review. J. Clean. Prod. **83**, 33–47 (2014)
3. Gajrani, K.K., Sankar, M.R.: Past and current status of eco-friendly vegetables oil based metal cutting fluids. Mater. Today Proc. Part A **4**(2), 3786–3795 (2017)
4. Raj, A., Leo, D., Varadajan, A.: Review on hard machining with minimal cutting fluids application. Int. J. Curr. Eng. Technol. **5**(6), 3717–3722 (2015)
5. Das, S.R., Panda, A., Dhupal, D.: Hard turning of AISI 4340 steel using coated carbide insert: surface roughness, tool wear, chip morphology and cost estimation. Mater. Today Proc. **5**(2), 6560–6569 (2018)
6. Noordin, N.Y., Venkatesh, V.C., Sharif, S.: Dry turning of tempered martensitic stainless tool steel using coated cermet and coated carbide tools. J. Mater. Process. Technol. **185**, 83–90 (2006)

Experimental Study of Single Pass Welding Parameter Using Robotic Metal Inert Gas (MIG) Welding Process

M. H. Osman(✉), N. F. Nasrudin, A. S. Shariff, M. K. Wahid, M. N. Ahmad, N. A. Maidin, R. Jumaidin, and M. H. Ab. Rahman

Fakulti Teknologi Kejuruteraan Mekanikal dan Pembuatan, Universiti Teknikal Malaysia Melaka, Hang Tuah Jaya, 76100 Durian Tunggal, Melaka, Malaysia
hairizal@utem.edu.my

Abstract. This paper presents the optimization of welding parameter for joining mild steel 1020 using ABB metal inert gas (MIG) robot welding. Mild Steel AISI 1020 with a thickness of 6 mm was selected in this experiment. The specimens cut into size 140 mm × 130 mm × 6 mm and then welded with ABB MIG robot welding used 1.0 mm electrode wire. Three welding parameters such as welding speed, Voltage, and welding pattern, each at three levels were considered. An L9 Orthogonal Array and signal-to-noise (S/N) ratio were employed to analyse the significant and percentage of each parameter for maximum tensile strength. The results revealed that the welding pattern gave significant main effects on the highest percentage distribution (61%), followed by the Welding Speed (11%) and Voltage (3%). Further, the results indicated that the combination of optimum parameter recorded as welding speed 5.5 mm/min, 22 V for Voltage and straight pattern travel capable to offer high tensile test strength. The observed data have been interpreted, discussed and analysed by using Taguchi method.

Keywords: Robot welding · MIG welding · Optimization · Taguchi

1 Introduction

Welding process can be divide into two major groups which is solid state and fusion welding. Fusion welding is a generic term for welding processes that rely upon melting to join materials of similar compositions and melting points. Due to the high-temperature phase transitions inherent to these processes, a heat-affected zone is created in the material. In contrast to fusion welding, solid-state welding does not involve the melting of materials. Solid state welding takes place by applied force or heat and force, it does not require any filler materials [1–6].

Welds can be geometrically arranged in a wide range of ways, the ordinarily in welding is butt joint, lap joint, corner joint, edge joint, and T-joint. The geometric is relying upon the procedure utilized and the thickness of the material, many pieces can be welded together in a lap joint geometry. Distinctive welding process required diverse joint plan. A few procedures can likewise be utilized to make multi-pass welds, in which

M. A. Zakaria et al. (Eds.): *Advances in Mechatronics, Manufacturing, and Mechanical Engineering*, LNME, pp. 10–21, 2021.
https://doi.org/10.1007/978-981-15-7309-5_2

one weld is permitted to cool, and after that another weld is performed over it. This takes into account the welding of thick areas masterminded in a solitary V planning joint [7–10].

The welding process is a combination of two parts, so the welding area must have strong strength. Many distinct factors influence the strength of welds and the material around them, including the welding method, the amount and concentration of energy input, the weldability of the base material, filler material, flux material, the design of the joint, and the interactions between all these factors. To test the quality of the welding, the non-destructive such as ultrasonic and color contrast or destructive test such as tensile test and fatigue test can be used [7, 11–13].

The impacts of welding on the material encompassing the weld can rely upon the materials utilized and the warmth contribution of the welding procedure utilized, the HAZ can be of shifting size and quality. The warm diffusivity of the base material assumes a substantial part if the diffusivity is high, the material cooling rate is high and the HAZ is generally little. On the other hand, a low diffusivity prompts slower cooling and a bigger HAZ. The measure of warmth infused by the welding procedure assumes a vital part too, as procedures prefer oxyacetylene welding have a concentrated warmth info and increment the span of the HAZ. Procedures like laser pillar welding give a profoundly focused, restricted measure of warmth, bringing about a little HAZ. Curve welding falls between these two extremes, with the individual procedures shifting fairly in warm [14, 15].

Since the welding parameter affected the welding quality, it is one of the factors that important to determine before the welding process. Optimization of the welding process parameters depends upon the ability to measure and control the process variables involved in the welding process. Three parameters of MIG welding such as current, voltage and welding pattern were taken for the experiment [16–18].

2 Experimental Process and Procedure

2.1 Design of Experiment

L_9 Orthogonal Array was used as the experimental setup for the welding parameter. The experimental array has three parameters and three levels as shown in Table 1. The factor of this experiment is Voltage, Welding Speed and Welding Pattern. For the Voltage and Welding Speed, there is three levels which is high, medium and low. Srirangan et al. used Taguchi L9 array with Grey relation analysis to optimize the process parameters in TIG welding of Incoloy 800HT with ultimate tensile strength, yield strength and impact toughness as performance characteristics. Meanwhile, R. Sathish et al. optimized.

The TIG welding parameters for dissimilar pipe joints using Taguchi method. Sudhir Kumar et al. used L9 Orthogonal Array to optimize process parameter of MIG welding process using AISI1018 mild steel plate. Anirban Tudu et al. optimized the welding process parameter based on larger is better response with current, gas flow and arc gap [19–22].

The experiments were carried out on MIG Robot Welding Machine as shown in Fig. 1. AISI 1020 mild steel was selected as a workpiece specimen which was prepared at 140 mm length × 130 mm width × 6 mm thickness size. Butt joint welding is selected

Table 1. The parameter at three levels and three factors

Parameters		Levels		
		1	2	3
Voltage (V)	A	21	22	23
Welding speed (m/mm)	B	4.5	5.5	6.5
Welding pattern	C	Straight	Triangle	Spiral

Fig. 1. ABB IRB1410 robotic mechanical arm

to executed this experiments and the programming for robotic welding has been done by using RobotWare Arc programming software and FlexPendant to teach it into the robot. The input parameter was generated using Minitab Software to achieve the optimization and these input values are fed in the robot programming.

Before start of the welding process, all specimens need to be groove on one side only each. The grooving bead geometry is thru in 30° merely and the process is essential to assure resilient connection due to complete penetration on the welded region. This geometry is considering one of important factor to achieve better quality. The robot welding will be setup from first parameter until the last parameter as show in Fig. 2. All of the nine samples will be test using tensile test machine to identify the strength of the joining.

The position of the plate must be setup according to American Standard Testing and Material (ASTM). After welding process, the welded plate must be cut into 'dogbone' shape using the ASTM E8 specifications as shown in Fig. 3 below. The cutting process

Fig. 2. The photographic view of welded sample

conducted using laser cut machine to get the precise dimensions. Soon after the cutting process done, tensile tensing was done on all 27 specimens using INSTRON 600Dx Tensile Machine. The data was collected and been analyze in Minitab Software. Lastly, the confirmation value was obtained from the Minitab Software and confirmation test was conducted.

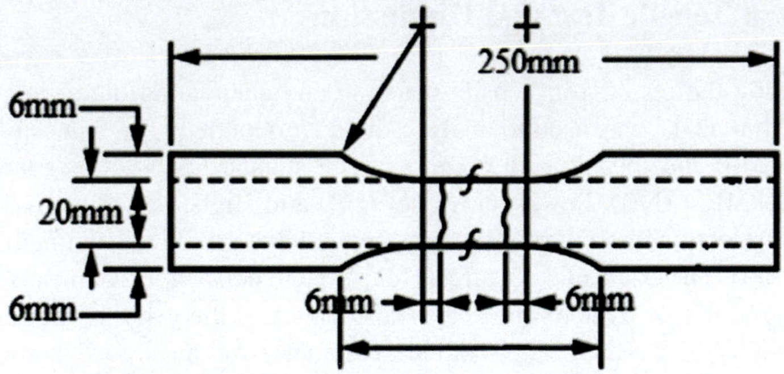

Fig. 3. Tensile test ASTM E8 diagram for metal inert gas welding

2.2 Tensile Test Procedure

A tensile test is also called as tension test is the most common type of test used to test the material behavior or the strength of welding part. In this experiment, the tensile test is used to get the value of strength for the welding part. By given a pulling force to a welding part until it breaks, the result will find its strength along with how much it will elongate. The point of failure is usually called its Ultimate Strength. The machine used for this process is Universal Tensile Testing Machine INSTRON 600DX. A specimen was placed at the center of the jig equally for top and bottom (Fig. 4).

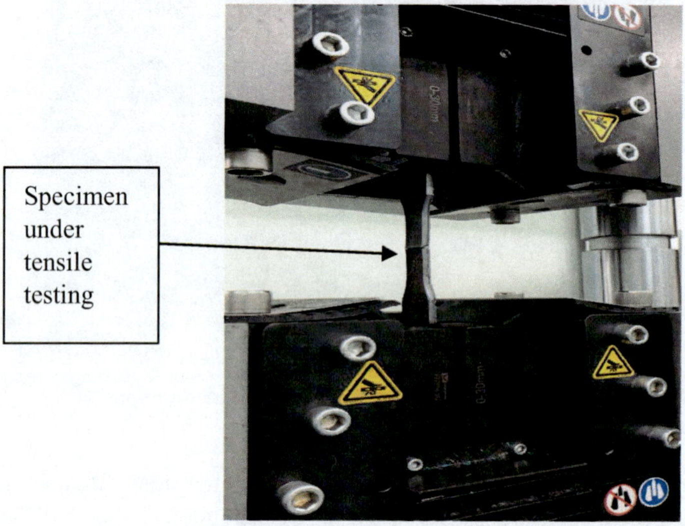

Fig. 4. The tensile testing process

3 Result of Tensile Test and Discussion

Upon obtaining the result from tensile machine and calculating the average for each parameter, Minitab 17 was used to analyze the data obtained. The values of Signal to ratio (S/N Ratio) and means were observed. The standard S/N ratios generally used are Nominal-is-Best (NB), Lower-the-Better (LB) and Higher-the-Better (HB). In this experiments, selection of Higher-is-Better is applied due to achieve optimum parameter of the tensile test. The graph of main effects for signal to noise ratio and means are plotted automatically as it's the type of characteristics. It means, the result of the higher value of tensile give the best output. The optimum parameter and most contributing factor in this experiment can be obtained from the graph [23].

Terms in Taguchi Method such as 'signal' represent the desirable value of mean for output characteristic while 'noise' means the undesirable value of standard deviation for output characteristics. Thus, the S/N ratio shows the ratio of mean to the standard deviation. The S/N ratio used to measure the deviation of a quality characteristic from the desired value. There are several types of S/N according to the characteristic such as lower is better, nominal is better and higher is better. For larger is better, the "Eq. (1)" is as shown below;

$$S/N = -10 \log \left(\frac{\sum_{i=1}^{n} \left[\frac{1}{y_i^2} \right]}{n} \right) \tag{1}$$

Where n is the number of observations and y is the observed data. The characteristic chosen in this experiment is larger is better which means the higher value of tensile strength is desirable. The number of experiments carried out is nine experiments based on Minitab Software that provide the optimization method and Orthogonal Array. Nine rows are corresponding to the number of tests with three columns at three levels. The output chosen to be studied is voltage, welding speed and welding pattern. Table 2 shows the S/N ratios and mean result using Larger-is-Better characteristics.

Table 2. Experimental design and the result of experiments

Exp no.	Designations	Means (MPa)	S/N ratio
1	A1B1C1	202.377	46.3059
2	A2B2C2	218.003	45.7691
3	A3B3C3	164.410	43.8664
4	A1B1C3	206.233	46.2832
5	A2B2C1	212.200	46.5271
6	A3B3C2	206.767	45.8277
7	A1B1C2	196.830	45.8764
8	A2B2C3	203.977	45.9894
9	A3B3C1	202.673	45.9532

The graph of Signal to Noise Ratio and Mean has been plotted. The analysis of graphs is made by observing the plotted value of each graph. The optimum parameter is observed by the highest level of plotted value in factor for both of the graphs since 'larger is better' were used. The S/N ratio identifies the factor that gives a higher value for the tensile strength, it means the highest value of S/N ratio shows the higher effect of the noise factor.

Meanwhile, the analysis of means graph is observed by the highest level plotted. It is because the graph of means shows that the average of three samples. The highest value gives the best result of tensile strength.

Main effect graph represents the value that shows the extent of influences of a factor on the response. Main Effect Plot represents the variation in the response variable with the variation in control factors and is used to examine the difference between level means for factors. The lines plotted has a high slope, it means the effect of each factor is significantly affect the tensile strength. The graph below shows the main effect Plots for S/N Ratio.

By referring to the Fig. 5, it shows the highest value plotted for welding speed is 5.5 mm/min, the highest value plotted for voltage is 22 V and for welding pattern, straight has the highest value plotted. All of the value plotted on the graph has a big difference from each other. It means the result for every level of factor gives difference effect for tensile strength. The variability in tensile strength must be maximized to meet the requirement that specimen with high tensile strength is the best. Figure 6 shows the

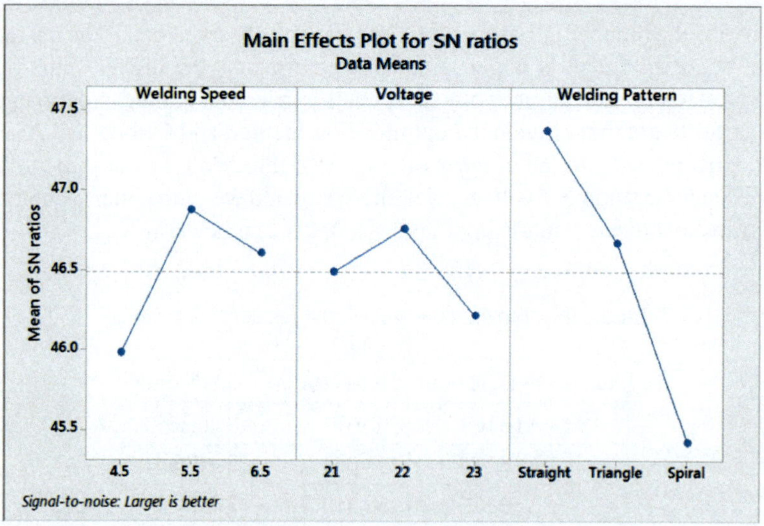

Fig. 5. Main effects plots for S/N ratios

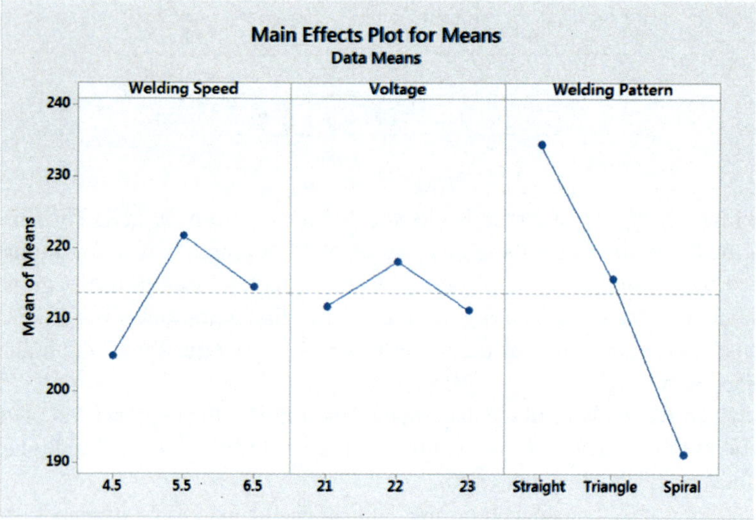

Fig. 6. Main effects plots for means

highest value plotted for welding speed is 5.5 mm/min, the highest value plotted for voltage is 22 V and the highest value plotted for Welding pattern is straight. As we can see from both of the graphs, the value that has the highest value plotted is the same. Therefore, the optimum parameter that gives the highest value of tensile that suggested by Taguchi Method is 5.5 mm/min, 22 V and Straight pattern.

Table 3 and 4 show the response table for the signal to noise ratio and means. For Table 5, the welding pattern is at the first rank with 1.81, followed by welding speed with

Table 3. The response table for signal to noise ratios (larger is better)

Level	Welding speed	Voltage	Welding pattern
1	46.14	46.5	47.39
2	46.89	46.77	46.67
3	46.61	46.37	45.58
Delta	0.75	0.40	1.81
Rank	2	3	1

0.75 and the last one is voltage with 0.40. For the Response Table for Means, the ranking for three factors is the same as the rank for a signal to noise ratio. Welding pattern is at the first rank with 43.2, followed by welding speed with 16.8 and the last one is voltage with 6.8. Throughout the observation, it demonstrates the value of delta of all factors is slightly different. It shows the three factor which is welding pattern, voltage and welding speed give the insignificant equivalent.

According to both of the table, the welding pattern is the first rank which means it is the most significant factor than welding speed and voltage. Three types of a pattern have been to choose to be examined such as straight pattern, triangle pattern and spiral pattern. The highest value plotted in the graph as Table 5 is straight pattern which has the value of means is 234.4 MPa and signal to ratio is 47.39, followed by triangle which has the value of means is 215.6 MPa and signal to ratio is 46.67 and lastly is spiral that has the value of means is 191.1 MPa and signal to noise ratio is 45.58. It might be due to the pattern itself such as how the welding does take place on the plate with that pattern and how the weld beads intersect to each other to fulfil the gap. It might be the triangle and spiral pattern do not fill up the gap with weld beads properly. Since the straight patterns do not move around while welding, the weld beads can flow smoothly through the gap that joins the two plates together. The second significant factor that influence in the tensile strength on Mild Steel 1020 is welding speed. Three levels chosen to be examined is 6.5 mm/min, followed by 5.5 mm/min and 4.5 mm/min from the table, the welding speed that give the highest tensile strength is 5.5 mm/min with the value of means is 221.7 MPa and signal to noise ratio is 46.89. The second highest welding speed that influence the strength is 6.5 mm/min with the value of means is 214.5 and signal to noise ratio is 46.61. Lastly, the welding speed that produces the lowest value of tensile strength is 4.5 mm/min with the value of means is 204.9 MPa and signal to noise ratio is 46.14. The linear rate (express in cm/min or mm/sec) at which the arc moves around along the joint, termed arc travel speed, affects weld bed size and penetration. With other variables kept constant, there is a certain value of travel speed at which the weld penetration is maximum. The value of welding speed that gives the best result of tensile strength might be because of that speed is at suitable rate which effect the weld bead size to formed structurally and gave more penetration to get to the bottom of the plate that join together. That cause the joining of two plates more strong and hard to be broke into two.

The last factor that influences the strength of welding is voltage. As usual, three levels of factors are used to be examined. 21 V, 22 V and 23 V are taken to be the level for voltage factor. From those three levels, 22 V gives the best mean with 218.1 MPa and the signal to noise ratio is 46.77. The second level that gives the strong tensile value is 21 V which give the means strength of 211.8 MPa and the signal to noise ratio is 46.50. The least value that is achieved for tensile strength is 23 V, it only gives means strength of 211.3 MPa and signal to noise ratio is 46.37. This is a very important variable in MIG welding, mainly because it determines the type of metal transfer by influencing the rate of drop late transfer across the arc. Arc voltage effect the arc length, the voltage with the smaller voltage get the shorter arc length and the highest voltage gets the longer arc length. The length of the arc determines the width and size of the arc cone. As arc length decreases, the arc cone becomes narrower and the arc is more focused. The weld bead that is more narrow and ropy and the level of weld penetration may decrease very slightly and as arc length increases, the arc cone becomes wider and the arc is broader. The weld bead that is wider and flatter and the level of weld penetration may increase very slightly.

Table 4. The response table for means

Level	Welding speed	Voltage	Welding pattern
1	204.9	211.8	234.4
2	221.7	218.1	215.6
3	241.5	211.3	191.1
Delta	16.8	6.8	43.2
Rank	2	3	1

4 ANOVA

Table 6 shows the analysis of Variance (ANOVA) of the recorded data, which specifies the percentage contribution significance of each factor. According to Table 5 below, the welding pattern recorded the highest percentage contribution which is 61%. This is followed by welding speed, recorded at 11%. The lowest percentage of contribution is a voltage which is 3%. It should be noted that the contribution from welding pattern is so much superior which indicates that dominant contribution to the whole tensile performance so much depended on the style of the pattern itself.

ANOVA is performed to test the centrality of the factor for the reaction which in this examination is elasticity. The term 'df' implies the degree of freedom which implies the number of terms that will add to the blunder in the forecast. The term 'Adj SS' alludes the balanced whole of squares which alludes to the entirety of squares got in the wake of evacuating insignificant term shape the model. The balanced mean square or 'Adj MS' shows the mean square acquired in the wake of expelling the inconsequential terms from

Table 5. ANOVA result of analysis

Source	DF	Seq SS	Adj SS	Adj MS	F	P	Percentage of contribution
Welding speed	2	0.8683	0.8683	0.4341	0.42	0.705	11%
Voltage	2	0.247	0.247	0.1235	0.12	0.894	3%
Welding pattern	2	4.9978	4.9978	2.4989	2.41	0.294	61%
Residual error	2	2.0766	2.0766	1.0383			

the reaction condition. The F-value is utilized to test the theory and it is computed as the proportion of balanced mean square an incentive to remaining mistake. The investigation is made by alluding the P-value and the rate of commitment computation. The outcome was translated by utilizing 95% of certainty level for all investigation of information.

ANOVA used to test the essentialness of every single primary factor and their collaborations by contrasting the Mean Square (MS) against a gauge of the exploratory mistakes at particular certainty levels. It is finished by isolating the aggregate inconstancy of S/N Ratios, which is measured by the entirety of the squared deviations from the aggregate mean S/N Ratios, into commitments by each of plan parameters and level.

5 Confirmation Test

The purpose of this confirmation test is to validate the prediction parameters of tensile strength at the optimum level which is recorded at A2B2C1 which is 22 V for voltage, 5.5 mm/min for welding speed and straight pattern. It is a crucial step which is highly recommended by Taguchi to verify the result obtained. The result shows that the percentage error for this comparison is around 4%, which is accepted for validation data.

6 Conclusion

This paper has presented the experimental Study of Single Pass Welding Parameter Using Robotic Metal Inert Gas (MIG) welding process. Taguchi method used to determine the main effect significant factors and optimum tensile strength parameters to the performance of robot welding process. Based on the result, some conclusion can be drawn:

1) The welding pattern has mainly affected the tensile strength based on the highest percentage distribution, followed by welding speed and voltage.
2) The optimum parameters are observed by using 22 V at 5.5 m/minspeed and straight welding pattern.
3) Welding pattern spiral gave the worst performance, leave highest significant percentage change.
4) The main effecting temperature on the specimens which resulting decrease of tensile strength and thermal softening caused by voltage and speed control efficiency of joining.

References

1. Zhang, H., Liu, J.: Microstructure characteristics and mechanical property of aluminum alloy/stainless steel lap joints fabricated by MIG welding–brazing process. Mater. Sci. Eng. A **528**(19–20), 6179–6185 (2011)
2. Frazier, W.E.: Metal additive manufacturing: a review. J. Mater. Eng. Perform. **23**(6), 1917–1928 (2014)
3. Cooper, D.R., Allwood, J.M.: The influence of deformation conditions in solid-state aluminium welding processes on the resulting weld strength. J. Mater. Process. Technol. **214**(11), 2576–2592 (2014)
4. Xu, G., Hu, J., Tsai, H.L.: Three-dimensional modeling of arc plasma and metal transfer in gas metal arc welding. Int. J. Heat Mass Transf. **52**(7–8), 1709–1724 (2009)
5. Maggiolino, S., Schmid, C.: Corrosion resistance in FSW and in MIG welding techniques of AA6XXX. J. Mater. Process. Technol. **197**(1–3), 237–240 (2008)
6. Taban, E., Kaluc, E.: Microstructural and mechanical properties of double-sided MIG, TIG and friction stir welded 5083-H321 aluminium alloy. Kovove Mater. **44**, 1–25 (2006)
7. Kah, P., Suoranta, R., Martikainen, J., Magnus, C.: Techniques for joining dissimilar materials: metals and polymers. Rev. Adv. Mater. Sci. **36**(2) (2014)
8. Martinsen, K., Hu, S.J., Carlson, B.E.: Joining of dissimilar materials. CIRP Ann. **64**(2), 679–699 (2015)
9. Verma, J., Taiwade, R.V.: Effect of welding processes and conditions on the microstructure, mechanical properties and corrosion resistance of duplex stainless steel weldments—a review. J. Manuf. Process. **25**, 134–152 (2017)
10. Eisazadeh, H., et al.: Effect of material properties and mechanical tensioning load on residual stress formation in GTA 304-A36 dissimilar weld. J. Mater. Process. Technol. **222**, 344–355 (2015)
11. Tamrin, K.F., Nukman, Y., Sheikh, N.A., Harizam, M.Z.: Determination of optimum parameters using grey relational analysis for multi-performance characteristics in CO2 laser joining of dissimilar materials. Opt. Lasers Eng. **57**, 40–47 (2014)
12. Harooni, M., Ma, J., Carlson, B., Kovacevic, R.: Two-pass laser welding of AZ31B magnesium alloy. J. Mater. Process. Technol. **216**, 114–122 (2015)
13. Carlson, J.S., Spensieri, D., Wärmefjord, K., Segeborn, J., Söderberg, R.: Minimizing dimensional variation and robot traveling time in welding stations. Procedia Cirp **23**, 77–82 (2014)
14. Zhang, L.J., Bai, Q.L., Ning, J., Wang, A., Yang, J.N., Yin, X.Q., Zhang, J.X.: A comparative study on the microstructure and properties of copper joint between MIG welding and laser-MIG hybrid welding. Mater. Des. **110**, 35–50 (2016)
15. Meng, X., Qin, G., Zhang, Y., Fu, B., Zou, Z.: High speed TIG–MAG hybrid arc welding of mild steel plate. J. Mater. Process. Technol. **214**(11), 2417–2424 (2014)
16. Dhas, J.E.R., Dhas, S.J.H.: A review on optimization of welding process. Procedia Eng. **38**, 544–554 (2012)
17. Karadeniz, E., Ozsarac, U., Yildiz, C.: The effect of process parameters on penetration in gas metal arc welding processes. Mater. Des. **28**(2), 649–656 (2007)
18. Kim, I.S., Son, J.S., Kim, I.G., Kim, J.Y., Kim, O.S.: A study on relationship between process variables and bead penetration for robotic CO2 arc welding. J. Mater. Process. Technol. **136**(1–3), 139–145 (2003)
19. Srirangan, A.K., Pulraj, S.: Multi response optimization of process parameter for TIG welding of Incoloy 800HT by Taguchi grey relational analysis. Eng. Sci. Int. J. **19**, 811–817 (2016)
20. Sathish, R., Naveen, B., Nijanthan, P., Geethan, K.A.V., Rao, V.S.: Weldability and process parameter optimization of dissimilar pipe joints using GTAW. Int. J. Eng. Res. Appl. **2**(3), 2525–2530 (2012)

21. Kumar, S., Singh, R.: Optimization of process parameters of metal inert gas welding with preheating on AISI 1018 mild steel using grey based Taguchi method. Measurement **1**(148), 106924 (2019)
22. Tudu, A., Pal, P.K.: Process parameter optimization of GTAW process on 1018 mild steel by using Taguchi method. J. Mater. Sci. Mech. Eng. **3**(2), 75–80 (2016)
23. Osman, M.H., Tamin, N.F., Ahmad, M.N., Rahman, M.A., Wahid, M.K., Maidin, N.A., Azahar, A.A.: Effect of cutting parameters on surface roughness in dry drilling of AISI D2 tool steel by using Taguchi method. J. Adv. Manuf. Technol. (JAMT) **121**(2), 535–546 (2018)

Real and Complex Wavelet Transform Using Singular Value Decomposition for Malaysian Speaker and Accent Recognition

Rokiah Abdullah[1](✉), Vikneswaran Vijean[1], Hariharan Muthusamy[2], Farah Nazlia Che Kassim[1], and Zulkapli Abdullah[3]

[1] School of Mechatronic Engineering, Universiti Malaysia Perlis (UniMAP), Kampus Pauh Putra, 02600 Arau, Perlis, Malaysia
rokiah@unimap.edu.my

[2] Department of Electronics Engineering, National Institute of Technology, Srinagar (Garhwal), Uttarakhand, India

[3] Pusat Kejuruteraan, Universiti Malaysia Perlis, Aras 1 Bangunan Pentadbiran, Kampus Tetap Pauh Putra, 02600 Arau, Perlis, Malaysia

Abstract. This paper presents a new approach for Malaysian speaker and accent recognition using wavelet feature extraction method, namely Wavelet Packet Transform (WPT), Discrete Wavelet Packet Transform (DWPT) and Dual Tree Complex Wavelet Packet Transform (DT-CWPT). Since Singular Value Decomposition (SVD) was based on factorization and summarization technique which reduces a rectangular matric, it is applied on those features to evaluate the performance for speaker and accent recognition. The features are derived from wavelets and SVD classified with three different classifiers namely *k*-Nearest Neighbors (*k*-NN), Support Vector Machine (SVM) and Extreme Learning Machine (ELM). In this work, English digits (0–9) and Malay words database uttered from 75 undergraduate students of Universiti Malaysia Perlis (UniMAP) which are Malays, Chinese and Indian. The Malay words had a combination of consonants and vowels in monosyllable and bi-syllable structure. The accuracy of file-based analysis achieved were above 81% while for frame-based analysis, 93.87% and above were obtained using three different classifiers (*k*-NN, SVM and ELM) for speaker and accent recognition. Through the experiments, it is observed that accent recognition achieved high recognition rate of 100% for both framed-based analysis and file-based analysis using SVM. The experimental results show the proposed features using SVD achieved high accuracy of 100% using SVM through English digits and Malay words in accent recognition. This indicated that feature extraction using wavelets (WPT, DWPT and DT-CWPT) with SVD can achieve a good performance for both English digits and Malay words.

Keywords: Discrete Wavelet Packet (DWPT) · Dual Tree Complex Wavelet Packet Transform (DT-CWPT) · Singular Value Decomposition (SVD)

M. A. Zakaria et al. (Eds.): *Advances in Mechatronics, Manufacturing, and Mechanical Engineering*, LNME, pp. 22–35, 2021.
https://doi.org/10.1007/978-981-15-7309-5_3

1 Introduction

Speech, speaker and accent recognition an important research topic in automatic speech recognition (ASR). In ASR, an important stage in determining the success of a system depends on the feature extraction and the classification [1, 2]. Previously various well-known feature extraction methods were employed to extract the characteristic from voice signal such as Mel Frequency Cepstral Coefficients (MFCC), Linear predictive Coding (LPC), Linear Predictive Cepstral Coefficients (LPCC) and etc. Wavelets is one alternative approach was employed in ASR and the recognition accuracy are very effective and promising [3, 4].

In this study, real and complex wavelet transform namely WPT, DWPT and DTCWPT using (SVD) is presented to improve the performance of speaker and accent recognition. A new database uttered from undergraduate students of University Malaysia Perlis with major races in Malaysia i.e. Malay, Chinese and Indians. The database consists of English digits (0–9) and selected Malay words consists six vowels of 'a, e, i, o, u, e'. It has a combination of consonants and vowels in monosyllable and bi- syllable structure. The following section will describe the real and complex wavelet transform, classifiers and the experimental results using the proposed approach.

2 Literature Review

Previous works investigated the feature extraction method employing SVD in their studies with different database in many field of ASR to evaluate and testing the performance of the proposed works. Some studies based on SVD in speech and speaker recognition are describes below.

Hariharan et al. [5] presents the Mel Frequency Band Energy Coefficients (MFBECs) combined with SVD technique for the classification of pathological or normal voice. Massachusetts Eye and Ear Infirmary (MEEI) database was used in the experiments with simple k-means nearest neighbourhood (*k*-NN) and Linear Discriminant Analysis (LDA) for testing the effectiveness of the (MFBECs) - SVD based feature vector. The classification accuracy was obtained 99.59% when the value of k is fixed as 2 using k-NN classifier and 98.48% for LDA classifier.

Feature selection using SVD and QR decomposition for text-independent speaker identification has been proposed by Chakroborty and Saha [6]. The performance after selection using a new feature of proposed work Gaussian Mel-Frequency Cepstral Coefficients (GMFCC) are compared with MFCC and Linear Frequency Cepstral Coefficients (LFCC). SVD was used to estimate number of features that has been used and QR decomposition with Column Pivoting (QRcp) provides of ranking for those features. The highest performance of speaker identification achieved 97.65% using GMFCC with 14 features employing YOHO database. Meanwhile, the accuracy using POLYCOST database obtained 81.75% with 15 GMFCC features.

The effect of SVD was investigated on speech recognition employing of vowels \a\, \e\, \u\ proposed by Balvinder et al. [7]. The effect of reduction in singular value was done and the experimental results show that the number of singular values drastically reduce without significantly affecting the perception of the vowels.

Speaker adaption of Deep Neural Network (DNN) in a large speaker code size and with SVD method has been researched by Xue et al. [8] to obtain a best performance in speech recognition. Large scale of 320-h Switchboard task was employed in the experiments containing 1540 speakers. The performance of using speaker-independent are evaluated in training models and the best models were used as the baselines. The composed matrices were used directly as the connection weights without training of them and the different speaker code size was used for testing. The random initialization for the connection weights are investigated to demonstrate the effect of SVD. The experimental results have shown that it is effective for providing well initializations and suitable in adapting large DNN models.

Muhammad Amirul and Norashikin [9] reported the combination of two techniques namely relative spectra-perceptual linear prediction (RASTA-PLP) and SVD in Malay language speech signals. Three Malay word "satu", "dua" and "sembilan" was employed and the results of the proposed works are compared with short-time fourier transform (STFT). The results show that for STFT, 99.98% of the energy is represented by 16 principal vectors for 2-syllable words and 30 principal vectors for 3-syllable words. Meanwhile for RASTA-PLP, this value appear at a much lower value, 9 principal vectors for 2-syllable words and 8 principal vectors for 3-syllable words. From the experiments, it can be concluded that the used of smaller number of principal vectors in recognizing words can cause less computation time.

Anggun et al. [10] researched on feature reduction data of MFCC using Principal Component Analysis (PCA) and SVD in speech recognition system. The data reduction process is designed in two version and classified using SVM classifier. The results showed that the MFCC + PCA version 2 and MFCC + SVD version 1 were able to provide the maximum accuracy from conventional MFCC method from 83.57% to 90.71%. The dimension of feature data decrease from 26 to 10 for MFCC + PCA version 2 and 26 into 14 for MFCC + SVD version 1. Combination of MFCC with PCA and SVD produced highest accuracy beside the computation time is faster than conventional MFCC with average of 7.819 s.

Suman et al. [11] works on robust automatic speaker identification using SVD in adverse conditions by applying Artificial Neural Network (ANN) with 25 speakers employing combination lock number "24-32-75". The experiments was done with different dB levels and the configuration of ANN are set up as 18 (number of input nodes-MFCC), 54 (hidden nodes) and output nodes are 25. For this work, the identification accuracy is same or better in case of proposed method than original method in all sorts of combinations between train-test Signal Noise Ratio (SNR) levels. The accuracy shows better recognition at lower SNR levels which is 100% achieved at 12 dB of SNR in training and testing environment while the original method achieved 98. 4% of accuracy at 12 dB of SNR.

Motivated by the previous studies, this study was undertaken to improve the performance of recognition rate of speaker and accent using real and complex wavelet transform namely WPT, DWPT and DT-CWPT. The studies have shown that the usage of wavelet gives a good result [12–14], however huge features set produced by wavelets affecting the processing time which is known as "curse of dimensionality". Number of smaller features are always preferred for the success of speaker/speech recognition system due to faster learning and improved performance. Then feature reduction/selection is one of the methods to solve this issue and many types of feature reduction employed in ASR. SVD is a popular and well-known method that was applied to test the performance of the recognition [15–17]. Therefore, SVD was employed to reduce number of features in this study.

3 Methodology

3.1 Database

The speech corpus was collected from 75 volunteers of undergraduate students of Universiti Malaysia Perlis (UniMAP). The speakers are from three main major races of Malaysia which are Malay, Chinese and Indians. They are individuals of both male (41) and female (34) genders aged from 19–24 years old. The purpose of sampling has been briefed to the participants and their consent is a must before recording session. Protocol of the study was examined and approved by local committee of experts. The Malay words are combination of consonants and vowels in monosyllable and bi- syllable structure represent the six vowels of 'a, e, i, o, u, e' and English digits (0–9) are pronounced for 15 sessions. Every session consists of predefined digit and Malay word organized randomly. The total speech samples are 23625 files. The recording was carried in a classroom sized semi-anechoic acoustic chamber using a handled condenser, supercardiod and unidirectional microphone. The background noise in this room was recorded approximately 22 dB which is this level is considered very quiet and controlled as compared to normal quiet room about 40 to 50 dB. Table 1 below summarizes the database.

Speech signals were recorded with a sampling rate of 44 kHz and down-sampled to 16 kHz. Based on the Shannon sampling theorem, the 16 kHz sampling was enough to reconstruct an 8 kHz bandwidth signal (telephony speech bandwidth) maximum frequency [18]. In Pre-processing stage, the speech signal was normalized and filtered, so that, only useful speech information was retained. The experiments are performed using two techniques (file-based analysis and frame-based analysis). In file-based analysis, speech signals are processed through feature extraction using different type of wavelets transform (WPT, DWPT and DT-CWPT) meanwhile in frame-based analysis, speech signals are segmented into frames of 32 ms long using Hamming window with 50% overlap before processed through feature extraction as described below in Fig. 1. After feature extraction, SVD was applied to reduce number of features from wavelets to improve the performance of speaker and accent recognition. Each audio samples file length averaged 2–3 s. Hence in frame-based analysis it generates 62–93 frame at 512 sample per frame. The accuracy was investigated individually using *k*-NN, SVM and ELM classifier. Figure 1 shows the block diagram of the proposed work.

3.2 Feature Extraction

Wavelet Packet Transform (WPT)

Wavelet Packet Transform (WPT) is an extension of Discrete Wavelet Transform (DWT)

Table 1. Database details

Item	Description		
Speakers	75		
Session/Speaker	15 times		
Wordlist	1. Digit English: (0–9) 2. Malay word: (Jam, Pas, Cap, Tol, Sen, Aku, Basi, Pulau, Rabu, Jalan, Muka)		
Age	19–24 years old		
Ethnic	Gender	No of speakers	Total
Malay	Male	14	25
	Female	11	
Chinese	Male	16	25
	Female	9	
Indian	Male	11	25
	Female	14	
Microphone	Stereo microphone		
Types of room	Controlled environment		
Sampling frequency	16 kHz		
Audio file format	Wav		

whereby the signal decomposed into two frequency bands namely low (approximation coefficients) and high (detail coefficients). Therefore, wavelet packet gives a balanced binary tree structure [19]. Considering Forth order Daubechies wavelets were used from previous works proposed by [20, 21], the particular wavelet family is best suited for analysis of speech signals. Daubechies wavelet are found to be time invariant, computationally fast and has sharp filter transition bands [21]. Five-level decomposition by WPT has a total 62 wavelet packet coefficients.

Discrete Wavelet Packet Transform (DWPT)
DWPT is an extension of DWT, which is both the detail and approximation coefficients are decomposed at each level of decomposition to create the full binary tree. Therefore, more information gained while features was generated based on approximation and detail coefficients at different levels applied [22]. The features were extracted from each sub-band for the analysis, which produced 62 feature vectors.

Dual Tree Complex Wavelet Packet Transform (DT-CWPT)
The DT-CWPT consists of two DWPT operating in parallel on an input signal. The second wavelet packet filter bank is obtained by replacing the first stage filter $h_i^{(1)}(n)$ by $h_i^{(1)}(n-1)$ and by replacing by $h_i^{'}(n-)$ for $i \in \{0, 1\}$. DT-CWPT's shift invariance

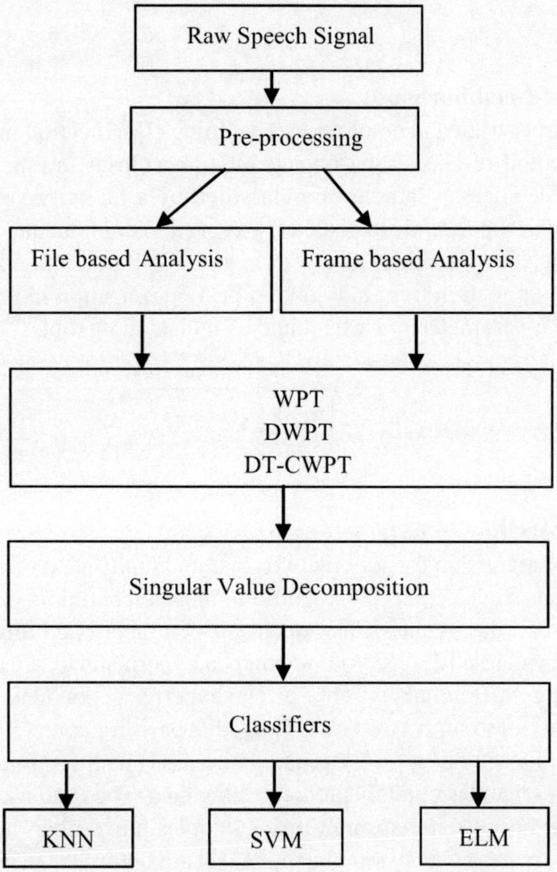

Fig. 1. Block diagram of the proposed work

and directional selectivity provides an accurate measure of spectral energy at a particular location in space, scale and orientation [13]. In this paper, input speech signals were decomposed into 5 levels using DT-CWPT which produced 124 wavelet packet coefficients.

Singular Value Decomposition (SVD)

Singular value decomposition (SVD) is a factorization technique which reduces feature matrix into smaller invertible and square matrix [23]. The theorem of SVD normally presented as given by

$$X = UDV^T \tag{1}$$

The column of U is called left singular vector, V are called right singular vector and D is singular value of the matrix. Column U and V are associated with the corresponding singular value. In this work, 32 singular value for WPT and DWPT, meanwhile 64 singular value for DT-CWPT gives a good summarization of the matrix of a signal.

3.3 Classifiers

K-NN (K-Nearest Neighborhood)

k-NN is simple, supervised and employs lazy learning classification method [24]. k-NN algorithm is a method of classifying objects based on closest training example in the training space [25]. The test samples are classified by a majority votes of the nearest neighbor's category. The Euclidean distance was used to compute the distance between training and testing. The Euclidean distance are given in Eq. 1. Suitable k value ranging from 1 to 10 were applied in this study due to best optimization in previous work [25, 26]. Best value of k-parameter = 1 was found as indicated in Tables 4, 5, 6 and 7.

$$d_E[x, y] = \sum_{i=1}^{N} \sqrt{x_i^2 - y_i^2} \tag{2}$$

Support Vector Machine (SVM)

SVM is supervised algorithm for solving two class and multi-class recognition problem in many applications such as speech recognition, image recognition, pattern recognition and etc. It is based on the Vapnic- Chervonenkis (VC) theory and Structure Risk Minimization (SRM) principle [27]. SVM constructs a hyperplane to separate the different classes by searching the optimal hyperplane. The hyperplane boundaries closest/nearest with the sample are called support vectors. The radial basis function (RBF) kernels function was considered to find best C (Cost) and gamma (G) with 10- fold cross validation to the training data set and accurately predict testing data. The RBF is chosen to demonstrate classification because it maps non-linear samples into higher dimensional space. The parameters of the best C (Cost) and gamma (G) are optimized using Lib SVM Tool [28] and are indicated in Table 4 until Table 7.

Extreme Learning Machine (ELM)

Extreme Learning Machine (ELM) is a supervised learning algorithm of single hidden layer feedforward (SLFN) proposed by Huang et al. [29]. The approach of ELM is the input weights and biases are randomly chosen and the output weights analytically determined by finding the least squares solution. The learning speed of ELM are very fast and provide good performance of accuracy. The best value of the regularization coefficients of ELM classifier were found between −10 and 10. The regularization coefficients of ELM classifier was found between at 1, 3 and 9 which is depicted in Tables 4, 5, 6 and 7.

Classifier Validation

For performance evaluation, ten-fold cross validation technique was used to evaluate k-NN, SVM and ELM algorithm. It evaluates predictive models by partitioning the original sample into a training set to train the model, and a test set to evaluate it. The data are divided equally into size 10 segment, which is 9 segments are used as training data and while the remaining is retained as the validation data for testing the model. The process of cross-validation repeated 10 times of training and testing in which, a different segment of data is held-out for validation in each repetition. Overlapping between training and validations set can be avoid after using this method because of all observations are used for both training and validation, and each observation is used for validation exactly once. As a result, it can provide accurate performance estimation of the algorithm.

4 Result and Discussion

For each database English digits (0–9) and Malay words, the performance of real and complex wavelet transform are evaluated using file-based analysis and frame-based analysis for speaker and accent recognition. Table 2 described performance of *k*-NN classifier using confusion matrix (CM) for accent recognition derives from WPT features. The total sample per class for English digits is 3750, total up to 11250 sample files for all classes (Malay, Chinese and Indian). Confusion matrices (CMs) characterize how often a presented utterance was recognized or confused with response alternatives.

Table 2. Confusion matrix (CM) for accent digit (WPT)

		Actual		
		Class 1	Class 2	Class 3
Predicted	Class 1	3525	174	30
	Class 2	194	3385	213
	Class 3	31	191	3507
	Total prediction	3750	3750	3750

True positives

Misclassified classes

The rows correspond to the predicted class (Class 1-Malay, Class 2-Chinese and Class 3-Indian) and the column corresponds to the actual class. The diagonal cells correspond to the observations that are correctly classified and the off-diagonal cells correspond to the misclassified classes.

Table 2 shows the results carried out for WPT feature using *k*-NN classifier for single run. The experimental results conducted 10 times for *k*-NN with average value is considered as maximum recognition accuracy of accent using WPT feature with standard deviation was shown in Table 3.

It is observed that from CM above, that Class 1 (Malay) is correctly predicted 3525 from 3750 predictions, meanwhile 174 is misclassified as Class 2 (Chinese) and 30 is misclassified as Class 3 (India). Accuracy can be calculated from sum of the element on the diagonal of the CM divided by the total number of predictions made. The equation is shown below.

$$Accuracy = \frac{Sum\ of\ diagonal}{Total\ number\ of\ predictions} \times 100 \quad (3)$$

Table 3. Confusion matrix (CM) for complete circle using *k*-NN

k-value	Accuracy	Class 1	Class 2	Class 3
1	92.60	94.00	90.27	93.52
1	92.56	94.05	90.51	93.12
1	92.40	94.03	90.19	92.99
1	92.30	93.60	90.03	93.28
1	92.47	94.16	90.13	93.12
1	92.30	93.36	90.35	93.20
1	92.36	93.76	90.16	93.15
1	92.57	94.27	90.29	93.15
1	92.69	94.37	90.48	93.23
1	92.35	94.11	89.79	93.15
Average	92.46	93.97	90.22	93.19
Standard deviation (std)	0.13	0.29	0.20	0.13

Accuracy measured is 10, 417 predictions (sum of diagonal) from 11, 250 (total prediction) is 92.60%. It is observed CM show that Class 1 (Malay) is higher for accent recognition and a little misclassified in Class 2 and Class 3. This is probably because of the way they uttered the English digit and their mother tongue affected the pronunciation.

The accuracies measured for the experiments are summarized in Tables 4, 5, 6 and 7. From the Tables 4, 5, 6 and 7, it is observed from both experiment (file-based analysis and frame-based analysis) that the maximum accuracy of 100% achieved using SVM classifier for accent and speaker recognition using English digits and Malay words. The accuracy of file-based analysis for both accent and speaker recognition achieved above 81.86% meanwhile for frame-based analysis, above 93.87% obtained using three different classifiers *k*-NN, SVM and ELM.

For accent recognition (file-based analysis) using English digits and Malay words from Table 4, it is observed that experiments using wavelets (DWPT, WPT and DT-CWPT) achieved 100% using SVM classifier. Meanwhile from Table 5 that 100% accuracy obtained employing non-linear feature namely DWPT using SVM classifier using the same database. In Table 6, the recognition rate for accent (frame-based analysis) achieved 100% using SVM classifier obtained from different wavelets (WPT, DWPT and DT-CWPT). Table 7 shows the maximum accuracy of speaker recognition (framed-based analysis) is 99.96% using English digits and 99.97% using Malay words derived from WPT features.

From Tables 4, 5, 6 and 7, it is shown that the accuracy of frame-based analysis is always better than the file-based analysis. The maximum accuracy yielded above 93.87% –100% using three different classifiers (*k*-NN, SVM and ELM). Since more samples are produced from frame-based analysis, and overlapping windows captures the missing information, it will generate a good result. Besides that, accent recognition employing English digits and Malay words always produced good result compared to speaker recognition. This is probably due to accent recognition has been trained with more data compared to speaker recognition. Moreover, the accent is affected of their articulation habit and differences in their mother tongue.

Table 4. Accuracy of accent recognition using English digits and Malay words (file-based)

Features extraction method (no of coefficient) after SVD	Accuracy (%) ± SD Accent (English digits)			Accuracy (%) ± SD Accent (Malay words)		
	KNN	SVM	ELM	KNN	SVM	ELM
WPT (32)	k = 1 92.46 ± 0.13	c = 1024 g = 0.001 100 ± 0.00	j = 9 99.99 ± 0.01	k = 1 92.85 ± 0.06	c = 1024 g = 0.002 100 ± 0.00	j = 9 99.99 ± 0.01
DWPT (32)	k = 1 92.46 ± 0.10	c = 1024 g = 0.001 100 ± 0.00	j = 9 99.99 ± 0.01	k = 1 92.69 ± 0.12	c = 1024 g = 0.002 100 ± 0.00	j = 9 99.99 ± 0.01
DT-CWPT (64)	k = 1 82.84 ± 0.16	c = 1024 g = 0.001 100 ± 0.00	j = 9 99.95 ± 0.01	k = 3 83.81 ± 0.22	c = 1024 g = 0.002 100 ± 0.00	j = 9 99.96 ± 0.01

Table 5. Accuracy of speaker recognition using English digits and Malay words (file-based)

Features extraction method (no of coefficient) after SVD	Accuracy (%) ± SD Speaker (English digits)			Accuracy (%) ± SD Speaker (Malay words)		
	KNN	SVM	ELM	KNN	SVM	ELM
WPT (32)	k = 1 86.22 ± 0.09	c = 1024 g = 0.0005 100 ± 0.00	j = 3 90.05 ± 0.07	k = 1 88.41 ± 0.09	c = 1024 g = 0.0005 99.99 ± 0.01	j = 3 92.63 ± 0.10
DWPT (32)	k = 1 86.28 ± 0.14	c = 1024 g = 0.0005 100 ± 0.00	j = 3 90.04 ± 0.14	k = 1 87.59 ± 0.12	c = 1024 g = 0.0005 100 ± 0.00	j = 3 92.08 ± 0.09
DT-CWPT (64)	k = 1 81.86 ± 0.15	c = 1024 g = 0.0005 9.97 ± 0.01	j = 5 86.74 ± 0.13	k = 1 83.25 ± 0.12	c = 1024 g = 0.0005 99.98 ± 0.01	j = 5 89.51 ± 0.13

From the previous work in speaker/speech and accent identification/recognition using SVD, Chakroborty and Saha [6] have adopted new feature selection method using SVD to estimate number of features that has been used and QRcp provides of ranking for those features. The highest performance of speaker identification achieved 97.65% using GMFCC with 14 features employing YOHO database. Meanwhile, the accuracy using

Table 6. Accuracy of accent recognition using English digits and Malay words (frame-based analysis)

Features extraction method (no of coefficient) after SVD	Accuracy (%) ± SD Accent (English digits)			Accuracy (%) ± SD Accent (Malay words)		
	KNN	SVM	ELM	KNN	SVM	ELM
WPT (32)	k = 1 97.26 ± 0.09	c = 1024 g = 0.002 100 ± 0.00	j = 9 99.98 ± 0.01	k = 1 97.39 ± 0.08	c = 1024 g = 0.002 100 ± 0.00	j = 9 99.98 ± 0.01
DWPT (32)	k = 1 97.13 ± 0.07	c = 1024 g = 0.002 100 ± 0.00	j = 9 99.99 ± 0.01	k = 1 97.34 ± 0.07	c = 1024 g = 0.002 100 ± 0.00	j = 9 99.99 ± 0.01
DT-CWPT (64)	k = 1 94.55 ± 0.03	c = 1024 g = 0.001 100 ± 0.00	j = 9 99.92 ± 0.01	k = 1 94.83 ± 0.06	c = 1024 g = 0.001 100 ± 0.00	j = 9 99.96 ± 0.01

Table 7. Accuracy of speaker recognition using English digits and Malay words (frame-based analysis)

Features extraction method (no of coefficient) after SVD	Accuracy (%) ± SD Speaker (English digits)			Accuracy (%) ± SD Speaker (Malay words)		
	KNN	SVM	ELM	KNN	SVM	ELM
WPT (32)	k = 1 95.11 ± 0.09	c = 1024 g = 0.0005 99.96 ± 0.004	j = 1 95.61± 0.07	k = 1 96.31 ± 0.09	c = 1024 g = 0.0005 99.97 ± 0.01	j = 3 96.99 ± 0.07
DWPT (32)	k = 1 95.22 ± 0.08	c = 1024 g = 0.0005 99.95 ± 0.01	j = 3 95.83 ± 0.07	k = 1 96.01 ± 0.06	c = 1024 g = 0.0005 99.96 ± 0.01	j = 3 97.09 ± 0.09
DT-CWPT (64)	k = 1 93.87 ± 0.12	c = 1024 g = 0.0005 99.93 ± 0.01	j = 3 95.09 ± 0.05	k = 1 94.85 ± 0.09	c = 1024 g = 0.0005 99.90 ± 0.01	j = 3 96.30 ± 0.06

POLYCOST database obtained 81.75% with 15 GMFCC features. Xue et al. [8] proposed speaker adaption of Deep Neural Network (DNN) in a large speaker code size and initiliazed with SVD method. The results have shown that the proposed SVD adaption method achieved up to 3–6% relative error reduction using a dozens of adaption utterances of speakers. Sim et al. [30], reported an investigation into the use of Factorized Hidden Layer (FHL) to achieve compact model adaptation to unseen domains. The authors found that the SVD to initialize the low-rank bases of an FHL model leads to a faster convergence and improved performance. The results show that SVD initialization consistently achieved better performance compared to random initialization by around 0.2–0.5% absolute.

In this study, the accuracy of speaker and accent recognition yielded 99.93%–100% using English digits and Malay words employing SVM classifier. Besides that, ELM contributes a good performance within range from 86.74%–99.99%. Meanwhile for each experiments study that using classifier k-NN, the accuracy recognition rate is from 81.86%–97.39%. SVM classifier achieved a good result in every experiment using all the datasets due to this algorithm maps training samples of classes into higher dimensional space through kernel function. It can predict the unknown data and classified it more accurately. Moreover, it has a better generalization (less overfitting) and robust to noise. Besides that, the performance of speaker and accent recognition using ELM are comparable with SVM classifier. The algorithm in ELM have a better performance in learning efficiency, universal approximation capability and fast learning speed.

From the experiments, it also observed that k-NN classifier give less performance compared to ELM and SVM classifier. The k-NN algorithm does not work well with high dimensional data because with large number of dimensions, it faced difficulties to calculate the distance in each dimension. It is found that feature reduction using SVD improved the accuracy of speaker and accent recognition. It is observed that the results are slightly better with previous works. This is because variety of information inside features in DWPT, DT-CWPT and WPT contributes the promising accuracy. The results proved that proposed feature extraction and classifier help to improve the recognition rate.

5 Conclusion

This paper reports the performance of new approach using different type of wavelet namely WPT, DWPT and DT-CWPT for Malaysian speaker and accent recognition. Feature reduction technique namely SVD was adopted with the proposed feature extraction to reduce large number of coefficients and classified with three different classifiers k-NN, SVM and ELM. The experiments were carried out in file-based analysis and frame- based analysis using English digits (0–9) and Malay words. The performance of the proposed works shows high accuracies in all experiments. The English digits and Malay words gives maximum accuracy of 100% for accent recognition (file- based and frame-based analysis). The accuracy of speaker and accent recognition obtained 99.93%–100% using English and Malay words with SVM classifier. Meanwhile the minimum recognition rate from file-based analysis, achieved 81.86% (speaker recognition) and 82.84% (accent recognition) using k-NN classifier derives from DT-CWPT features. On the other hand, for framed-based analysis, the minimum rate is obtained 93.87% (speaker recognition) and 94.55% (accent recognition) derives from DT-CWPT features. It is observed that the performance of different type of wavelet are not much different each other either in file-based analysis and framed based analysis. This is because of the features extracted from all wavelet derived from various level of frequency filters sub-band. Therefore, features extracted from various wavelet transform leads to no significant differences in accuracy recognition rate. From the results presented in this paper, it can be concluded that proposed method using SVD can obtain best performance in ASR where is smaller size of feature and less computation time. The proposed method can be extended in other applications in ASR to detect stuttered speech and pathological disease. In future work, non-linear feature extraction method will also be investigated with different feature extraction and feature reduction/selection to enhance the recognition rate.

Acknowledgements. The authors would like to acknowledge the support and appreciation to Dr. Noriha Basir, Senior Lecturer of Center for International Languages (CIL), Universiti Malaysia Perlis (UniMAP) as language consultant for the wordlist used in the study.

References

1. Dahake, P.P., Shaw, K., Malathi, P.: Speaker dependent speech emotion recognition using MFCC and support vector machine. In: International Conference on Automatic Control and Dynamic Optimization Techniques (ICACDOT), pp. 1080–1084 (2016)
2. Srivastava, S., Bhardwaj, S., Bhandari, A., Gupta, K., Bahl, H., Gupta, J.R.P.: Wavelet packet based mel frequency cepstral features for text independent speaker identification. In: Intelligent Informatics, pp. 237–247. Springer, Heidelberg (2013)
3. Mnassri, A., Bennasr, M., Adnane, C.: A robust feature extraction method for real-time speech recognition system on a raspberry Pi 3 board. Eng. Technol. Appl. Sci. Res. **9**(2), 4066–4070 (2019)
4. Abdullah, R., Muthusamy, H., Vijean, V., Abdullah, Z., Kassim, F.N.C.: Real and complex wavelet transform approaches for Malaysian speaker and accent recognition. Pertanika. J. Sci. Technol. **27**(2), 737–752 (2019)
5. Hariharan, M., Paulraj, M.P., Yaacob, S.: Identification of vocal fold pathology based on mel frequency band energy coefficients and singular value decomposition. In: IEEE International Conference on Signal and Image Processing Applications, pp. 514–517 (2009)
6. Chakroborty, S., Saha, G.: Feature selection using singular value decomposition and QR factorization with column pivoting for text-independent speaker identification. Speech Commun. **52**(9), 693–709 (2010)
7. Kour, B., Singh, R., Lehana, P., Rajput, P.: Effect of singular value decomposition-based processing on speech perception. Int. J. Nat. Lang. Comput. **2**(1), 9–16 (2013)
8. Xue, S., Jiang, H., Dai, L., Liu, Q.: Unsupervised speaker adaptation of deep neural network based on the combination of speaker codes and singular value decomposition for speech recognition. In: IEEE International Conference on Acoustics, Speech and Signal Processing (ICASSP), pp. 4555–4559 (2015)
9. Zulkifly, M.A.A., Yahya, N.: Relative spectral-perceptual linear prediction (RASTA-PLP) speech signals analysis using singular value decomposition (SVD). In: IEEE 3rd International Symposium in Robotics and Manufacturing Automation (ROMA), pp. 1–5 (2017)
10. Winursito, A., Hidayat, R., Bejo, A., Utomo, M.N.Y.: Feature data reduction of MFCC using PCA and SVD in speech recognition system. In: International Conference on Smart Computing and Electronic Enterprise (ICSCEE), pp. 1–6 (2018)
11. Senapati, S., Chakroborty, S., Saha, G.: Robust automatic speaker identification based on singular value decomposition technique in adverse conditions. CiteSeer X (2018)
12. Paulraj, M.P., Yaacob, S., Hariharan, M.: Diagnosis of voice disorders using MEL scaled WPT and functional link neural network. Int. J. Biomed. Soft Comput. Hum. Sci. Spec. Issue Biosens. Data Acquis. Process. Control. **14**(2), 55–60 (2009)
13. Lim, W.J., Muthusamy, H., Vijean, V., Yazid, H., Nadarajaw, T., Yaacob, S.: Dual-tree complex wavelet packet transform and feature selection techniques for infant cry classification. J. Telecommun. Electron. Comput. Eng. (JTEC) **10**(1–16), 75–79 (2018)

14. Almaadeed, N., Aggoun, A., Amira, A.: Speaker identification using multimodal neural networks and wavelet analysis. IET Biometrics **4**(1), 18–28 (2015)
15. Mishra, B.P., Chakroborty, S., Saha, G.: Improving speaker identification via singular value decomposition-based feature transformer. In: TENCON 2008-2008 IEEE Region 10 Conference, pp. 1–6 (2008)
16. Bhamre, P.D., Memane, S.B.: Face recognition using singular value decomposition and hidden Markov model. Int. J. Mod. Trends Eng. Res. **2**(10), 323–332 (2015)
17. Seki, H., Yamamoto, K., Akiba, T., Nakagawa, S.: Rapid speaker adaptation of neural network based filterbank layer for automatic speech recognition. In: IEEE Spoken Language Technology Workshop (SLT), pp. 574–580 (2018)
18. Gruhn, R.E., Minker, W., Nakamura, S.: Statistical Pronunciation Modeling for Non-Native Speech Processing. Springer, Heidelberg (2011)
19. Johari, N.A.A.B., Hariharan, M., Saidatul, A., Yaacob, S.: Multistyle classification of speech under stress using wavelet packet energy and entropy features. In: IEEE Conference on Sustainable Utilization and Development in Engineering and Technology (STUDENT), pp. 74–78 (2011)
20. Bong, S.Z., Wan, K., Murugappan, M., Ibrahim, N.M., Rajamanickam, Y., Mohamad, K.: Implementation of wavelet packet transform and nonlinear analysis for emotion classification in stroke patient using brain signals. Biomed. Signal Process. Control **36**, 102–112 (2017)
21. Lei, L., Kun, S.: Speaker recognition using wavelet packet entropy, i-vector, and cosine distance scoring. J. Electr. Comput. Eng. **2017**, 1–9 (2017)
22. Zhang, Y., Dong, Z., Wang, S., Ji, G., Yang, J.: Preclinical diagnosis of magnetic resonance (MR) brain images via discrete wavelet packet transform with Tsallis entropy and generalized eigenvalue proximal support vector machine (GEPSVM). Entropy **17**(4), 1795–1813 (2015)
23. Haitsma, J., Kalker, T., Oostveen, J.: Robust audio hashing for content identification. In: International Workshop on Content-Based Multimedia Indexing, pp. 117–124 (2001)
24. Hariharan, M., Chee, L.S., Ai, O.C., Yaacob, S.: Classification of speech dysfluencies using LPC based parameterization techniques. J. Med. Syst. **36**(3), 1821–1830 (2012)
25. Oung, Q.W., Muthusamy, H., Basah, S.N., Lee, H., Vijean, V.: Empirical wavelet transforms-based features for classification of Parkinson's disease severity. J. Med. Syst. **42**(2), 29 (2018)
26. Hariharan, M., Polat, K., Yaacob, S.: A new feature constituting approach to detection of vocal fold pathology. Int. J. Syst. Sci. **45**(8), 1622–1634 (2014)
27. Amami, R., Ayed, D.B., Ellouze, N.: Practical selection of SVM supervised parameters with different feature representations for vowel recognition. Int. J. Digit. Content Technol. Appl. (JDCTA) **7**(9), 418–424 (2013)
28. Chang, C.C., Lin, C.J., Yaacob, S.: LIBSVM: a library for support vector machines. J. ACM Trans. Intell. Syst. Technol. (TIST) **2**(3), 1–27 (2011)
29. Huang, G.B., Zhu, Q.Y., Siew, C.K.: Extreme learning machine: a new learning scheme of feedforward neural networks. In: 2004 IEEE International Joint Conference on Neural Networks, vol. 2, pp. 985–990 (2004)
30. Sim, K.C., Narayanan, A., Misra, A., Tripathi, A., Pundak, G., Sainath, T.N., Bacchiani, M.: Domain adaptation using factorized hidden layer for robust automatic speech recognition. In: Interspeech, pp. 892–896 (2018)

A Comparison Study of Font Reconstruction Using Differential Evolution

N. Roslan(✉), Z. R. Yahya, W. Z. A. W. Muhamad, and N. A. Rusdi

Institute of Engineering Mathematics, Universiti Malaysia Perlis, Kampus Pauh Putra, 02600 Arau, Perlis, Malaysia
nurshazneem@unimap.edu.my

Abstract. The use of Differential Evolution (DE) to simultaneously optimize parameter t and middle control points (P_1 and P_2) of cubic Bézier curve is presented in this paper. The main objective of this research is to reduce the error of the numerical result. In addition, the numerical result from this research also being compared with the previous article [4] and [5]. The main steps involved are boundary extraction and corner point detection of the images. Then, followed by the process of DE in optimizing the parameter t and middle control points. Finally, by using the optimized value, the piecewise cubic Bézier curve is fitted to each segment. In addition, the Sum Square Error (SSE) has been used as an objective function to calculate the distance of the fitted Cubic Bézier curve with the boundary of the original images. The numerical result for this research produced more errors because of two parameter value that were optimized has been run simulataneously. The result that has been obtained does not guarantee that this algorithm is not good because the use of DE is subjective. DE is an interesting algorithm that can be explored in more detail and can be applied in various problems. The new research of DE can be extended more to its various mutation strategies.

Keywords: Bézier curve · Curve fitting · Differential Evolution · Optimization · Sum Square Error

1 Introduction

In reverse engineering problems, curve reconstruction has been widely used. There are various aspect of curve reconstruction such as font designing, medical images reconstruction and data visualization. The main focused in curve reconstruction is the smoothness and the accurateness of the curve. In this research, Differential Evolution (DE) has been used to optimize parameter value t and middle control points of cubic Bézier curve. The performance of the DE algorithm is sensitive to the mutation strategy, crossover strategy and control parameters such as the population size (NP), crossover rate (CR) and the mutation factor (F). The best settings for the control parameters can be different for different optimization problems [1]. A suitable choice of mutation operator and control parameter settings can help the algorithm lead the population toward the global optimum [2]. A lot of research has been done related to the population size of DE [3].

M. A. Zakaria et al. (Eds.): *Advances in Mechatronics, Manufacturing, and Mechanical Engineering*, LNME, pp. 36–44, 2021.
https://doi.org/10.1007/978-981-15-7309-5_4

Previously, two research has been done which is optimizing the middle control points (P_1 and P_2), [4] and optimizing the parameter t [5]. Therefore, in this research the optimization of the middle control point (P_1 and P_2) and parameter t is done simultaneously with hope that it can reduce the error of the numerical result. Besides that, the numerical result obtained from this research also will be compared with the previous article [4] and [5]. The Sum Square Error (SSE) has been used as an objective function in order to get the optimized parameter value.

The organization of the paper is as follows. Section 2 discussed about Cubic Bézier curve. Then, it was followed by the explanation of the process involved in Differential Evolution algorithm. Next, explanation on the adopted procedure on how algorithm works for images reconstruction. In Sect. 4, the numerical result of the new approach of curve reconstruction using Differential Evolution, compared with the previous research is been illustrated. Lastly, the paper has been concluded in Conclusion section.

2 Bézier Curve

Bézier curve was developed by Paul de Casteljau in 1959 and by Pierre Étienne Bézier in about 1962. A parametric Bézier curve at parameter t is defined as:

$$Q(t) = \sum_{i=0}^{n} B_i^n(t)P_i \quad 0 \leq t \leq 1 \tag{1}$$

with P_i is a control point, B_i^n is the Bernstein polynomial basis function and t is the parameter. The numeric value of t is normally between 0 and 1. The definition of Bernstein basis function is as follow:

$$B_i^n(t) = {}^nC_i(1-t)^{n-i}t^i \tag{2}$$

Figure 1 shows the example of the cubic Bézier curve. Cubic Bézier curve have four control points ($n = 3$). Therefore from Eq. (1) the cubic Bézier is illustrated as:

$$\begin{aligned} Q(t) &= B_0^3(t)P_0 + B_1^3(t)P_1 + B_2^3(t)P_2 + B_3^3(t)P_3 \\ &= (1-t)^3P_0 + 3t(1-t)^2P_1 + 3t^2(1-t)P_2 + t^3P_3 \\ &= t^3[-P_0 + 3P_1 - 3P_2 + P_3] + t^2[3P_0 - 6P_1 + 3P_2] + t[-3P_0 + 3P_1] + P_0 \end{aligned} \tag{3}$$

As shown in Fig. 1, P_0 and P_3 are the end points while P_1 and P_2 are the middle control points. This study focuses on how to optimize the parameter t and estimate those middle control point (P_1 and P_2) by using DE algorithm. In general, the optimization and estimation is defined as a process of searching the most suitable value of P_1 and P_2, as well as the parameter t which will help to produced a better curve.

In this research, the cubic Bézier has been used because of their computationally efficient and having high level of accuracy to approximate character shapes [6]. Besides that, cubic Bézier gives reasonable design flexibility while avoiding the increased calculations needed with higher polynomials [7]. Therefore, in this study, cubic Bézier curve will be applied to reconstruct the Japanese font, where the same font has been use in previous research [4] and [5].

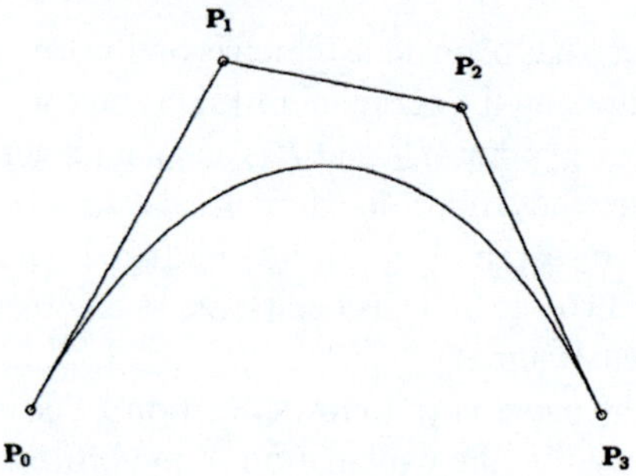

Fig. 1. Cubic Bézier curve.

3 Differential Evolution

Differential Evolution was proposed by Rainer Storn in 1995. The purpose of DE is to solve the optimization problem, which is to get the best result based on its parameter value. DE algorithm able to find the acceptable solution with reasonable time demand, but efficiency of the search is sensitive to the setting of control parameters [19]. In addition, it also performs better than the Genetic Algorithm (GA) or the Particle Swarm Optimization (PSO) over several numerical benchmarks [20]. The main steps of the DE algorithm are presented in Fig. 2.

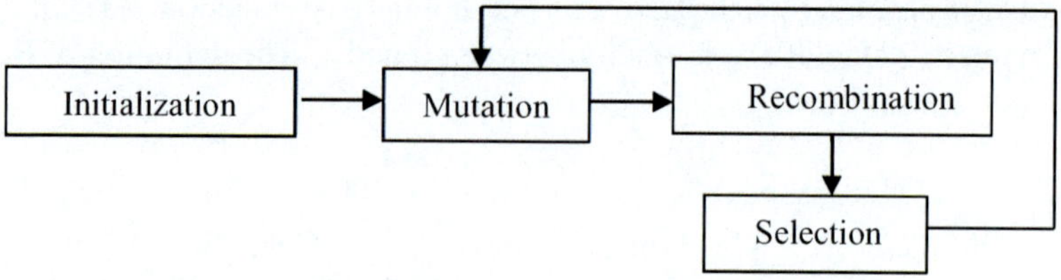

Fig. 2. Process on DE algorithm.

In this research, Differential Evolution (DE) has been used to optimize parameter t and middle control point of cubic Bézier curve. The process of DE is started by generate the initial solution based on the population size which is set by the author. Formula for the initialization is as follows:

$$X_{i,j} = X_j^L + rand(0, 1) \bullet \left(X_j^U - X_j^L\right) \quad (4)$$

where $rand(0, 1)$ is a uniformly distributed random number lying between 0 and 1; while X_j^U and X_j^L is an upper boundary and lower boundary of data points for each segments. Since this research optimize parameter t and middle control point, therefore the initial solution is assign differently for parameter t and middle control point (P_1 and P_2).

In the next step, the target vector $X_{i,G}$ of the population members will be change into donor vector $V_{i,G}$ according to this mutation strategy, namely DE/rand/1.

$$V_{i,G} = X_{r1,G} + F\left(X_{r2,G} - X_{r3,G}\right) \quad (5)$$

There are various type of DE mutation strategy describe in [15, 21]. The notation DE/rand/1 specifies that the vector to be perturbed is randomly chosen and the perturbation consist of one weighted difference vector. The given parameter vector $X_{r1,G}$, $X_{r2,G}$ and $X_{r3,G}$ are selected randomly such that they are distinct. The mutation factor that's normally used is between [0.5, 1], ([15, 16]).

The recombination process incorporates successful solution from previous generation where the trial vector $U_{j,i,G}$ is produced from the target vector $X_{i,G}$ and elements of the donor vector $V_{i,G}$ with the probability of CR [22]. This process is illustrated as follows:

$$U_{j,i,G} = \begin{cases} V_{j,i,G} & if \;\; rand_{j,i}(0,1) \leq CR, \\ & or \; j = j_{rand} \\ X_{j,i,G} & else \end{cases}$$
$$i = 1, 2, \ldots, N; \quad j = 1, 2, \ldots, D \tag{6}$$

From Eq. 6, CR is a known as a parameter value like mutation factor, F. The range value of CR is between [0, 1]. The next step is a selection process. This step determines whether the target vector or the trial vector will survive to the next generation. The selection process is as follows:

$$X_{i,G+1} = \begin{cases} U_{i,G}, & if \; f(U_{i,G}) \leq f(X_{i,G}) \\ X_{i,G}, & else \end{cases} \tag{7}$$

where $f(X)$ is the function to be minimized. So, the best parameter that gives the (lowest) fitness value will be included in the next generation.

Mutation process, recombination, and the selection process will be repeated until some stopping criterion is reached. Commonly, stopping criteria are satisfied when a predefined small number of fitness value or a fixed number of generation have been reached [22]. In this research, our objective function is to optimize the value of parameter t and estimate the middle control point, such that the there is a small error between the original images and the new approximate curve.

4 Adopted Procedure

4.1 Boundary Extraction and Corner Detection

The first phase of curve fitting is called boundary extraction and corner point detections. All images were obtained from electronic device and scanning. In this research, the built in MATLAB function called "Boundaries" has been used in order to extract the boundary of the images. There are few research that used the same technique for the boundary extraction [5, 11–13, 18]. Some of the researcher used Avarahmi algorithm for the purpose of boundary extraction [8, 9] and Chain Code [10]. The version of MATLAB that has been used in this research is MATLAB version 9.2, R2017a.

After boundary extraction had completed, the next process is to find the corner point of the images. In this research SAM06 algorithm [14] been used in order to detect the corner of the images. SAM06 algorithm also has been used [5, 11, 14, 18]. SAM06 algorithm is robust, simple to implement, efficient and performs well on noisy shapes, [14]. Detected corners by SAM06 were precise and without detecting any wrong ones. The corner point were used to divide the boundary of the images into smaller segments.

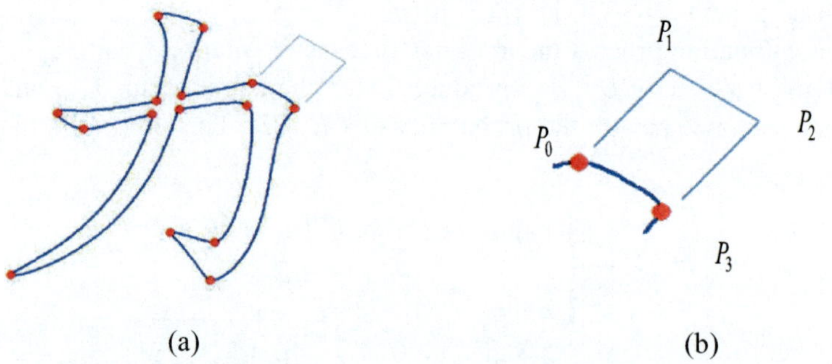

Fig. 3. Boundary and corner points of the images.

The red dots in Fig. 3(a) represent the corner points of the images. While Fig. 3(b) shows the segmentation of the images. At this phase, Douglas-Peucker algorithm [17] been used in order to reduce the data points of the curve. The purpose of this algorithm is to use a small number of data points to construct a curve, as well as to reduce the computational time [5, 11–13, 18]. Boundary of the images and corner detection of the boundary is represented in Fig. 4.

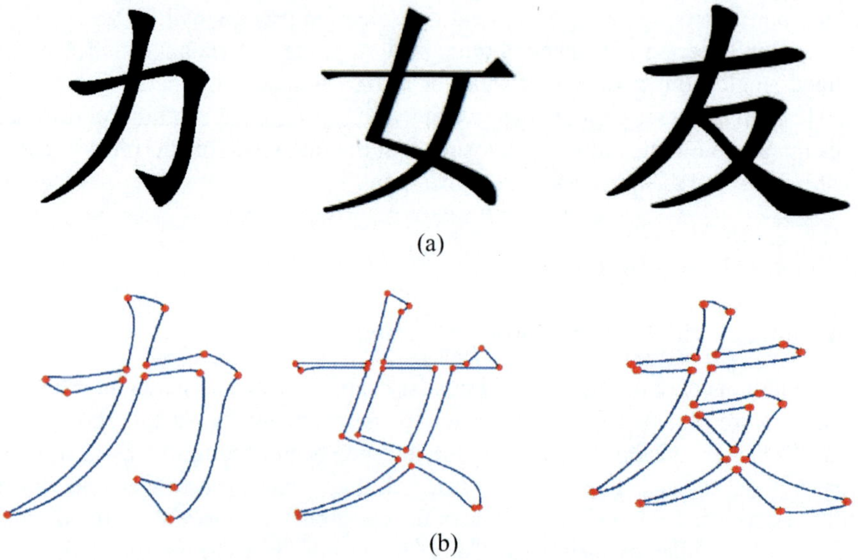

Fig. 4. The image of: (a) original image and (b) the boundary and corner point detection.

4.2 Curve Fitting

Process that involved in curve fitting is called parameterization. The expansion in Eq. (3) can be represented in matrix form as follows:

$$Q(t) = \begin{bmatrix} t^3 & t^2 & t & 1 \end{bmatrix} \begin{bmatrix} -1 & 3 & -3 & 1 \\ 3 & -6 & 3 & 0 \\ -3 & 3 & 0 & 0 \\ 1 & 0 & 0 & 0 \end{bmatrix} \begin{bmatrix} P_0 \\ P_1 \\ P_2 \\ P_3 \end{bmatrix} \tag{8}$$

In this study, each segment will be approximated by using cubic Bézier curve. The parameter t and middle control points (P_1 and P_2) in Eq. (8) will be optimized by using Differential Evolution algorithm. There are four free parameters such as population size (NP), number of iteration (NI), crossover rate (CR) and mutation factor (F) involved in DE algorithm process. For the comparison purposes, the same parameter as research paper [4] and [5] has been used. The parameter value of population size, $NP = 100$, number of iteration, $NI = 50$, crossover rate, $CR = 0.5$ and mutation factor, $F = 0.8$.

4.3 Sum Square Error

After having the best value of middle control points (P_1 and P_2), then fitted the cubic Bézier curve. Bear in mind that the main objective of this study is to minimize the distance between boundaries of the original images and fitted cubic Bézier curve. Therefore, to calculate the error between those two curves, Sum Square Error (SSE) has been applied as the objective function, $f(x)$. In addition, for this study, the stopping criterion has been decided in which every process will be repeated ten times and the mean error will be recorded. Errors of all segments of the image is calculated by using Sum Square Error (SSE). The sum square error is computed as follows:

$$SSE = \sum_{i=1}^{n} \left[Q_i(t) - p_i\right]^2 \tag{9}$$

where p_i is the original data point and $Q_i(t)$ denotes the point on the fitting curve that is related to the optimized parameter value of t and middle control point. Result and discussion is discussed in the next section. Table 1 shows the numerical result of the new approach, compared with the previous research [4] and [5].

5 Result and Discussion

This section will illustrate the result of the new approach of curve reconstruction using Differential Evolution. As referred to Table 1, the performance of the proposed method is not good since it produced more errors compared to the previous research. By comparing all the numerical result of three images, the best result is from research [5], where DE algorithm has been used to optimize the parameter t.

Figure 5 shows the comparison of the images, between the best result that is obtained from research [5] and the proposed method. We can see clearly in Fig. 5(b), there is big distance between the original and the new approximate curve (refer to the arrow). A big distance between the curves will contribute to a big values of SSE.

Table 1. Comparison of numerical result based on: optimizing the middle control points (P_1 and P_2) [4], optimizing the parameter t [5] and proposed method.

Image	Optimizing (P_1 and P_2) [4]	Optimizing parameter t [5]	Proposed method
力	569.1344	**463.5338**	6598.50
女	158.1562	**151.1479**	2355.90
友	387.6141	**192.2468**	2838.40

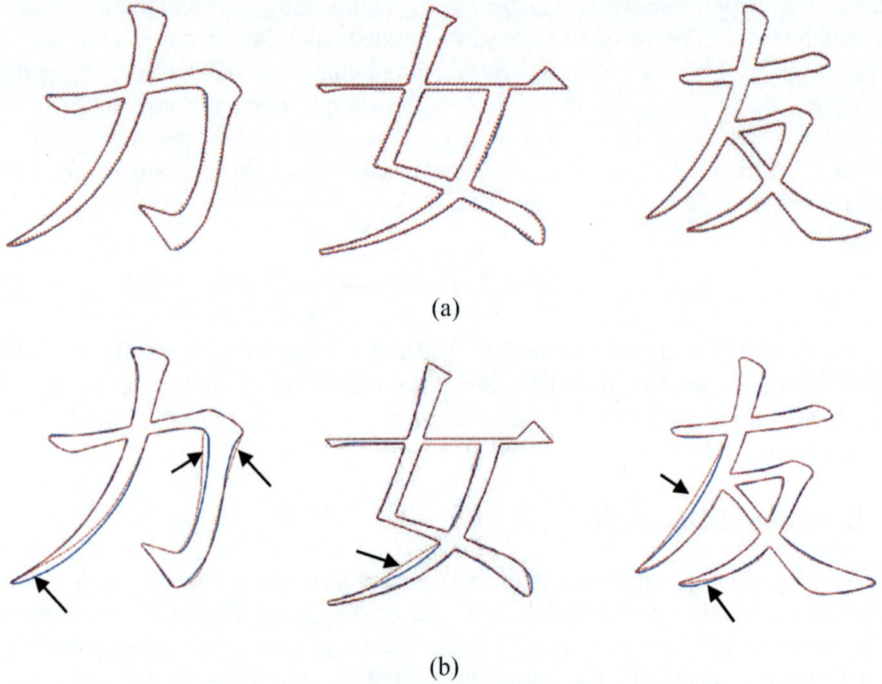

Fig. 5. Image of the original data (solid line) and the new parametric curve (dotted line) for: (a) previous research for optimizing the parameter t [5] and (b) new proposed method.

6 Conclusion

In this research, the optimization of the middle control point (P_1 and P_2) and parameter t is done simultaneously with hope that it can reduce the error of the numerical result. The comparison of the numerical result (refer Table 1) shows that DE algorithm is usefull to optimize parameter t, but it is not suitable to optimize parameter t and middle control points (P_1 and P_2) simultaneously. By referring to Table 1, the best result obtained by DE algorithm is by optimizing the parameter t [5] since it produce a smallest error. Therefore, this research concluded that the use of DE is subjective, because some of the problem is suitable to use DE but some maybe not. Besides that, further research can explore the usage of DE to optimize the parameter t and middle control point (P_1 and P_2) by using a different mutation strategies.

Acknowledgments. This research was supported by University Malaysia Perlis and Fundamental Research Grant Scheme (FRGS) under a grant number of FRGS/1/2017/STG06/UNIMAP/03/2 funded by Ministry of Education Malaysia.

References

1. Mallipeddi, R., Suganthan, P.N., Pan, Q.K., Tasgetiren, M.F.: Differential evolution algorithm with ensemble of parameters and mutation strategies. Appl. Soft Comput. **11**(2), 1679–1696 (2011)
2. Yi, W., Gao, L., Li, X., Zhou, Y.: A new differential evolution algorithm with a hybrid mutation operator and self adapting control parameters for global optimization problems. Appl. Intell. **42**(4), 642–660 (2015)
3. Piotrowski, A.P.: Review of differential evolution population size. Swarm Evol. Comput. **32**, 124 (2017)
4. Roslan, N., Yahya, Z.R.: Reconstruction of font with Cubic Bézier using differential evolution. Sains Malays. **44**(8), 1203–1208 (2015)
5. Roslan, N., Yahya, Z.R.: Parameterization method on cubic Bézier curve fitting using differential evolution. In: International Conference on Mathematics, Engineering & Industrial Applications 2016 (ICoMEIA 2016), Songkhla, Thailand, 10–12 August 2016, p. 030075. https://doi.org/10.1063/1.4965195
6. Sarfraz, M., Khan, M.A.: An automatic algorithm for approximating boundary of bitmap characters. Future Gener. Comput. Syst. **20**(8), 1327–1336 (2004)
7. Masood, A., Sarfraz, M.: Capturing outlines of 2D objects with Bézier cubic approximation. Image Vis. Comput. **27**(6), 704–712 (2009)
8. Avrahami, G., Pratt, V.: Sub-pixel edge detection in character digitization. In: Raster Imaging and Digital Typography II, pp. 54–64 (1991)
9. Itoh, K., Ohno, Y.: A curve fitting algorithm for character fonts. Electron. Publ. **6**(3), 195–205 (1993)
10. Sarfraz, M., Khan, M.A.: Automatic outline capture of Arabic fonts. Inf. Sci. **140**(3), 269–281 (2002)
11. Rusdi, N.A., Yahya, Z.R.: Reconstruction of Arabic font with quartic Bézier curve. Sains Malays. **44**(8), 1209–1216 (2015)
12. Roslan, N., Yahya, Z.R.: Different mutation strategies for reconstruction of Japanese character. In: Malaysian Technical Universities Conference on Engineering and Technology 2015 (MUCET 2015), Johor Bharu, Malaysia, 11–13 October 2015

13. Rusdi, N.A., Yahya, Z.R.: Reconstruction of Arabic font using artificial bee colony algorithm. ARPN J. Eng. Appl. Sci. **11**(8), 10761–10767 (2016)
14. Sarfraz, M.: Corner detection for curve segmentation. In: Interactive Curve Modeling: With Applications to Computer Graphics, Vision and Image Processing, pp. 209–240. Springer, London (2008)
15. Storn, R.: On the usage of differential evolution for function optimization. In: Proceedings of Fuzzy Information Processing Society, 1996. NAFIPS, 1996 Biennial Conf. North American, pp. 519–523 (1996)
16. Arunachalam, V.: Optimization using differential evolution 2008, pp. 3–18 (2008)
17. MATLAB: Ramer-Douglas Peucker algorithm demo by Ligong Han. http://www.mathworks.com/matlabcentral/fileexchange/41986-ramer-douglas-peucker-algorithm. Accessed 16 Aug 2014
18. Rusdi, N.A., Yahya, Z.R., Roslan, N., Muhamad, W.Z.W.: Reconstruction of medical images using artificial bee colony algorithm. Math. Probl. Eng. **2018,** 1–7 (2018)
19. Josef, T.: Adaptive differential evolution and exponential crossover. In: International Multiconference on Computer Science and Information Technology, pp. 927–931 (2008)
20. Das, S., Abraham, A., Chakraborty, U.K., Konar, A.: Differential evolution using a neighborhood-based mutation operator. IEEE Trans. Evol. Comput. **13**(3), 526–553 (2009)
21. Das, S., et al.: Particle swarm optimization and differential evolution algorithms: technical analysis, applications and hybridization perspectives. In: Advances of Computational Intelligence in Industrial Systems, pp. 1–38. Springer, Heidelberg (2008)
22. Priza, P., Siti, M.H.S.: Differential evolution optimization for Bézier curve fitting. In: 2010 Seventh International Conference on Computer Graphics, Imaging and Visualization (CGIV), pp. 68–72. IEEE (2010)

Numerical Analysis of Vibration for Flexible Frame of a Lightweight Electric Vehicle

J. Md. Sah[1(✉)], K. A. Ismail[1], and Z. Taha[2]

[1] Structural Mechanics and Dynamics Research Group, School of Manufacturing Engineering, Universiti Malaysia Perlis, 02600 Arau, Perlis, Malaysia
jamali@unimap.edu.my

[2] Dzuki Consultancy and Training, 26600 Pekan, Pahang, Malaysia

Abstract. Vehicle vibration is one of the main factors for driving fatigue, discomfort and health problems. Most studies on ride dynamics and suspension systems of vehicles used lumped-mass model to represent the vehicle body and the wheels. However, a lightweight design of an electric vehicle frame may deteriorate the rigidity of the structure and, based on a previously performed finite element modal analysis, induces a flexible vibration. The flexible vibration affects ride performance and study on dynamic characteristics of the vehicle helps to understand the main parameters required to improve the ride comfort especially when using passive type suspension system. This paper presents comparison of numerical analysis of rigid-body and flexible full-vehicle models for lightweight aluminium frame vehicle with experimental results. The MATLAB Simulink was employed to simulate vibration responses at four areas namely front unsprung mass, rear unsprung mass, floor and seat. Sinusoidal input signals to the three wheels at low frequency of 4 Hz were applied in such a way that the vibration accelerations at the front and rear unsprung masses are equal to the experimental RMS accelerations that had been measured on the vehicle driven on road surface. The measured time-domain signal and PSD for front unsprung mass and seat were plotted to represent the actual vibration conditions of the vehicle. The RMS acceleration were computed for the four areas and compared with simulated data. It was found that the accelerations of the floor and seat using a rigid vehicle body model were not comparable with the experimental data. However the flexible vehicle model managed to generate data that corresponds well with the experiment. Thus it can be said that the flexibility model allows accurate analysis and hence better understanding of vehicle dynamics in order to improve ride comfort of the driver.

Keywords: Vehicle vibration · Full-vehicle model · Frame flexibility · Ride comfort · Electric vehicle

1 Introduction

A three-wheeled electric vehicle which uses solar energy to charge its batteries and power the electric motor was designed and fabricated. The frame was made using aluminium hollow pipes specifically to reduce the vehicle weight in order to improve performances

M. A. Zakaria et al. (Eds.): *Advances in Mechatronics, Manufacturing, and Mechanical Engineering*, LNME, pp. 45–55, 2021.
https://doi.org/10.1007/978-981-15-7309-5_5

in terms of travel distances and battery energy consumption [1, 2]. However, lightweighting the structure generally results in higher noise, vibration and harshness and vehicle vibration is one of the main factors for driving fatigue, discomfort and health problems [3]. It is a common phenomenon in vehicles where vibration due to road surface condition results to whole-body vibration (WBV) of the occupant inside the vehicle. Most studies on ride dynamics and suspension systems of vehicles used lumped-mass model to represent the vehicle body and the wheels. However, a lightweight design of vehicle frame may deteriorate the rigidity of the structure.

The rigid-body models representing dynamic system for vehicles are no longer suitable to predict the ride performances accurately [4]. The influences of rail vehicle carbody flexibility on ride quality and vehicle dynamics has been studied by Carlbom [5, 6] through measurements and simulations. Diana et al. [7] has included the carbody flexibility function in the modal superposition method for analysing the ride comfort of railway vehicles. A total of 33 modes within 5 Hz to 20 Hz were considered and it was found that the first bending mode significantly influenced to the dynamic behaviour of the carbody. Evaluation of vertical ride comfort for railway vehicles was carried out by Dumitriu [8] using a uniform Euler-Bernoulli beam principle to model the carbody supported by a passive type suspension system to reduce the bending vertical vibration and limiting the rotation of the cross-sections. Anti-bending bars effectively reduced the vertical bending vibration in the carbody due to the increment of the carbody bending frequencies. A finite element model of an electric multiple unit train was used by Wang et al. [9] to analyse vibration modal using ANSYS. A three-dimensional rigid-flexible coupling dynamics model considering the carbody flexible vibration and underneath equipment was developed to investigate the influence of rigidity and elasticity connection on carbody elastic vibration. It was found that suspension frequency, suspended position and the equipment mass significantly affect the vibration of the carbody while the damping ratio had little influence on the vibration. These studies have shown the importance of applying flexibility function for analysing vertical ride quality for rail vehicles. However, the influence of bending vibration for lightweight electric vehicle has not been addressed considering the importance of ride performances evaluation especially for aluminium constructed frame. Furthermore, most of the studies employing the Euler-Bernoulli simple beam method developed half-vehicle models that considered bounce, pitch and flexible bending modes. The purpose of this work is to compare the numerical analysis of rigid-body and flexible full-vehicle models for lightweight aluminium frame vehicle with experimental results measured when the vehicle was driven on an urban road.

A preliminary modal analysis was conducted using computer-aided three-dimensional interactive application (CATIA) software in order to understand the bending characteristics of the frame. Several of the natural frequencies are shown in Table 1 that resulted to maximum displacement of the frame in the X-axis (Tx), Y-axis (Ty) and vertical Z-axis (Tz). The first mode was induced at frequency below 10 Hz that resulted in significant vertical bending of the frame as well as the third mode at 10.21 Hz. The modal shapes of the two frequencies are shown in Fig. 1 where these conditions of the frame may have contributed to the vibration measurements and ride performances [9].

Therefore, it is important to consider the flexibility of the frame especially for vertical bending frequency below the 10 Hz to improve the analysis.

Table 1. Modal analysis results.

Mode	Frequency (Hz)	*Tx* (%)	*Ty* (%)	*Tz* (%)
1	6.41	0.61	1.48	42.53
2	7.07	71.58	5.83	0.53
3	10.21	1.32	8.94	17.45
4	16.51	9.30	44.93	0
5	21.38	0.07	4.43	6.61

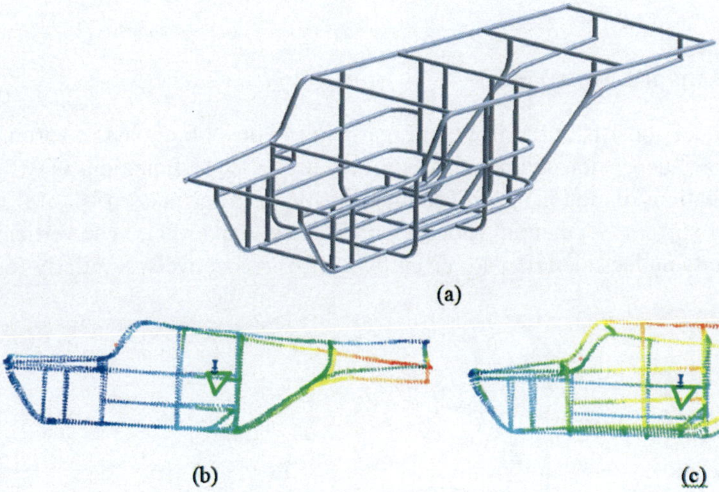

Fig. 1. (a) The aluminium frame significant vertical Z-axis displacement of vibration in the (b) First mode shape, and (c) Third mode shape.

2 Method

A linear six DOFs full vehicle lumped-mass model equations incorporating a seated four DOFs driver model [10] were developed using Newton's second law. The equations were then modified to consider flexibility of the aluminium frame based on Euler-Bernoulli simple uniform supported beam and compared with experimental data. The MATLAB (R2015a) and Simulink (8.5) were employed to simulate vibration responses at four areas namely front unsprung mass, rear unsprung mass, floor and seat. Sinusoidal input signals to the three wheels at low frequency of 4 Hz were applied [11, 12] in such a way that the vibration accelerations at the front and rear unsprung masses are equal to the experimental RMS accelerations that had been measured when the vehicle was driven on an urban road surface condition. These equations account for three rigid body modes with three different bending modes. The parameters representing mass, stiffness and damping for the models are based on a previous work [1].

Vibration measurements were done when the vehicle was driven at 40 km/h for approximately 10 min on an 8 km urban road stretch at Putrajaya. Data were taken at four selected points of the vehicle, namely front wheel (FW), rear wheel (RW), vehicle floor (FL) and seat (SE). Data acquisition and analysis were performed using National Instrument NI9233 twelve channels, ±5 V, 50 kS/s per channel, 24-Bit IEPE dynamic signal acquisition module and NI CompactDAQ. The analogue inputs from Kistler 8690C piezobeam accelerometers were analysed using NI Sound and Vibration software with LabView SignalExpress interface. The measurement rate and sample size were set to 50k Hz and 500k respectively using continuous samples acquisition mode. The signals were processed using Butterworth lowpass filter with 100 Hz cut-off frequency. The measurement on the seat was done using SAE pad which was placed at the centre of the seat cushion.

2.1 Full-Vehicle Lumped-Mass Model

The full-vehicle model consists of the front and rear wheels, unsprung masses, sprung mass and passive type suspension components as shown in Fig. 2 [1]. Equations (1), (2) and (3) show the equations of motion in vertical direction for unsprung masses M_{ufl}, M_{ufr} and M_{ur} in terms of stiffness K, damping coefficient C, pitch θ and roll φ. The vertical displacement, velocity and acceleration are given by Z, $\dot{Z}$, $\ddot{Z}$ respectively. Similarly for

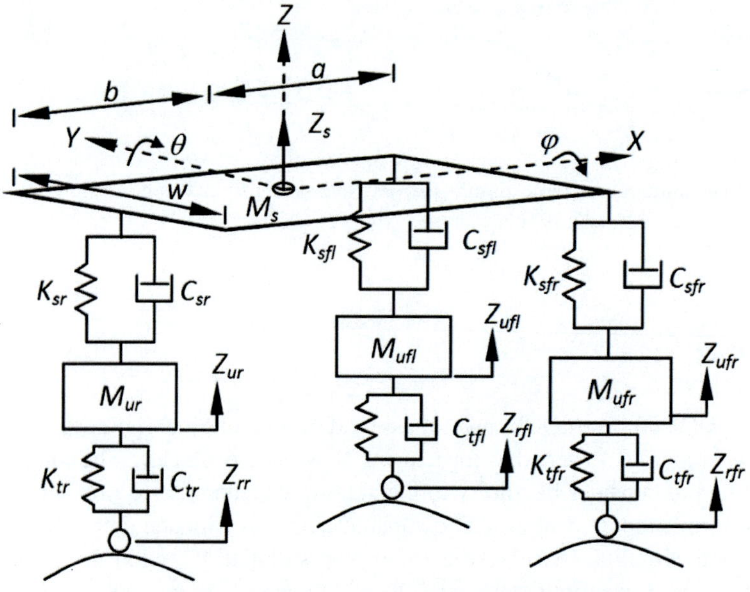

Fig. 2. Six DOFs full-vehicle model [1].

pitch and roll, the 'dot' and 'double dot' above the symbols represent the velocity and acceleration respectively.

$$\begin{aligned} M_{ufl}\ddot{Z}_{ufl} = {} & K_{sfl}Z_s + C_{sfl}\dot{Z}_s - aK_{sfl}\theta - aC_{sfl}\dot{\theta} + cK_{sfl}\varphi + cC_{sfl}\dot{\varphi} \\ & - K_{sfl}Z_{ufl} - C_{sfl}\dot{Z}_{ufl} - K_{tfl}Z_{ufl} - C_{tfl}\dot{Z}_{ufl} + K_{tfl}Z_{rfl} + C_{tfl}\dot{Z}_{rfl} \end{aligned} \quad (1)$$

$$\begin{aligned} M_{ufr}\ddot{Z}_{ufr} = {} & K_{sfr}Z_s + C_{sfr}\dot{Z}_s - aK_{sfr}\theta - aC_{sfr}\dot{\theta} - dK_{sfr}\varphi - dC_{sfr}\dot{\varphi} \\ & - K_{sfr}Z_{ufr} - C_{sfl}\dot{Z}_{ufr} - K_{tfr}Z_{ufr} - C_{tfr}\dot{Z}_{ufr} + K_{tfr}Z_{rfr} + C_{tfr}\dot{Z}_{rfr} \end{aligned} \quad (2)$$

$$\begin{aligned} M_{ur}\ddot{Z}_{ur} = {} & K_{sr}Z_s + C_{sr}\dot{Z}_s + bK_{sr}\theta + bC_{sr}\dot{\theta} - K_{sr}Z_{ur} - C_{sr}\dot{Z}_{ur} - K_{tr}Z_{ur} \\ & - C_{tr}\dot{Z}_{ur} + K_{tr}Z_{rr} + C_{tr}\dot{Z}_{rr} \end{aligned} \quad (3)$$

Equations (4), (5) and (6) show the equations of motion for the sprung mass M_s, in vertical, pitch and roll motions respectively.

$$M_s\ddot{Z}_s = F_1^{'} + F_2^{'} + F_3^{'} \quad (4)$$

where,

$$F_1^{'} = -K_{sfl}\left[Z_s - a\theta + c\varphi - Z_{ufl}\right] - C_{sfl}\left[\dot{Z}_s - a\dot{\theta} + c\dot{\varphi} - \dot{Z}_{ufl}\right]$$

$$F_2^{'} = -K_{sfr}\left[Z_s - a\theta - d\varphi - Z_{ufr}\right] - C_{sfr}\left[\dot{Z}_s - a\dot{\theta} - d\dot{\varphi} - \dot{Z}_{ufr}\right]$$

$$F_3^{'} = -K_{sr}[Z_s + b\theta - Z_{ur}] - C_{sr}\left[\dot{Z}_s + b\dot{\theta} - \dot{Z}_{ur}\right]$$

$$\begin{aligned} I_{yy}\ddot{\theta} = {} & aK_{sfl}\left[Z_s - a\theta + c\varphi - Z_{ufl}\right] + aC_{sfl}\left[\dot{Z}_s - a\dot{\theta} + c\dot{\varphi} - \dot{Z}_{ufl}\right] \\ & + aK_{sfr}\left[Z_s - a\theta - d\varphi - Z_{ufr}\right] + aC_{sfr}\left[\dot{Z}_s - a\dot{\theta} - d\dot{\varphi} - \dot{Z}_{ufr}\right] - bK_{sr}[Z_s + b\theta - Z_{ur}] \\ & - bC_{sr}\left[\dot{Z}_s + b\dot{\theta} - \dot{Z}_{ur}\right] \end{aligned} \quad (5)$$

$$\begin{aligned} I_{xx}\ddot{\varphi} = {} & - cK_{sfl}\left[Z_s - a\theta + c\varphi - Z_{ufl}\right] - cC_{sfl}\left[\dot{Z}_s - a\dot{\theta} + c\dot{\varphi} - \dot{Z}_{ufl}\right] \\ & + dK_{sfr}\left[Z_s - a\theta - d\varphi - Z_{ufr}\right] + dC_{sfr}\left[\dot{Z}_s - a\dot{\theta} - d\dot{\varphi} - \dot{Z}_{ufr}\right] \end{aligned} \quad (6)$$

2.2 Full-Vehicle Model with Flexible Bending Modes

The aluminium frame was modelled as a simple uniform Euler-Bernoulli beam supported on passive type suspension systems in which bounce, pitch, roll and flexible bending modes vertical vibrations were considered. When n modes are considered, the vertical displacement of the aluminium frame can be written as [4].

$$z(x, y, t) = Z_s(t) + (-x)\theta_s(t) + \sum_{i=3}^{n} P_i(x)q_i(t) + (y)\varphi_s(t) + \sum_{i=3}^{n} R_i(y)u_i(t) \quad (7)$$

For $i > 2$, the shape functions are taken as

$$P_i(x) = \cosh\beta_i x + \cos\beta_i x - \frac{\cosh\lambda_i - \cos\lambda_i}{\sinh\lambda_i - \sin\lambda_i} \times (\sinh\beta_i x + \sin\beta_i x) \quad (8)$$

$$R_i(y) = \cosh\beta_i y + \cos\beta_i y - \frac{\cosh\lambda_i - \cos\lambda_i}{\sinh\lambda_i - \sin\lambda_i} \times (\sinh\beta_i y + \sin\beta_i y) \quad (9)$$

where λ_i and β_i satisfy

$$1 - \cosh\lambda_i \cos\lambda_i = 0,\ \beta_i = \frac{\lambda_i}{L} \quad (10)$$

Equations (11), (12) and (13) represent the equations of motion for the M_s, considering the flexibility of the aluminium frame.

$$M_s\ddot{Z}_s = F_1 + F_2 + F_3 \quad (11)$$

where,

$$F_1 = -K_{sfl}\left[Z_s - a\theta + \sum_{m=3}^{n} P_m(a)q_m(t) - c\varphi + \sum_{m=3}^{n} R_m(c)u_m(t) - Z_{ufl}\right] - C_{sfl}\left[\dot{Z}_s - a\dot{\theta} + \sum_{m=3}^{n} P_m(a)q_m(t) - c\dot{\varphi} + \sum_{m=3}^{n} R_m(c)u_m(t) - \dot{Z}_{ufl}\right]$$

$$F_2 = -K_{sfr}\left[Z_s - a\theta + \sum_{m=3}^{n} P_m(a)q_m(t) - d\varphi + \sum_{m=3}^{n} R_m(d)u_m(t) - Z_{ufr}\right] - C_{sfr}\left[\dot{Z}_s - a\dot{\theta} + \sum_{m=3}^{n} P_m(a)q_m(t) - d\dot{\varphi} + \sum_{m=3}^{n} R_m(d)u_m(t) - \dot{Z}_{ufr}\right]$$

$$F_3 = -K_{sr}\left[Z_s + b\theta + \sum_{m=3}^{n} P_m(b)q_m(t) - Z_{ur}\right] - C_{sr}\left[\dot{Z}_s + b\dot{\theta} + \sum_{m=3}^{n} P_m(b)q_m(t) - \dot{Z}_{ur}\right]$$

$$I_{yy}\ddot{\theta} = F_1 \times (-a) + F_2 \times (-a) + F_3 \times (b) \quad (12)$$

$$I_{xx}\ddot{\varphi} = F_1 \times (c) + F_2 \times (-d) \quad (13)$$

The modal coordinate equations for the vehicle are given by Eq. (14) and Eq. (15).

$$\ddot{q}_m(t) = -2\xi_m\omega_m\dot{q}_m(t) - \omega_m^2 q_m(t) + \frac{P_m(a)}{M_s}F_1 + \frac{P_m(a)}{M_s}F_2 + \frac{P_m(b)}{M_s}F_3 \quad (14)$$

where,

$$P_m(x) = \cosh\beta_m x + \cos\beta_m x - \frac{\cosh\lambda_m - \cos\lambda_m}{\sinh\lambda_m - \sin\lambda_m} \times (\sinh\beta_m x + \sin\beta_m x)$$

$$1 - \cosh\lambda_m \cos\lambda_m = 0;\ \beta_m = \frac{\lambda_m}{(a+b)};\ \frac{EI_{yy}\beta_m^4}{\rho} = \omega_m^2;\ \frac{\mu I_{yy}\beta_m^4}{\rho} = 2\xi_m\omega_m$$

$$\ddot{u}_m(t) = -2\psi_m\tau_m\dot{u}_m(t) - \tau_m^2 u_m(t) + \frac{R_m(c)}{M_s}F_1 + \frac{R_m(d)}{M_s}F_2 \tag{15}$$

where,

$$R_m(y) = \cosh\alpha_m y + \cos\alpha_m y - \frac{\cosh\gamma_m - \cos\gamma_m}{\sinh\gamma_m - \sin\gamma_m} \times (\sinh\alpha_m y + \sin\alpha_m y)$$

$$1 - \cosh\gamma_m \cos\gamma_m = 0;\ \alpha_m = \frac{\gamma_m}{(c+d)};\ \frac{EI_{xx}\alpha_m^4}{\rho} = \omega_m^2;\ \frac{\mu I_{xx}\alpha_m^4}{\rho} = 2\psi_m\tau_m$$

2.3 Seat and Driver's Model

Behaviour of human body seated in a car can be modelled as a 4 DOFs system [10]. It comprised of four masses which were coupled with spring and damping elements. It was assumed that the model was linear in which the driving point mechanical impedance did not vary significantly over the range of excitation type and levels in the frequency range. The masses of the lower legs and feet, and the hand and arm are not included based on the assumption of their negligible contributions to the biodynamic response of the seated body. The addition of the driver and seat models resulted in modifications to Eq. (11) that resulted in Eq. (16).

$$M_s\ddot{Z}_s = F_1 + F_2 + F_3 + K_{se}Z_{se} + C_{se}\dot{Z}_{se} - K_{se}Z_s - C_{se}\dot{Z}_s \tag{16}$$

Vehicle's seat and the lumped-parameter model of seated driver equations of motion:

$$M_{se}\ddot{Z}_{se} = K_{se}Z_s - K_{se}Z_{se} + K_{bl}Z_{bl} - K_{bl}Z_{se} + C_{se}\dot{Z}_s - C_{se}\dot{Z}_{se} + C_{bl}\dot{Z}_{bl} - C_{bl}\dot{Z}_{se} \tag{17}$$

$$M_{bl}\ddot{Z}_{bl} = K_{bl}Z_{se} - K_{bl}Z_{bl} + K_{bo}Z_{bo} - K_{bo}Z_{bl} + C_{bl}\dot{Z}_{se} - C_{bl}\dot{Z}_{bl} + C_{bo}\dot{Z}_{bo} - C_{bo}\dot{Z}_{bl} \tag{18}$$

$$M_{bo}\ddot{Z}_{bo} = K_{bo}Z_{bl} - K_{bo}Z_{bo} + C_{bo}\dot{Z}_{bl} - C_{bo}\dot{Z}_{bo} \tag{19}$$

$$M_{he}\ddot{Z}_{he} = K_{he}Z_{bl} - K_{he}Z_{he} + C_{he}\dot{Z}_{bl} - C_{he}\dot{Z}_{he} \tag{20}$$

Simulink block diagrams for the equations were developed [13] as in Fig. 3 to solve and compare the vibration responses at the four areas, M_{ufl}, M_{ur}, M_s and M_{se}. A MATLAB function allows sinusoidal excitation to be developed and used as input to the model. The mathematical expression of the excitation is given by:

$$\left.\begin{array}{lll} y = 0 & \text{for} & 0 \le u < 10 \\ y = A\sin(4 \times 2\pi \times u) & \text{for} & 10 \le u \le 30 \\ y = 0 & \text{for} & u > 30 \end{array}\right\} \tag{21}$$

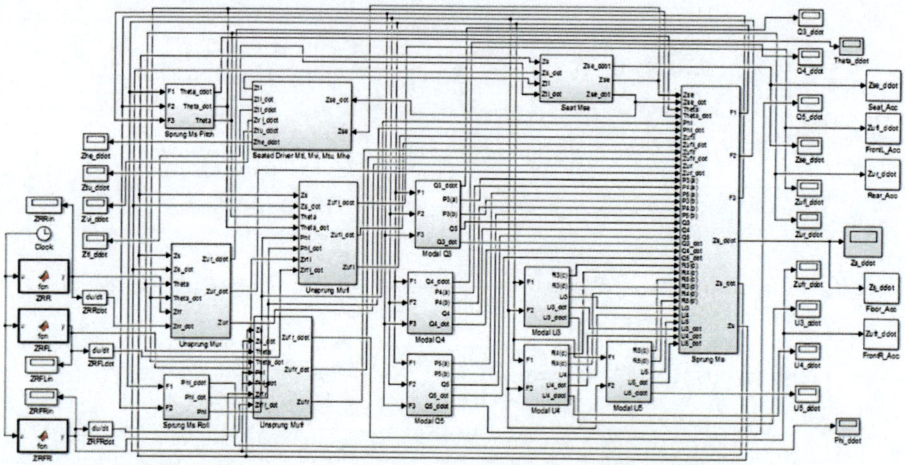

Fig. 3. Simulink block diagram of the three-wheeled full-vehicle model for rigid-body and flexible frame analysis.

3 Results and Discussion

Experimental data showed that vibrations in the vertical Z-axis have produced the highest RMS acceleration compared to other axes and therefore can be considered as the 'worst axis'. The worst axis of vibration is usually considered for ride comfort evaluation and health risk assessments [12]. The graph as in Fig. 4 shows the frequency weighted time-domain vibration signal for a period of 10 s and PSD of the signal at the FL in vertical Z-axis direction. The vibration signal oscillates within ±5 m/s^2 with several values of the time responses going beyond the acceleration range. It was recorded that the RMS and peak accelerations were 1.45 m/s^2 and 6.35 m/s^2 respectively and the dominant frequencies between 4 Hz and 50 Hz.

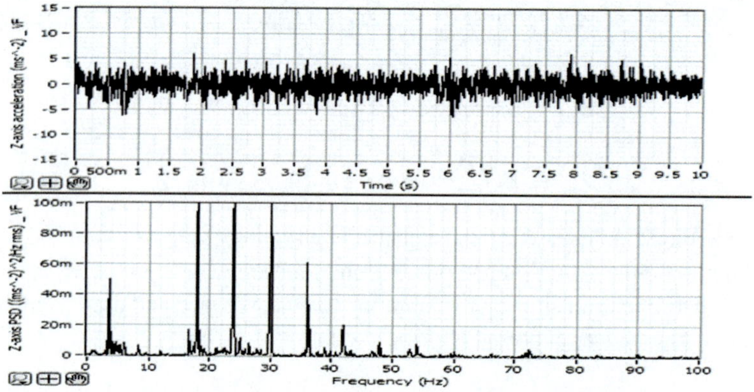

Fig. 4. Z-axis vibration acceleration and PSD on FL.

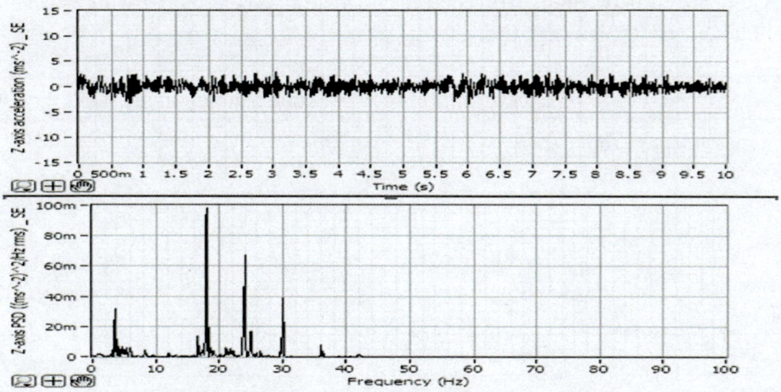

Fig. 5. Z-axis vibration acceleration and PSD on SE.

The vibration signal measured on the SE is shown in Fig. 5. The acceleration can be seen oscillated within ±3 m/s^2 for the vertical Z-axis direction. The RMS and peak accelerations were 0.98 m/s^2 and 3.34 m/s^2 respectively. The dominant frequencies were observed between 2 Hz to 30 Hz. Several peak frequencies can be seen from both graphs of the FL and SE namely at 4, 18, 24 and 30 Hz. The 4 Hz has been suggested as the resonance frequency for conventional car seats [12, 13] and therefore it was used as the simulation frequency for the input signals. The full-vehicle lumped mass model was simulated with sinusoidal input signal as given in Eq. 21 where the amplitude A was set to 0.0265 m for the front wheels (Z_{rfl} and Z_{rfr}) and A was set to 0.0328 m for the rear wheel (Z_{rr}). The results of the vibration acceleration for the M_{ufl}, M_{ur}, M_s and M_{se} are shown in Table 2.

Table 2. Comparison between experimental data and lumped-mass model simulation results.

	Experiment (m/s^2)	Lumped-mass model (m/s^2)	Diff (%)
Unsprung mass (M_{ufl})/Front wheel (FW)	2.46	2.46	0
Unsprung mass (M_{ur})/ Rear wheel (RW)	2.18	2.20	0.9
Sprung mass (M_s)/ Floor (FL)	1.45	1.70	17.2
Seat (M_{se})/(SE)	0.98	1.17	19.4

The full-vehicle model with flexible bending modes simulation was done using the same sinusoidal input signals as for the lumped-mass model. As mentioned earlier, these equations account for three rigid body modes with three different bending modes where $n = 4$, 5 and 6. The results of the vibration acceleration for the M_{ufl}, M_{ur}, M_s and M_{se} are shown in Table 3.

Table 3. Comparison between experimental data and full-vehicle model with flexible bending modes simulation results.

	Experiment (m/s^2)	Flexible bending modes (m/s^2)	Diff (%)
Unsprung mass (M_{ufl})/Front wheel (FW)	2.46	2.49	1.2
Unsprung mass (M_{ur})/ Rear wheel (RW)	2.18	2.17	0.5
Sprung mass (M_s)/ Floor (FL)	1.45	1.43	1.4
Seat (M_{se})/(SE)	0.98	0.99	1.0

An important improvement in vertical bending vibration can be seen from the analysis as the flexible model reduces the differences significantly compared to the lumped-mass model. This is crucial for the prediction of structural dynamics of the vehicle frame and how it interacts with the driver. The full-vehicle model has considered up to three vertical bending modes [14] for eigenfunction $P_m(x)$ and $R_m(y)$ as used in Eq. 14 and Eq. 15 respectively. As the damping factor is small, these eigenfunctions for the flexible frame can be assumed orthogonal for the three bending modes. As a result, the flexible frame consumes vibration energy imparted to the driver-seat dynamics system. The human body exerts damping effect on the flexural vibration of the vehicle frame, thus reducing vibration level experienced by the driver at the floor and seat [15].

4 Conclusion

Vibration model for lightweight vehicle needs to consider the effects of frame flexibility on ride performances. Flexibility is introduced by employing a simple uniform Euler-Bernoulli beam supported on suspension systems into the coupled full-vehicle with seated driver model. The full-vehicle with flexible bending modes produced better prediction for simulation of the lightweight vehicle with error of less than 1.5% especially for the vertical vibration of the floor and seat. In future, the validated model can be extended for analysis at different road conditions and improvement for comfort level of the driver can be suggested via parametric studies.

References

1. MdSah, J., Taha, Z., Ismail, K.A.: Lightweight vehicle and driver's whole-body models for vibration analysis. IOP Conf. Ser. Mater. Sci. Eng. **318**, 012069 (2018)
2. Delogu, M., Zanchi, L., Dattilo, C.A., Pierini, M.: Innovative composites and hybrid materials for electric vehicles lightweight design in a sustainability perspective. Mater. Today Commun. **13**, 192–209 (2017)
3. Doke, P., Fard, M., Jazar, R.: Vehicle concept modeling: a new technology for structures weight reduction. Procedia Eng. **49**, 287–293 (2012)

4. Zhou, J., Goodall, R., Ren, L., Zhang, H.: Influences of car body vertical flexibility on ride quality of passengers railway vehicles. Proc. IMechE Part F J. Rail Rapid Transit **223**(5), 461–471 (2009)
5. Carlbom, P.: Carbody and passengers in rail vehicle dynamics. Doctoral Dissertation, Royal Institute of Technology, Sweden (2000)
6. Carlbom, P.: Combining MBS with FEM for rail vehicle dynamics analysis. Multibody Syst. Dyn. **6**, 291–300 (2001)
7. Diana, G., Cheli, F., Andrea, C., Corradi, R., Melzi, S.: The development of numerical model for railway vehicles comfort assessment through comparison with experimental measurements. Veh. Syst. Dyn. **38**(3), 165–183 (2002)
8. Dumitriu, M.: A new passive approach to reducing the carbody vertical bending vibration of railway vehicles. Veh. Syst. Dyn. **55**(11), 1787–1806 (2017)
9. Wang, Q., Zeng, J., Wei, L., Zhang, C.: Study on coupled vibration performance of flexible carbody and under-frame suspended equipment of inter-city EMU. J. Adv. Veh. Eng. **3**(1), 40–45 (2017)
10. Bai, X.X., Xu, S.X., Cheng, W., Qian, L.J.: On 4-degree-of-freedom biodynamic models of seated occupants: lumped-parameter modeling. J. Sound Vib. **402**, 122–141 (2017)
11. Mansfield, N.J.: Human Response to Vibration. CRC Press LLC, Boca Raton (2005)
12. Carlbom, P., Berg, M.: Passengers, seats and carbody in rail vehicle dynamics. Veh. Syst. Dyn. Suppl. **37**, 290–300 (2002)
13. Jamali, M.S., Ismail, K.A., Taha, Z., Aiman, M.F.: Development of Matlab Simulink model for dynamics analysis of passive suspension system for lightweight vehicle. J. Phys. Conf. Ser. **908**, 012066 (2017)
14. Nagai, M., Sawada, Y.: Active suspension for flexible structure control of high speed ground vehicles. IFAC Proc. Vol. **20**(5), 197–202 (1987)
15. Tomioka, T., Takigami, T.: Experimental and numerical study on the effect due to passengers on flexural vibrations in railway vehicle carbodies. J. Sound Vib. **343**, 1–19 (2015)

Utilization of Fuzzy Analytical Hierarchy Process (FAHP) and TOPSIS-AHP for Selecting the Best Plant Maintenance Strategy

K. N. Kamaludin[1], L. Abdullah[1(✉)], L. Y. Sheng[1], M. N. Maslan[1], R. Zamri[1], M. Mat Ali[1], M. S. Syed Mohamed[1], M. Zainon[2], and M. S. Noorazizi[3]

[1] Faculty of Manufacturing Engineering, Universiti Teknikal Malaysia Melaka, 76100 Durian Tunggal, Melaka, Malaysia
lokman@utem.edu.my

[2] Faculty of Electrical and Electronic Engineering Technology, Universiti Teknikal Malaysia Melaka, 76100 Durian Tunggal, Melaka, Malaysia

[3] Department of Engineering, Razak Faculty of Technology and Informatics, UTM Kuala Lumpur, Level 6 Razak Tower, Jalan Sultan Yahaya Petra, 54100 Kuala Lumpur, Malaysia

Abstract. Determination of a plant maintenance strategy in a manufacturing plant is a crucial process. The appropriateness of the adopted maintenance strategy will directly and indirectly affect the plant manufacturing cost. Analytical Hierarchy Process (AHP) offers as an alternative instrument to cater the decision-making problem. However, in a higher critical condition, AHP analysis as only one method is insufficient for a concrete decision making. This paper explores and extends the methodology and results of a research study which utilizes Fuzzy-AHP and TOPSIS-AHP. The objective of this research is to select the best maintenance management method between corrective maintenance (CM), predictive maintenance (PDM) and condition-based maintenance (CBM) via methodologies. Fuzzy-AHP is one popular decision-making instrument which has the ability to capture the uncertainty in judgement. TOPSIS-AHP is another simple and powerful alternative instrument to solve multi-criteria decision making (MCDM) problem. It is concluded that condition-based maintenance (CBM) strategy is the best plant maintenance strategy using all the MCDM methods. For future study, other MCDM instrument such as Preference Ranking Organization Method (PROMETHEE) can be explored, and compared.

Keywords: Multi-criteria decision making · Fuzzy analytical hierarchy process · Maintenance strategy selection · Pairwise comparison · Plant maintenance strategy development

1 Introduction

Maintenance is one of the pillars in a manufacturing plant, Palucha [1]. In a complex scenario manufacturing plant, the management needs a well arranged system to execute the activity, so called strategies. A few selections of maintenance strategies can be

M. A. Zakaria et al. (Eds.): *Advances in Mechatronics, Manufacturing, and Mechanical Engineering*, LNME, pp. 56–67, 2021.
https://doi.org/10.1007/978-981-15-7309-5_6

adopted, corrective maintenance (CM), predictive maintenance (PDM) and condition-based maintenance (CBM). The importance of precise strategy is very important, as directly and indirectly accords to the manufacturing cost. According to Salonen [2], maintenance is a relatively a huge part of the operating budget in a manufacturing plant with heavy investment in machinery and equipment, which accounts from 15%, to up to 70% of the total manufacturing cost, hence the crucial factor.

In this research, Fuzzy Analytic Hierarchy Process (FAHP), a decision making instrument which integrates artificial intelligence to the Traditional Analytic Hierarchy Process (AHP), and Technique for Order Preference by Similarity to Ideal Solution (TOPSIS) with AHP will be explored and adopted to select the best plant maintenance strategy for the wire manufacturing industry. The criteria judgement in this research is based by a research by Parmar [3].

2 Literature Review

According to Parmar [3], AHP was utilized to decide on the best plant maintenance strategy of a wire manufacturing industry in India. The result showed that condition-based maintenance was the best maintenance strategy compared to corrective maintenance and predictive maintenance. Figure 1 shows the hierarchy model of the maintenance strategy criteria selection.

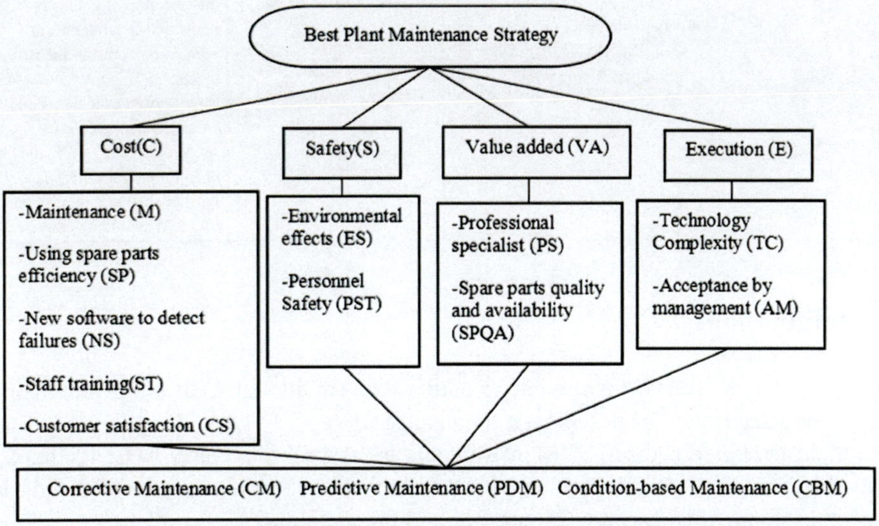

Fig. 1. Hierarchy model, Parmar [3].

Through a thorough gap analysis, TOPSIS-AHP and Fuzzy-AHP has shown to be an excellent method to support and compliment AHP analysis. A short research gap in Table 1 shows the recent studies applying MC-DM techniques in application with the summary

Table 1. Research gap analysis for MCDM approaches.

Author	Title	MC-DM type	Summary
Özcan, Ünlüsoy and Eren [4]	A Combined Goal Programming AHP approach supported with TOPSIS for Maintenance Strategies Selection in Hydroelectric Power Plant	AHP, TOPSIS and Goal Programming (GP)	This paper used GP-AHP supported by TOPSIS to select the best maintenance strategy for hydroelectric power plant. GP is a extension of Linear Programming (LP) to solve multi criteria decision making problem
Ayhan [5]	A Fuzzy AHP Approach for Supplier Selection Problem: A Case Study in a Gear motor Company	Triangular Fuzzy AHP	This paper used Triangular Fuzzy AHP to determine best supplier for a gear-motor company
Pun, Tsang, Choy, Tang and Lam [6]	Selection of Material Handling Equipment by using Analytical Hierarchy Process	Analytic Hierarchy Process (AHP)	AHP was applied to select the best material handling equipment from industrial truck, conveyor, hoist and forklift in pump industry
Hanine, Boutkhoum, Tikniouine and Agouti [7]	Application of an integrated multi–criteria decision making AHP–TOPSIS methodology for ETL software selection	AHP-TOPSIS	AHP-TOPSIS method applied to select the best ETL software. Serious limitations in imprecise judgments. Fuzzy environment should be included for further study
Parmar, Soni and Patida [3]	Selection of Plant Maintenance Strategies and Performance Enhancement of a Wire Manufacturing Industry Using AHP	Analytic Hierarchy Process (AHP)	Traditional AHP was used in the selection of plant maintenance strategies in wire manufacturing industry by considering the factors of safety, costs, executions and values added. The result showed that safety was the most important criteria. Further supplement of other MC-DM should be extended

3 Methodology

In this research, 3 MCDM instrument, in addition to Traditional AHP, are explored and integrated to determine the best plant maintenance strategy. To keep the application subject simple and flexible, the MCDM instruments are Trapezoidal Fuzzy AHP, Triangular Fuzzy AHP and TOPSIS AHP. Other MCDM is always in the area of interested and can be an extension to this research. Figure 2 shows the methodology flowchart.

As Fig. 2 shows, the methodology is straight forward. Once the criteria for the subject has been defined, the data collected will be analyzed using the four MCDM. The result obtained from each MCDM will be compares for consistency. The consistency will result in the decision making for the applicants. Upon inconsistency, the data will need a review and re-analyzed.

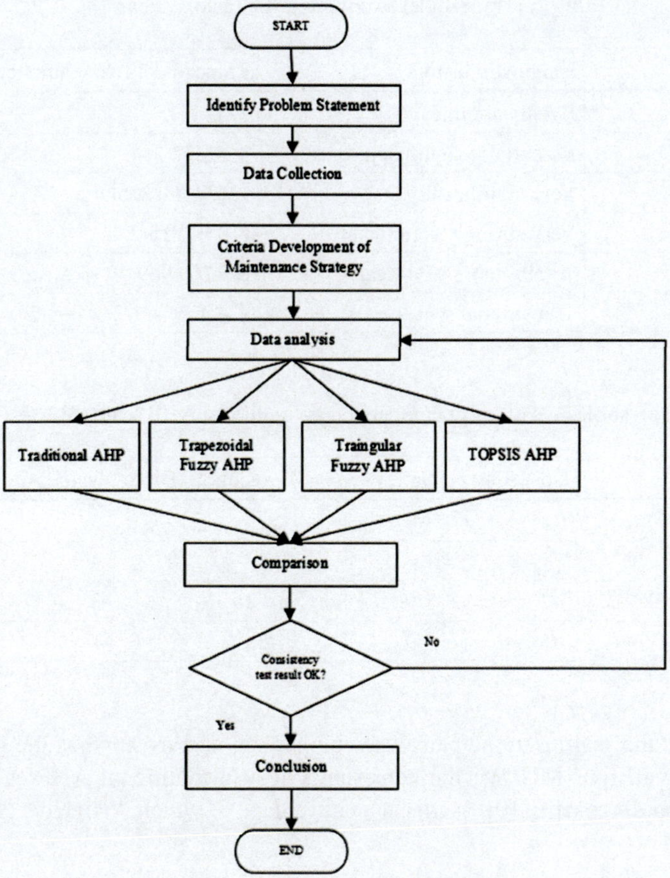

Fig. 2. Methodology flow chart.

The judgement of importance of criteria are referred to Parmar [3]. Two MCDM methods in detail will be referenced to other research (Trapezoidal and Triangular FAHP), whilst TOPSIS AHP steps will be presented in detail in this research. The results of the final selection of best plant maintenance strategy from four different methods including Traditional AHP will finally be analyzed and compared.

3.1 Trapezoidal Fuzzy AHP

Steps of Trapezoidal FAHP, are referenced to Zheng [8]. There are mainly four steps in utilizing Trapezoidal FAHP. Based on the pair-wise comparison in traditional AHP scale, as shown in Table 2, and comparison to Trapezoidal Fuzzy Linguistic factors in Table 1, pair-wise comparison in trapezoidal Fuzzy numbers is generated, as in Table 3. The methods of implementing traditional AHP can be referred to Kamaludin [9]. The accountability of the scaling process will be by the expert of the process or field, in this case, are the maintenance engineers.

Table 2. Trapezoidal fuzzy linguistic factor, Zheng [8].

AHP scale	Linguistic factors	Trapezoidal fuzzy number
1	Evenly significant	1, 1, 1, 1
3	Essentially significant	2, 5/2, 7/2, 4
5	Very significant	2, 9/2, 11/2, 6
7	Very strongly significant	6, 13/2, 15/2, 8
9	Totally more significant	8, 17/2, 9, 9
X = 2, 4, 6, 8	Transitional scales	$x - 1, x - 1/2, x + 1/2, x + 1$

Table 3. Pair-wise comparison in traditional AHP scale [3].

Criteria	Cost (C)	Safety (S)	Execution (E)	Value Added (VA)
Cost (C)	1	1/7	2	5
Safety (S)	7	1	9	9
Execution (E)	1/2	1/9	1	3
Value Added (VA)	1/5	1/9	1/3	1

To ensure the scaling in the pair-wise comparison is consistent, as the scaling will be applied to all four MCDM, the consistency test in traditional AHP calculation is sufficient according Ariff [10], as this is an advantage of which AHP have compared to other MCDM.

Table 4. Pair-wise comparison in trapezoidal fuzzy numbers.

Criteria	Cost (C)				Safety (S)				Execution (E)				Value Added (VA)			
Cost (C)	1	1	1	1	1/8	2/15	2/13	1/6	1	3/2	5/2	3	4	9/2	11/2	6
Safety (S)	6	13/2	15/2	8	1	1	1	1	8	17/2	9	9	8	17/2	9	9
Execution (E)	1/3	2/5	2/3	1	1/9	1/9	2/17	1/8	1	1	1	1	2	5/2	7/2	4
Value Added (VA)	1/6	2/11	2/9	1/4	1/9	1/9	2/17	1/8	1/4	2/7	2/5	1/2	1	1	1	1

Following the detail steps as in Zheng [8], the final overall priority or overall ranking are summarized in Table 5.

Table 5. Overall priority ranking by trapezoidal fuzzy AHP.

Criteria	C	S	VA	E	
	0.22516	0.28028	0.27113	0.22343	Overall priority
Alternatives					
CM	0.35279	0.34808	0.33983	0.34574	0.34638
PDM	0.29522	0.28411	0.30558	0.34306	0.30560
CBM	0.35199	0.36781	0.35459	0.31120	0.34802

3.2 Triangular Fuzzy AHP

In Triangular Fuzzy AHP, the traditional AHP scale is replaced with triangular fuzzy numbers as shown in Table 6. Details of the methodology of Triangular AHP are referenced to Ayhan [5]. Altogether there are eight main steps in completing Triangular Fuzzy AHP method.

Table 6. Triangular fuzzy numbers, Ayhan [5].

Saaty scale	Definition	Fuzzy triangular scale
1	Evenly significant	(1, 1, 1)
3	Essentially significant	(2, 3, 4)
5	Very significant	(4, 5, 6)
7	Very strongly significant	(6, 7, 8)
9	Totally more significant	(9, 9, 9)
2	Transitional scales between two values	(1, 2, 3)
4		(3, 4, 5)
6		(5, 6, 7)
8		(7, 8, 9)

From the tradition AHP pair-wise comparison matrix, as in Table 2, triangular fuzzy pair wise comparison matrix is generated as in Table 7.

Table 7. Pair-wise comparison matrix for criteria, triangular fuzzy numbers.

Criteria	Cost			Safety			Execution			Value Added		
Cost	1	1	1	1/8	1/7	1/6	1	2	3	4	5	6
Safety	6	7	8	1	1	1	9	9	9	9	9	9
Execution	1/3	1/2	1	1/9	1/9	1/9	1	1	1	2	3	4
Value Added	1/6	1/5	1/4	1/9	1/9	1/9	1/4	1/3	1/2	1	1	1

Following the detail steps as in Ayhan [5], the final overall rankings are summarized as in Table 8.

Table 8. Weights of alternatives relative to criteria and overall ranking matrix.

Criteria	C	S	VA	E	
	0.15787	0.70268	0.04339	0.09606	Overall priority
Alternatives					
CM	0.38512	0.06648	0.09223	0.35085	0.14522
PDM	0.24509	0.15774	0.22368	0.45323	0.20278
CBM	0.36979	0.77578	0.68409	0.19591	0.65201

3.3 TOPSIS-AHP

TOPSIS-AHP is the last instrument to calculate the best plant maintenance strategy. The steps in TOPSIS-AHP method will be demonstrated in the following sub-topics. Technique for Order Preference by Similarity to Ideal Solution (TOPSIS) was developed by Yoon [11] and the steps below are referred to the paper and Behzadian [12].

Step 1: Global Weight for Sub-Criteria

$$W_j = (Sub - criteria) \times (corresponding - criteria) \quad (1)$$

Table 9. Global weight of sub-criteria.

Criteria	Level 1	Sub-criteria	Level 2	Global weight
Cost	0.164833	Maintenance (M)	0.164849	0.027180
		Spare Part (SP)	0.075650	0.012473
		New software (NS)	0.034143	0.005629
		Staff Training (ST)	0.418529	0.069008
		Customer (C)	0.306829	0.050591
Value add	0.046742	Professional Specialist (PS)	0.1	0.004674
		Parts Quality & Availability (SPQA)	0.9	0.042067
Execution	0.096654	Tech Complex(TC)	0.5	0.048326
		Accept by management (AM)	0.5	0.048326
Safety	0.691721	Environment (E)	0.5	0.345860
		Personal Safety (PST)	0.5	0.345860
		Total	4	1

Step 2: Decision Matrix of Alternatives Relative to Sub-Criteria

Table 10. Decision matrix of alternatives relative to sub-criteria.

Alternatives	M	SP	SN	ST	CS	PS	SPQA	TC	AM	EE	PST
CM	1	7	9	5	1	1	1	5	1	1	1
PDM	5	4	3	3	3	3	3	3	7	5	3
CBM	7	1	1	1	7	8	5	1	5	9	9
	0.02718	0.01247	0.00563	0.06901	0.05059	0.00467	0.04207	0.04833	0.04833	0.34586	0.34586

Step 3: Normalization of Decision Matrix

$$\text{Normalized Matrix } R_{ij} = \frac{X_{ij}}{\sqrt{\sum_{i=1}^{m} X_{ij}^2}} \tag{2}$$

Table 11. Normalized decision matrix.

Alternatives	M	SP	SN	ST	CS	PS	SPQA	TC	AM	EE	PST
CM	0.11625	1.69775	2.84605	1.58114	0.13131	0.11704	0.17150	1.58114	0.11625	0.09713	0.10541
PDM	0.70711	0.56569	0.33129	0.58835	0.42426	0.37210	0.58835	0.58835	1.37281	0.55216	0.33129
CBM	1.37281	0.12403	0.10541	0.17150	2.21359	2.52982	1.58114	0.17150	0.70711	1.76505	2.84605
W_j	0.02718	0.01247	0.00563	0.06901	0.05059	0.00467	0.04207	0.04833	0.04833	0.34586	0.34586

Step 4: Weighted Decision Matrix

$$V = V_{ij} = R_{ij} \times W_j \tag{3}$$

Table 12. Weighted decision matrix.

Alternatives	M	SP	SN	ST	CS	PS	SPQA	TC	AM	EE	PST
CM	0.00316	0.02118	0.01602	0.10911	0.00664	0.00055	0.00721	0.07641	0.00562	0.03359	0.03646
PDM	0.01922	0.00706	0.00187	0.04060	0.02146	0.00174	0.02475	0.02843	0.06634	0.19097	0.11458
CBM	0.03731	0.00155	0.00059	0.01183	0.11199	0.01182	0.06651	0.00829	0.03417	0.61046	0.98434

Step 5: Identification of the Positive and Negative Ideal Solution

The Positive Ideal Solution (A+) is the maximum weight relatives to sub-criteria. The Negative Ideal Solution (A−) is the minimum weight relatives to sub-criteria. The steps are applied to all sub-criteria.

Table 13. Positive and negative ideal solutions.

Sub-criteria	CM	PDM	CBM	A+	A−
M	0.00316	0.01922	0.03731	0.03731	0.00316
SP	0.02118	0.00706	0.00155	0.02118	0.00155
⋮	⋮	⋮	⋮	⋮	⋮
PST	0.03646	0.11458	0.98434	0.98434	0.03646

Step 6: Calculating the Separation Distance of Each Alternative from A+ and A−
The separation distance of alternatives from Positive Ideal Solution (A+) can be calculated by using the Eq. 4 and Eq. 5 for Negative Ideal Solution (A−).

$$S^{+} = \sqrt{\sum\nolimits_{j=1}^{n} (A_{ij}^{+} - A_{ij})^2} \tag{4}$$

$$S^{-} = \sqrt{\sum\nolimits_{j=1}^{n} (A_{ij}^{n} - A_{ij})^2} \tag{5}$$

Step 7: Selective Closeness to Ideal Solution

Table 14. Separation distances.

Separation distance			
	CM	PDM	CBM
S+	1.11841	0.97479	0.12555
S−	0.12135	0.19134	1.11713

$$C_i = \frac{S_i^{-}}{S_i^{+} + S_i^{-}} \tag{6}$$

Table 15. Selective closeness to ideal solution.

Selective closeness to ideal solution			
	CM	PDM	CBM
	0.09789	0.16408	0.89897
Rank	3	2	1

4 Discussion

Comparing each result from all the four methods shows that the best plant maintenance strategy for the wire manufacturing industry are as shown in Table 16, which is CBM.

Table 16. Overall rankings.

Method	Overall ranking
Traditional AHP	CBM > PDM > CM
Trapezoidal Fuzzy AHP	CBM > CM > PDM
Triangular Fuzzy AHP	CBM > PDM > CM
TOPSIS-AHP	CBM > PDM > CM

Overall ranking by Trapezoidal Fuzzy AHP method has different order of ranking for second and third as compared to other 3 methods. The difference in overall ranking is due to the scales in Trapezoidal Fuzzy numbers which overlaps with other classes, as also explained by Zaki [13]. This is the ability of the Fuzzy function to overcome uncertainty between two sets.

The consistency of the four methods shows that the methodology is sufficient to select CBM as the ultimate priority for best plant maintenance. Though the rankings for the highest priority are the same, the final weight (or percentage) differs (AHP = 0.64446, Trapezoidal FAHP = 0.34802, Triangular Fuzzy AHP = 0.65201 and AHP-TOPSIS = 0.89897). This is due to the different methodology of calculation for each MC-DM.

CBM as a strategy, should be the best practice of maintenance. The justification for CBM (which falls under preventive or pro-active maintenance) above other strategy is the ability to optimize maintenance cost, improvement in the system reliability, reduction in a number of maintenance operation therefore, reduce human errors, as explained by Niu [14]. Even though the initial cost of setting the system up is higher (due to the requirement of normal or intelligent monitoring system), the benefits mentioned overtakes this hurdle.

PDM, which also falls under preventive maintenance category is the secondary choice. Due to the lack of monitoring system, the maintenance method can sometimes be less optimized. In some organization, PDM is sufficient, but with enough resources, CBM should be implemented. CM fall under re-active maintenance where maintenance activities are performed upon equipment failure, which has the least preferences in as suggested through the methodologies. This method contains a lot inefficiency. A serious damage or failure to the facilities, personnel and equipment is likely to occur through this strategy as explained by Gandhare [15] and application on system modelling by [16].

5 Conclusion

This research demonstrates four alternative instruments to solve multi-criteria decision making (MCDM) problems. AHP can easily be explored and applied by engineers or users, to determine the best decision, with reference to a hierarchy model. Condition based maintenance (CBM) is selected as the best plant maintenance strategy for wire manufacturing industry which had been proven by the four methods. Utilization of MCDM also reduces the chances of errors in decision making.

Acknowledgement. The authors would like to acknowledge the financial support by Universiti Teknikal Malaysia Melaka (UTeM) under the special budget by the Centre for Research and Innovation Management (CRIM) for the publication in MUCET 2019.

References

1. Pałucha, K.: World class manufacturing model in production management. Arch. Mater. Sci. Eng. **58**(2), 227–234 (2012)
2. Salonen, A., Deleryd, M.: Cost of poor maintenance: a concept for maintenance performance improvement. J. Qual. Maint. Eng. **17**(1), 63–73 (2011)
3. Parmar, M., Soni, M., Patidar, S.: Selection of plant maintenance strategy and performance enhancement of a wire manufacturing industry using AHP. Int. J. Res. Aeronaut. Mech. Eng. **4**(1), 81–85 (2016)
4. Özcan, E.C., Ünlüsoy, S., Eren, T.: A combined goal programming – AHP approach supported with TOPSIS for maintenance strategy selection in hydroelectric power plants. Renew. Sustain. Energy Rev. **78**(2016), 1410–1423 (2017)
5. Ayhan, M.B.: A fuzzy AHP approach for supplier selection problem: a case study in a gearmotor company. Int. J. Manag. Value Supply Chain. **4**(3), 11–23 (2013)
6. Pun, K.P., Tsang, Y.P., Choy, K.L., Tang, V., Lam, H.Y.: A fuzzy-AHP-based decision support system for maintenance strategy selection in facility management. In: 2017 Portland International Conference on Management of Engineering and Technology (PICMET), pp. 1–7 (2017)
7. Hanine, M., Boutkhoum, O., Tikniouine, A., Agouti, T.: Application of an integrated multi-criteria decision making AHP-TOPSIS methodology for ETL software selection. SpringerPlus **5**(1), 1–17 (2016)
8. Zheng, G., Zhu, N., Tian, Z., Chen, Y., Sun, B.: Application of a trapezoidal fuzzy AHP method for work safety evaluation and early warning rating of hot and humid environments. Saf. Sci. **50**(2), 228–239 (2012)
9. Kamaludin, K.N., Abdullah, L., Maslan, M.N., Zamri, R., Ali, M.M., Syed Mohamad, MS.: Utilization of Analytical Hierarchy Process (AHP) for selecting the best design concept of conveyor system. In: Lecture Notes in Mechanical Engineering, pp. 265–279 (2019)
10. Ariff, H., Salit. M.S., Ismail, N. Nukman, Y.: Use of analytical hierarchy process (AHP) for selecting the best design concept. Jurnal Teknologi **49**(A), 9–11 (2008)
11. Yoon, K., Hwang, C.: TOPSIS (Technique for order preference by similarity to ideal solution)– a multiple attribute decision making, W: Multiple attribute decision making–methods and applications, a State-of-the-Art Survey. Springer, Berlin (1981)
12. Behzadian, M., Khanmohammadi Otaghsara, S., Yazdani, M., Ignatius, J.: A state-of the-art survey of TOPSIS applications. Expert Syst. Appl. **39**(17), 13051–13069 (2012)

13. Zaki, A., Noor, M., Jafar, F.A., Zainudin, S.F.: Fusion of fuzzy AHP in selecting material for drinking water bottle based on customer needs. ARPN J. Eng. Appl. Sci. **12**(14), 4248–4249 (2017)
14. Niu, G., Yang, B.S., Pecht, M.: Development of an optimized condition-based maintenance system by data fusion and reliability-centered maintenance. Reliab. Eng. Syst. Saf. **95**(7), 786–796 (2010)
15. Gandhare, B.S., Akarte, M.: Maintenance strategy selection. In: Ninth AIMS International Conference on Management, pp. 1330–1336 (2012)
16. Fong, T.T., Jamaludin, Z., Abdullah, L.: System identification and modelling of rotary inverted pendulum. Int. J. Adv. Eng. Technol. **6**(66), 2342 (2014)

Effect of Tool Engagement on Cutting Force for Different Step Over in Milling AISI P20 Tool Steel

R. Hamidon(✉), N. I. Mohamed, R. Saravanan, H. Azmi, Z. A. Zailani, and M. Fathullah

School of Manufacturing, Universiti Malaysia Perlis, 02600 Arau, Perlis, Malaysia
roshaliza@unimap.edu.my

Abstract. In mold production, end milling with tool path strategies is required for the process known as pocket operation. Different step overs involve depending on the type of tool path strategy used. Thus, different engagement will occur and leads to fluctuation of cutting force due to different step over during the process. However, most of study before focused on the effect of cutting speed, feed rate and depth of cut only in machining AISI P20. Thus, in this study, step over will be considered as one of the factor to improve machining force. The objective of this study are to evaluate the effect of cutting parameters and step over on cutting force and to study the behavior of cutting force for different tool engagement. A series of milling operation was carried out by varying cutting speed and feed rate. However, the depth of cut was set to 0.25 mm for each run. Step over with 100%, 75% and 50% were selected in this study. L27 Taguchi and S/N ratio were used to determine the significant factors that influence the result. Within the range of cutting parameters selected, feed rate were found to be the most significant parameters that influence cutting force. The highest cutting force found for 100% step over compared to 75% and 50% step over. According to the result, cutting force increased as the step over increased. In can be concluded that, step over is one of the important cutting parameter that affected machining output.

Keywords: Step over · Tool engagement · Cutting force · AISI P20

1 Introduction

Cutting force is very important component to control surface quality and dimensional accuracy of machined part. It could affect cutting temperature, tool deflection, tool wear, vibration and so on [1, 2]. Most of studies focus on the effect of cutting parameters such as cutting speed, feed rate and depth of cut on cutting force. However, machining a pocket to form a mold is more complex process compared to normal cutting process. The other cutting parameters such as step over also need to be considered. In machining a pocket, cutting tool needs to travel in straight and corner cutting depending on the shape of the pocket. Fluctuation of cutting force is the main problem in machining a pocket. Due to the problem, it somehow affected the cutting temperature and surface

M. A. Zakaria et al. (Eds.): *Advances in Mechatronics, Manufacturing, and Mechanical Engineering*, LNME, pp. 68–76, 2021.
https://doi.org/10.1007/978-981-15-7309-5_7

quality of the mold. The fluctuation of force occurs due to different step over involves in the tool path strategy in machining a pocket. According to Topal [3], step over ratio may affect the final surface roughness by determining the fact as to how many times the tool is being passed and scraped again on a finished surface.

Previous studies focus on modification of tool path strategy to achieve a constant tool engagement and indirectly maintaining the cutting force [4–6]. Most of the method is suitable for rough cutting only. According to Hamamci [7], step over have significant effect on surface roughness. From the result, the significant effect of step over on surface roughness almost equal as cutting speed.

Moreover, the selection of step over and cutting tool diameter will determine the engagement between the cutting tool and workpiece. The possible engagement angle between cutting tool and work material is 180° where the value of step over is equal to diameter of cutting tool. Adesta [8] stated, the highest cutting force occurs when full engagement occurs between cutting tool and workpieces. Besides that, during the engagement, tool will experience up milling and down milling which also could affect the cutting force. They also obtained higher cutting force obtained for up milling.

Thus in this study, the most significant cutting parameter that affect cutting force will be investigated by considering step over as one of the cutting parameter. In addition the behavior of cutting force for different step over and tool engagement also will be studied. The behavior of cutting force is one of the reasons that determine the surface quality of machined surface.

2 Methodology

2.1 Selection of Material

Normally, hard to cut material is used in mould and die industry. This is due to the properties of material. This type of material owned greater strength and high resistant to deformation. It can be classified into low-alloy steels, stainless steel, and lightweight metal and high temperature alloy. The workpiece used in this study is P20 (100 mm × 90 mm × 50 mm) tool steel. P20 tool steel is generally used in a heat-treated state. P20 tool steels are also used for special applications like injection molding because the resistance to abrasion is an important criterion for a mold that will be used to produce hundreds of thousands of parts. The presence of chromium and nickel enhances the toughness and hardness of P20 steels. To perform machining operation, CoroMill 490 cutting tool with coated carbide insert was used. The tool diameter is 20 mm with two indexable insert positions.

2.2 Experimental Procedure

The selected cutting parameters used for the experiment listed in Table 1. Three factors with three level of cutting speed, federate and step over was chosen to run the experiment. In this study the depth of cut kept constant which the value is 0.25 mm. According to Hamidon et al. [10], depth of cut has least influence on cutting force in milling AISI H13 steel. The illustration of step over (50%, 75% and 100%) used in this study shown in Fig. 1.

Table 1. Experimental parameters and their levels

Cutting parameter	Level		
	1	2	3
Cutting speed (m/min)	50	75	100
Feed rate (mm/min)	100	200	300
Step over (%)	10	75	100

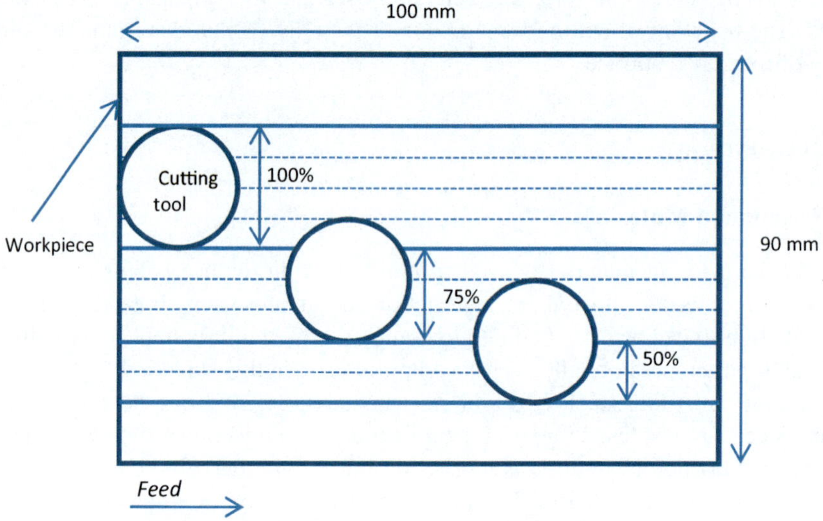

Fig. 1. Illustration of step over during milling process

Experiment was conducted by using CNC milling machine, AKIRA SEIKI Performa SR3 XP. To measure the cutting force, Kistler 9265B dynamometer was attached to the workpiece as shown in Fig. 2. The dynamometer was supported by charge amplifier and signal conditioning to measure cutting forces in X-, Y- and Z-direction. The force was applied to dynamometer when the cutting tool starts to cut the material. Electric charge was generated through pressure detected from highly sensitive quartzes plate installed in the dynamometer. Then, it was transmitted to 3-channel charge amplifier and converted to voltage. The analogue signal converted to digital signal at sampling rate of 1000 Hz.

The digital signal can be read and analyzed through Dynoware Software. The schematic diagram of experimental setup was illustrated in Fig. 3.

L27 full factorial design was employed to run the experiment. For each run, fresh tool was used. For S/N ratio, smaller the better characteristic was chosen. The analysis of the result to determine the most significant factor was performed in Minitab 16. Then, the behavior of cutting force was analyzed through excel. The resultant force (Fr) was obtained by using equation the following equation:

$$F_r = \sqrt{F_x^2 + F_y^y + F_z^2} \tag{1}$$

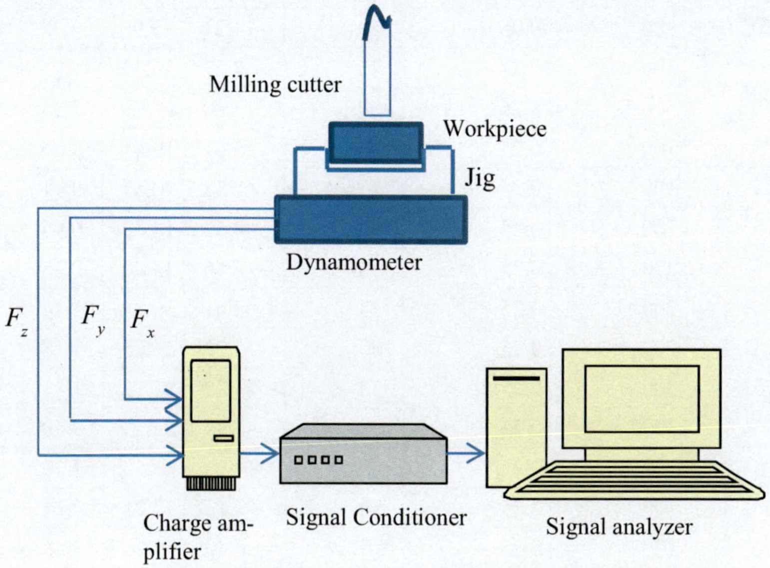

Fig. 2. Schematic diagram of experimental setup

3 Result and Discussion

Taguchi's method is widely used in various engineering fields for experimental planning design. Taguchi's method is an efficient and incomplex technique for optimization especially to improve quality, cost and performance design. Table 2 shows the average value of cutting force acting on X (Fx), Y (Fy) and Z (Fz) directions and the resultant force (Fr), for all the 27 runs of experiment. The lowest cutting force recorded was 36.15N which obtained from the cutting parameters combination of cutting speed of 50 m/min, feed rate of 100 mm/min and step over ratio of 50%. While, the highest cutting force obtained was 102.57 N for 75 m/min, 300 mm/min and 75% step over.

The effect of cutting parameters on machining output was examined through Signal-to-Noise S/N response. The S/N ratio used to transform the experimental data on the

Table 2. Response for L27 orthogonal array Taguchi method

No	Cutting speed, V (m/min)	Feed rate, f (mm/min)	Depth of cut (mm)	Step over (%)	Average cutting force (N)			
					Fx	Fy	Fz	Fr
1.	50	100	0.25	50	−9.28	26.90	22.29	36.15
2.	50	100	0.25	75	2.64	39.62	31.49	50.68
3.	50	100	0.25	100	21.91	44.96	43.63	66.37
4.	50	200	0.25	50	−12.73	34.86	31.39	48.61
5.	50	200	0.25	75	−1.59	49.43	41.54	64.59
6.	50	200	0.25	100	13.42	52.46	50.20	73.84
7.	50	300	0.25	50	−14.31	35.98	33.78	51.39
8.	50	300	0.25	75	−2.23	56.49	44.95	72.23
9.	50	300	0.25	100	17.80	69.99	62.25	95.34
10.	75	100	0.25	50	−5.61	19.44	19.06	27.79
11.	75	100	0.25	75	2.88	30.51	29.11	42.27
12.	75	100	0.25	100	−0.52	18.73	20.99	28.14
13.	75	200	0.25	50	−15.16	33.90	30.38	47.98
14.	75	200	0.25	75	−2.61	36.34	35.74	51.0
15.	75	200	0.25	100	5.22	31.75	37.18	49.17
16.	75	300	0.25	50	−21.79	82.75	51.41	99.83
17.	75	300	0.25	75	−1.99	83.53	59.49	102.57
18.	75	300	0.25	100	1.97	56.52	52.29	77.02
19.	100	100	0.25	50	−7.12	27.21	27.18	39.11
20.	100	100	0.25	75	0.40	30.24	33.88	45.41
21.	100	100	0.25	100	4.59	21.93	27.87	35.76
22.	100	200	0.25	50	−9.25	25.11	30.54	40.60
23.	100	200	0.25	75	2.60	39.94	48.07	62.55
24.	100	200	0.25	100	30.40	53.73	66.64	90.84
25.	100	300	0.25	50	−13.97	31.52	33.91	48.35
26.	100	300	0.25	75	−1.31	43.49	48.15	64.89
27.	100	300	0.25	100	18.44	50.35	64.71	84.03

machining performance before the optimization which it can also categorized as special kind of data summary. S/N ratio is the most common method to analyze data regarding the effect of the cutting parameters. Hamidon et al. [10] mentioned that the significant factor can be obtained by calculating the average S/N ratio without further Analysis of Variance (ANOVA). For 'the lower the better' following expression in Eq. (2) was used to calculate the S/N ratio:

$$\frac{S}{N} = -10\log\frac{1}{n}\left(\sum y^2\right) \tag{2}$$

The influence of cutting parameters on cutting force displayed in response table and response graph as shown in Table 3 and Fig. 3 respectively. From the graph, feed rate has the highest influence on cutting force and followed by step over. The least influence factor is cutting speed. The best combinations of cutting parameters to obtain smaller cutting force are A2B1C1. A study on influence of tool path strategies on cutting force by Shajari [7] also claimed that the most significant factor that influences the cutting force is step over and followed by feed rate and cutting speed. However, ball end milling was used in the study. Figure 4 shows the cutting force increased as the percentage of step over increased. Cutting force value gets higher as the value of feed rate and step over increases due to the chip load increased. For the tool engagement, for the three cases of step over the highest of cutting force occurred when the tool reach 100° tool engagement. At this stage, up milling occurred and the chip thickness is maximum. As mentioned before, higher cutting force obtained in up milling compared to down milling.

Figure 5 shows the engagement angle of the cutting tool and workpiece. In this study, cutting tool with two indexable insert were used. For 50% step over, the cutting tool engaged from 90° to 180° with the workpiece. Meanwhile, for 75% step over, the engagement angle will engage from 60° to 180°. For full engagement (100% step over), the tool engaged from 0° to 180°. For 50% step over, the tool experience down milling only. In contrast, for 75% and 100% step over, the tool experience both up milling and down milling. From the result in Fig. 6, the cutting force increased when the first insert start to cut the workpiece. The highest cutting force obtained for 100% step over compared to 50% and 75%. The highest cutting force obtained when the cutting tools reached 90° engagement with the workpiece. The maximum chip thickness occurs at 90°. The cutting force reduces when the tool engagement approach 180°. At this stage, the chip thickness reduces and approaches to zero. The cutting force rises again when the second insert start to cut. Theoretically, the cutting forces in up milling are higher compared to down milling. In up-milling, the cutting tool rotation is in the opposite direction of the feed direction which results in a higher cutting force. However, in down-milling, the cutting rotation is in the same direction as the feed direction which results in a lower magnitude of the cutting forces.

Table 3. Response table for cutting force

Level	Cutting speed [A]	Feed rate [B]	Step over [C]
1	−35.55	−32.02	−33.26
2	−34.44	−35.13	−35.52
3	−34.64	−37.48	−35.85
Delta	1.12	5.47	2.60
Rank	3	1	2

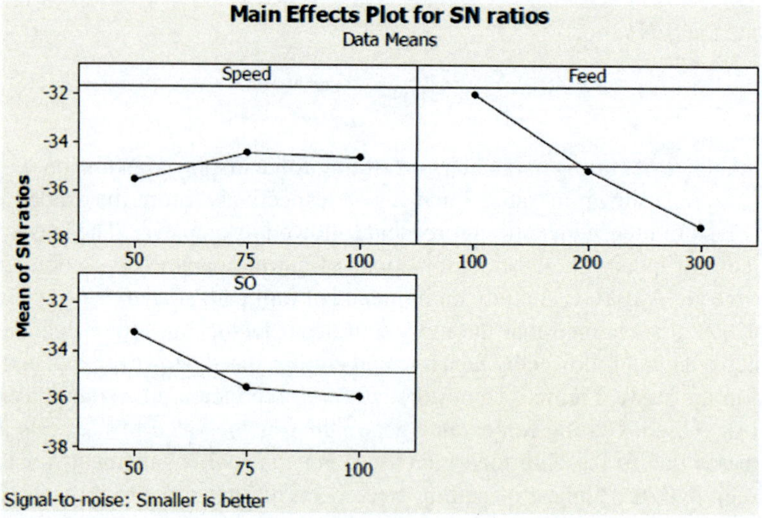

Fig. 3. Response graph for cutting force

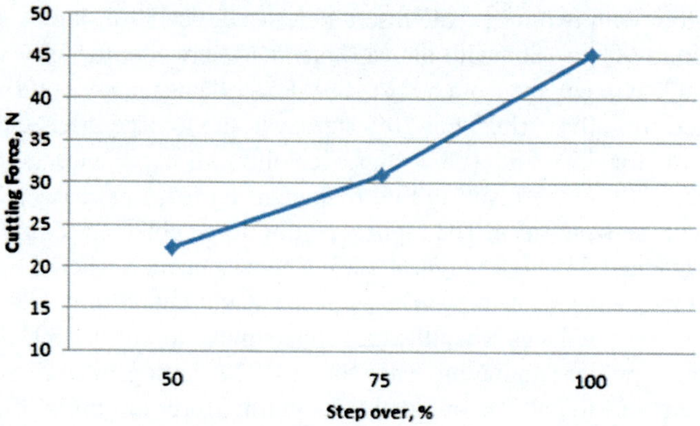

Fig. 4. Cutting force for different step over

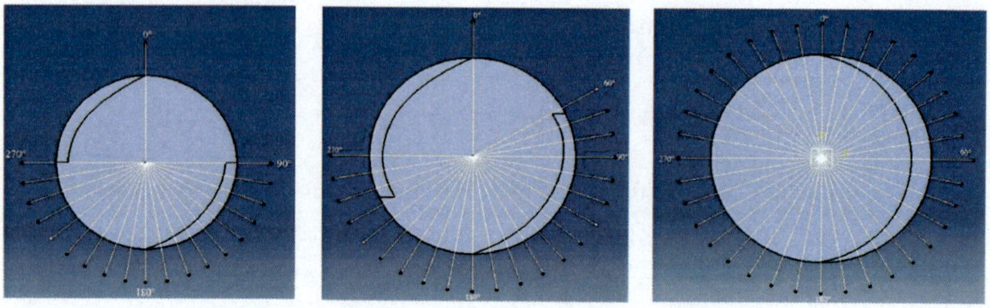

Fig. 5. Tool engagement angle for (a) 50% step over (b) 75% step over (c) 100% step over

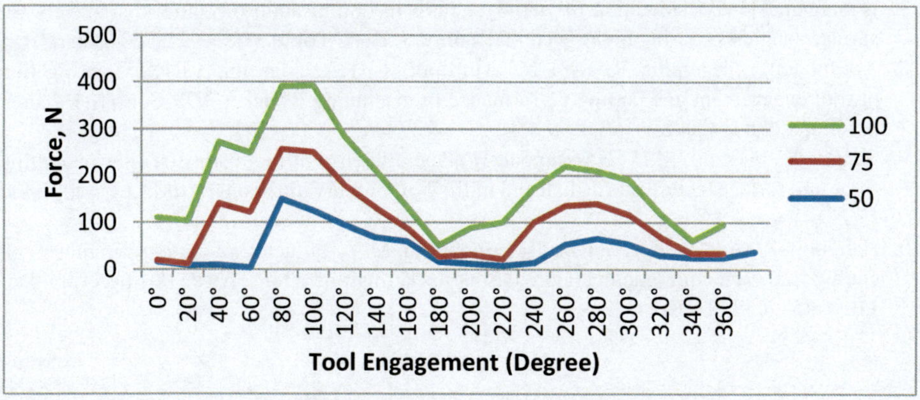

Fig. 6. Behavior of cutting for different tool engagement

4 Conclusion

Step over is important in designing tool path strategy to form a mold. From the result obtained, feedrate was the most significant factor followed by step over and cutting speed. The steps over shows promising result that affect the cutting force. Furthermore, it shows that by increasing the step over, cutting force also increased. Highest cutting force obtained for 100% step over when the tool makes full contact with the workpiece. The tool engagement is much related to the step over that affected cutting force. By analyzing the cutting force behavior towards tool engagement, the highest cutting force obtained when the tool 90° in contact with the workpiece. This is due to the maximum chip thickness at this stage. Higher cutting speed should be considered in future study so that it can be applied in related industry. In addition, detail analysis on chip thickness should be done as it contributes to more accuracy of the result.

References

1. Ding, T., Zhang, S., Wang, Y., Zhu, X.: Empirical models and optimal cutting parameters for cutting forces and surface roughness in hard milling of AISI H13 steel. Int. J. Adv. Manuf. Technol. **51**(1–4), 45–55 (2010)
2. Kaneko, J., Horio, K.: Fast cutter workpiece engagement estimation method for prediction of instantaneous cutting force in continuous multi-axis controlled machining. Procedia CIRP **4**, 151–156 (2012)
3. Topal, E.S.: The role of stepover ratio in prediction of surface roughness in flat end milling. Int. J. Mech. Sci. **51**, 782–789 (2009)
4. Kim, H.-C.: Optimum tool path generation for 2.5D direction-parallel milling with incomplete mesh model. J. Mech. Sci. Technol. **24**(5), 1019–1027 (2010)
5. Stori, J., Wright, P.: Constant engagement tool path generation for convex geometries. J. Manuf. Syst. **19**(3), 172–184 (2000)
6. Uddin, M.S., Ibaraki, S., Matsubara, A., Nishida, S., Kakino, Y.: Constant engagement tool path generation to enhance machining accuracy in end milling. JSME Int J., Ser. C **49**(1), 43–49 (2006)

7. Hamamci, M.: Consideration of step-over ratio in optimisation of cutting parameters for surface roughness during high speed machining. J. Balk. Tribol. Assoc. **21**, 753–766 (2015)
8. Adesta, E.Y.T., Hamidon, R., Riza, M., Alrashidi, R.F.F.A., Alazemi, A.F.F.S.: Investigation of tool engagement and cutting performance in machining a pocket. IOP Conf. Ser. Mater. Sci. Eng. **290**(1) (2018)
9. Shajari, S., Sadeghi, M.H., Hassanpour, H.: The influence of tool path strategies on cutting force and surface texture during ball end milling of low curvature convex surfaces. Sci. World J. (2014)
10. Hamidon, R., Adesta, E.Y.T., Riza, M., Suprianto, M.Y.: Influence of cutting parameters on cutting force and cutting temperature during pocketing operations. ARPN J. Eng. Appl. Sci. **11**(1), 453–459 (2016)

Progressive Tool Wear in Machining of Aluminum Alloy: The Influence of Solid Lubricant Nanoparticles

Z. A. Zailani(✉), N. S. Jaaffar, R. Hamidon, A. Harun, and H. Jaafar

School of Manufacturing Engineering, Universiti Malaysia Perlis, 02600 Arau, Perlis, Malaysia
zailani@unimap.edu.my

Abstract. The abstract should summarize the contents of the paper in short terms, i.e. 150–250 words. Aluminum alloy 7075 are used in a variety of applications particularly in automotive and aerospace industry owing to their features of lightweight, high-strength and corrosion resistant properties. However, build up edge (BUE) and material adhesion either on cutting tool or workpiece render these materials difficult to machine. Their machining is associated with rapid tool wear and poor workpiece quality. Cutting fluids are currently the common solution to these problems although there are concerns on their use in terms of health footprint and environmental effects. Thus, new innovations are crucial to enhance the machinability as well as diminishing hazards through encouraging greener machining techniques. In this research, the use of solid lubricants; graphene and hexagonal boron nitride nanoparticles to augment minimum quantity lubricant were researched in macro drilling. Effects of four different machining conditions namely dry, minimum quantity lubricant, minimum quantity lubricant dispersed with graphene and hexagonal boron nitride nanoparticles were investigated on their progressive tool wear behavior. A notable finding is that the nanoparticles of solid lubricants had a significant factor in improving machinability of aluminum alloy 7075 compared to dry and minimum quantity lubricant alone. It was observed that the use of minimum quantity lubricant dispersed with hexagonal boron nitride demonstrated desirable tool life enhancement, tool wear reduction and number of holes drilled increment.

Keywords: Aluminum alloy · Solid lubricants · Nanoparticles · Minimum quantity lubricant · Tool wear

1 Introduction

Machining of aluminum alloy 7075 is difficult owing to the properties of its ductility that allow the formation of large chip tool and chip thickness high ratio that contributes to increasing the cutting force, heat generation as well as poor surface finish [1]. Moreover, excessive heat leads to build up edge (BUE) and material sticky that contribute to tool wear and decrease the tool life. Cutting fluids have been used extensively to alleviate this machinability issues through lubricating, cooling and flushing action with around 85%

M. A. Zakaria et al. (Eds.): *Advances in Mechatronics, Manufacturing, and Mechanical Engineering*, LNME, pp. 77–84, 2021.
https://doi.org/10.1007/978-981-15-7309-5_8

of these fluids either petroleum-based or mineral-based oil [2]. Unfortunately, since the fluids are not biodegradable, they could potentially cause major environmental pollution throughout their life cycle [3].

Recent improvement strategies have included applying different types of cutting fluids in the machining process. Effective unconventional methods such as dry cutting, minimum quantity lubrication (MQL), cryogenic cooling, chilled air, and the use of solid lubricants have been found to be viable substitutes of cutting fluids to optimise machining performance and minimise risks [4]. The development of solid lubricants nanoparticles is one of the alternatives to overcome these issues. The morphological of solid lubricants are positively improve the wear resistance that could be produced between contact surface in friction owing to lamellar structure of these materials [5].

When dealing with nanoparticles, there are four possible lubrication mechanisms of nanoparticles, i.e.; (a) rolling effect, (b) protective film, (c) tribo-film or mending effect, and (d) polishing effect as shown in Fig. 1. In addition, the usefulness of this lubrication is dependent on quantity, the morphology and crystal structure, and the delivery method to the tool–workpiece interface [6].

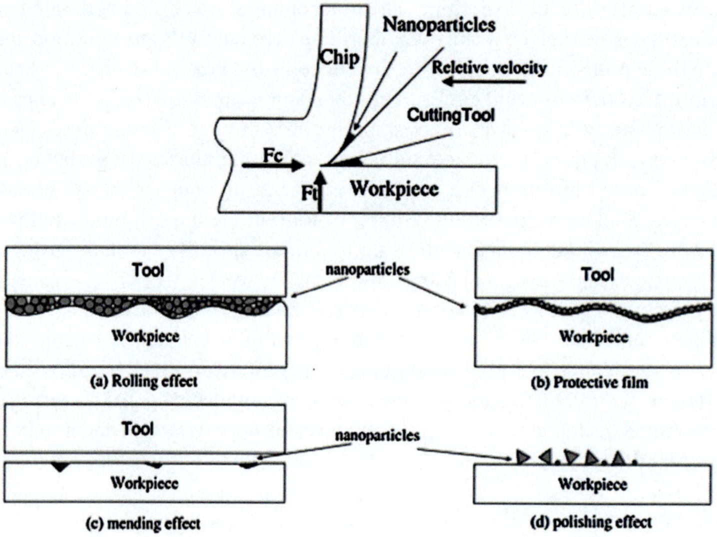

Fig. 1. The presence of nanoparticle reduces the tool and workpiece contact [6].

On the other hand, solid lubricants in machining are reported to be effective in prolonging tool life [7], and improving surface finish [8]. Their effectiveness in improving the machinability of Al 7075 however has limited scientific inquiry as documented support. This lack of inquiry and evidence-based study is therefore the motivating factor and driver for this research pursuit.

2 Methodology

In this research, a series of drilling tests were carried out using a Tongtai EZ – 5A CNC Milling Machine on a workpiece of aluminium alloy with of dimensions of 100 mm length by 100 mm width and 10 mm height. The cutting tools selected were 2-flutes uncoated High-Speed Steel (HSS) with 6.0 mm diameter tool (Fig. 2). Before cutting tests, all micro tools were imaged under optical microscope to inspect their geometry and cutting edge radius quality. The average cutting-edge radii of cutting tools were measured to be 0.16 mm.

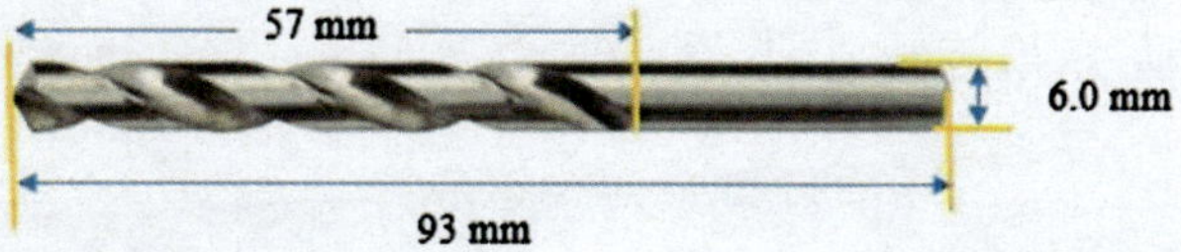

Fig. 2. 2-flutes uncoated High-Speed Steel (HSS) tool.

Four different modes of cooling/lubricant were studied for this experiment, i.e., dry, minimum quantity lubricant, minimum quantity lubricant dispersed with graphene and hexagonal boron nitride (hBN) nanoparticles as shown in Table 1. For MQL, 100% biodegradable vegetable-based oil, karanja oil cutting lubricant was selected. A constant cutting velocity of 105 m/min and feedrate of 0.15 mm/rev was used. Each test was repeated with new cutting tools. The experimental setup for drilling aluminum alloys 7075 has been shown in Fig. 3.

Table 1. Machining condition

Cutting condition	Cutting speed	Feed rate
Dry cutting	105 m/min	0.15 mm/rev
MQL	105 m/min	0.15 mm/rev
MQL + Graphene nanoparticles	105 m/min	0.15 mm/rev
MQL + hBN nanoparticles	105 m/min	0.15 mm/rev

For the nanoparticles, graphene and hexagonal boron nitride were supplied by US Research Nanomaterials, Inc. SEM images of the nanoparticles at 5 μm scale are shown in Fig. 4.

The nanofluids were prepared by mixing the nanoparticles with the non-edible vegetable oil; karanja oil. Before the addition, the nanoparticles were dried in a drying oven at 120 °C for 2 h. In this experiments, 30 mL of karanja oils was used as a based lubricant for each mixture. 0.15 g nanoparticles were added to the base oil. As a surfactant, sodium dodecyl sulfate with 0.1% weight of the nanoparticles was added to stabilize the particles and enhance the uniformity. Weighing process was undertaken using a Electronic Densimeter MD-3005. Then, the mixtures were sonicated in a Branson Digital

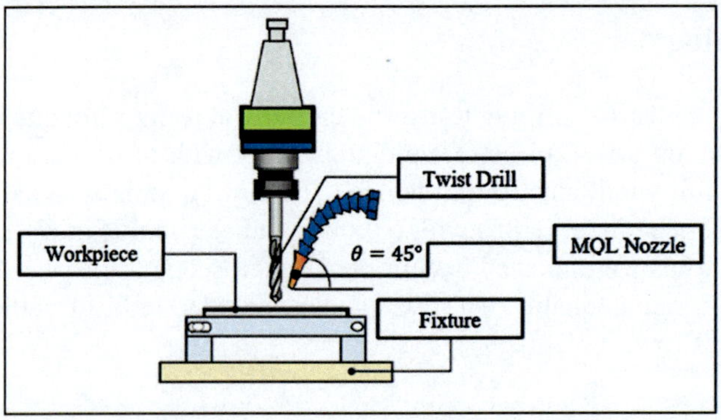

Fig. 3. Experimental setup for drilling aluminum alloys 7075.

Fig. 4. SEM nanoparticle images of graphene (left), and hexagonal boron nitride (right).

Sonifier Ultrasonic Liquid Processor for 60 min to suspend the particles homogeneously as shown in Fig. 5. Tool wear rate was periodically measured using Digital Microscopy Image Analyser which in conjunction with *i*Solution-Lite computer software. For the tool wear measurement, the procedure was repeated until the tool wear met one of these following rejection criteria: (i) average flank wear of 0.3 mm; or (ii) excessive chipping/flanking or fracture of the cutting edge. It was measured after every two holes and the patterns of wear on the high-speed steel (HSS) cutting tool were observed.

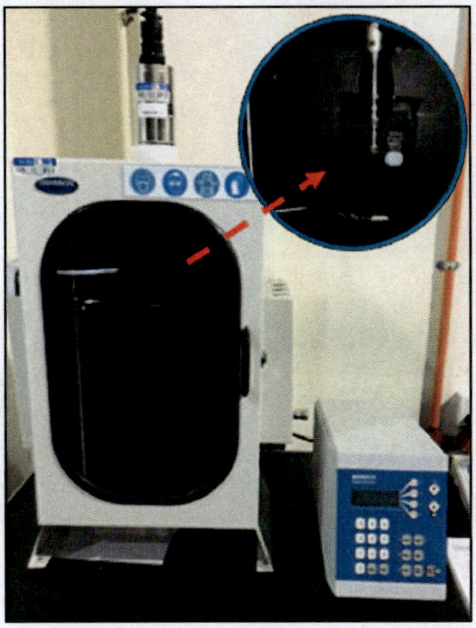

Fig. 5. Sonication process using a Branson Digital Sonifier Ultrasonic Liquid Processor.

3 Results

3.1 Tool Wear Modes

Several modes of wear were found after drilling process as shown in Fig. 6, for instances, (a) outer corner wear, (b) chisel edge wear, (c) crater wear, (d) margin wear, (e) flank wear and (f) chipping wear. The wear zone is the area where the coating had vanished and the substrate material visible. The examples pointed to flank wear as the major mode of wear found in the experiment. This is crucial since it has direct impact to the effective diameter and dimensional accuracy of the drilled holes. The effectiveness of hole diameter is influenced by the outer corner wear, where one of the factors outer corner wear occurred is caused by built up edge (BUE) or adhesion on the flank face.

3.2 Flank Wear

Figure 7 shows average flank wear and number of drilled holes and machining time under different cutting conditions. The magnitude of flank wear was determined according to ISO 8688-2 [9]. The tools were imaged from the flat bottom face to focus on each cutting edge. Lines were drawn from the cutting edge to the end of the worn land and the average of these lines used as the average flank wear.

From the result, it clearly shows that the application of minimum quantity lubricant concurrently with nanoparticles was effective in improving the wear performance and number of holes drilled. This could be attributed to the impact of nanoparticles that could form a protective coating and thus resulted in reduced friction [10]. In addition, the small size of hBN nanoparticles (80–100 nm) let them to simply penetrate the contact interface for lubricating. That is the reason why they yielded better wear resistant compared to other conditions. As expected, without CFs during machining lead to high friction

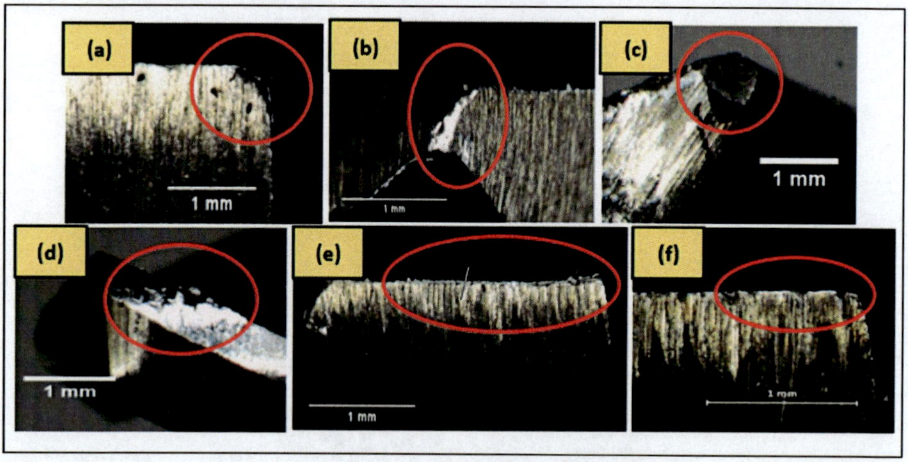

Fig. 6. Types of wear during machining process.

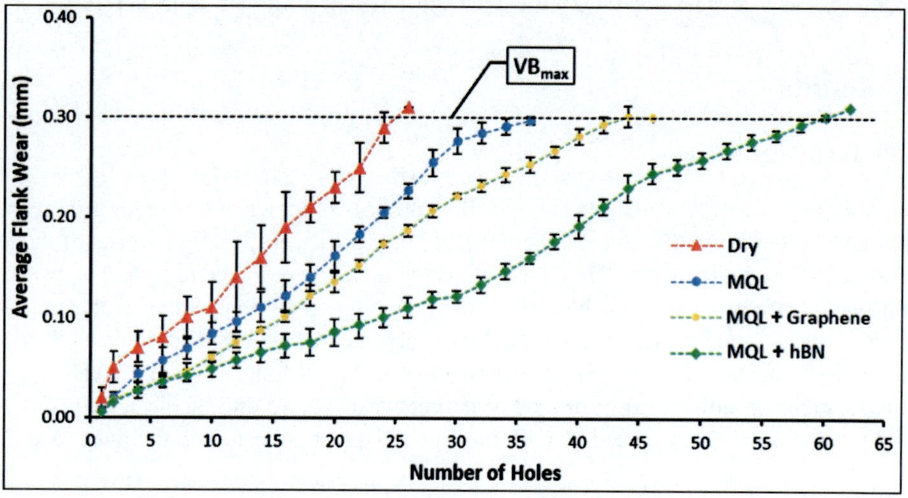

Fig. 7. Number of holes drilled under different cutting conditions.

occurred between tool-workpiece interfaces that contributed to heat generation. Thus, the tool wear rate increased and tool life shortened.

The number of holes drilled under dry cutting, MQL, MQL dispersed with graphene and MQL dispersed with hBN were 26, 36, 44 and 60, respectively. From the results, it can be noticed that rapid tool wear was happened under dry cutting where it was surpassed the tool criterion just upon the 26th hole drilled after 18 min of machining times. On the other hand, under MQL dispersed with hBN, the wear line increased gradually and can withstand up to 41 min of cutting process. This contributed to approximately 130% holes increment as compared to dry cutting. This clearly shows that the effectiveness of hBN nanoparticles can reduced friction at tool-workpiece interface due to tribolochemical reaction. Besides, the repeatability of the results was also improved when the solid lubricants were applied.

3.3 Tool Edge Radius

One of the indicators in evaluating tool wear is tool edge radius. Figure 8 shows the edge radius under different cutting condition. From the results, the significant positive correlation between the effects of hBN can be seen after 1st hole drilled. The edge radius value was almost similar with the as received value. This might be due to the morphology of the smaller nanoparticles size of hBN (80–100 nm) which are able to penetrate with ease into two contact surfaces and improve tribological behaviour and reduce friction through the rolling, filling and polishing action of nanoparticles [11]. The effectiveness of hBN in maintaining the edge radius was still promising even after 60th holes as compared to other conditions with lesser holes.

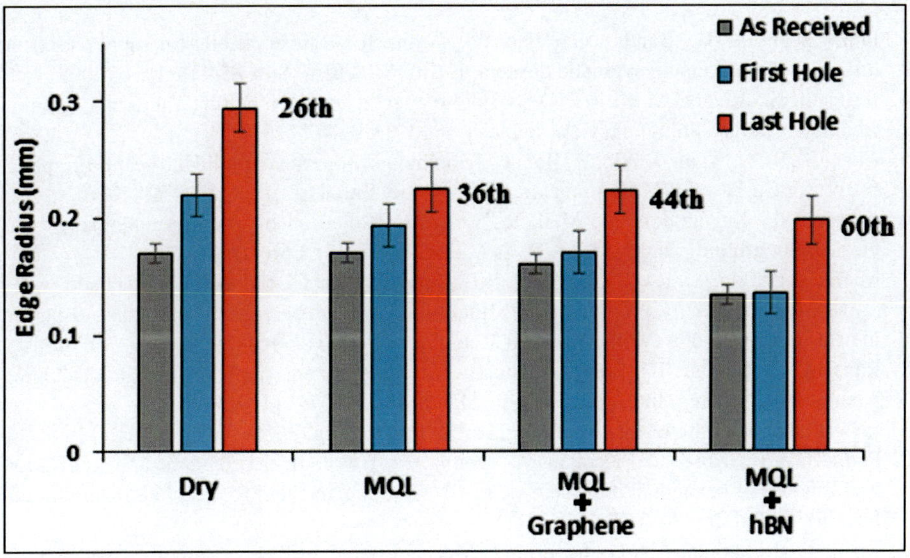

Fig. 8. Tool edge radius under different cutting conditions.

4 Conclusion

The use of graphene and hBN nanoparticles to augment minimum quantity lubrication were found to be significant in reducing flank wear and increasing number of holes drilled through their impressive lubrication mechanism even with a small amount of nanoparticles compared to dry cutting and minimum quantity lubricant alone. Approximately 130% increment in terms of total holes drilled was achieved under simultaneous use of MQL and hBN nanoparticles as compared to dry cutting. The presence of hBN also reduces the average wear and slowing down the wear rate. Besides, not only the wear reduced, the edge radius also could be maintained in good shape. This strongly recommends that the combination of both MQL as lubricant and nanoparticles as friction reducer is vital for reducing tool wear and prolonging tool life.

Acknowledgement. This research was undertaken at the School of Manufacturing Engineering and Faculty of Engineering Technology, Universiti Malaysia Perlis. Their support on machines and laboratory facilities are gratefully acknowledged.

References

1. Singh, G., Gupta, M.K., Sharma, V.S.: Sustainable drilling of aluminium 6061-T6 alloy by using nano-fluids and Ranque-Hilsch vortex tube assisted by MQL: an optimisation approach. Int. J. Mach. Mach. Mater. **20**(3), 252 (2018)
2. Shokrani, A., Dhokia, V., Newman, S.T.: Environmentally conscious machining of difficult-to-machine materials with regard to cutting fluids. Int. J. Mach. Tools Manuf **57**, 83–101 (2012)
3. Pop, L., Puscas, C., Bandur, G., Vlase, G., Nutiu, R.: Basestock oils for lubricants from mixtures of corn oil and synthetic diesters. J. Am. Oil Chem. Soc. **85**(1), 71–76 (2008)
4. Debnath, S., Reddy, M.M., Yi, Q.S.: Environmental friendly cutting fluids and cooling techniques in machining: a review. J. Clean. Prod. **83**, 33–47 (2014)
5. Chen, B., Bi, Q., Yang, J., Xia, Y., Hao, J.: Tribological properties of solid lubricants (graphite, h-BN) for Cu-based P/M friction composites. Tribol. Int. **41**(12), 1145–1152 (2008)
6. Alberts, M., Kalaitzidou, K., Melkote, S.: An investigation of graphite nanoplatelets as lubricant in grinding. Int. J. Mach. Tools Manuf **49**(12–13), 966–970 (2009)
7. Marques, A., Guimaraes, C., Da Silva, R.B., Fonseca, M.P.C., Sales, F.W., Machado, A.R.: Surface Integrity Analysis of Inconel 718 after turning with different solid lubricants dispersed in neat oil delivered by MQL. Procedia Manuf. **5**, 609–620 (2016)
8. Rao, D.N., Krishna, P.V.: The influence of solid lubricant particle size on machining parameters in turning. Int. J. Mach. Tools Manuf **48**(1), 107–111 (2008)
9. ISO 8688-2, Tool life testing in milling, Part 2: End milling (1989)
10. Podgornik, B., Kosec, T., Kocijan, A., Donik, C.: Tribological behaviour and lubrication performance of hexagonal boron nitride (h-BN) as a replacement for graphite in aluminium forming. Tribol. Int. **81**, 267–275 (2015)
11. Rahmati, B., Sarhan, A.A.D., Sayuti, M.: Morphology of surface generated by end milling AL6061-T6 using molybdenum disulfide (MoS_2) nanolubrication in end milling machining. J. Clean. Prod. **66**, 685–691 (2014)

Effect of Milling Parameter and Fiber Pull-Out on Machinability Kenaf Fiber Reinforced Plastic Composite Materials

H. Azmi[1](✉), C. H. C. Haron[2], R. Hamidon[1], Z. A. Zailani[1], T. C. Lih[1], A. R. Yuzairi[1], and H. Sanusi[1]

[1] School of Manufacturing Engineering, Universiti Malaysia Perlis, 02600 Arau, Perlis, Malaysia
azmiharun@unimap.edu.my

[2] Department of Mechanical and Materials, Faculty of Engineering and Built Environment, Universiti Kebangsaan Malaysia, 43650 Bangi, Selangor, Malaysia

Abstract. Milling surface quality normally depends on the value of surface roughness and delamination factor. The milling parameters, which are cutting tool geometry and fiber pull-out, are the major factors affecting the value of surface roughness and delamination factor in milling kenaf fiber reinforced plastic composite. The objectives of this research are to study the effects of milling parameters, to evaluate the fiber behavior, and to determine the optimum conditions for a range of milling parameters in order to minimize surface roughness (R_a) and delamination factor (F_d) using response surface methodology (RSM). RSM with central composite design (CCD) approach was used to conduct a non-sequential experiment and analyzed the data from the measurements of surface roughness and delamination factor. This study focused on the investigation of relationship between the milling parameters and their effects on kenaf reinforced plastic composite materials during cutting process. Kenaf composite panels were fabricated using vacuum assisted resin transfer molding (VARTM) method that was pressurized below 15 psi using a vacuum pressure. The results showed that the optimum parameters for better surface roughness and delamination factor were cutting speed of 16 Vm/min, feed rate of 0.1 mm/tooth, and depth of cut of 2.0 mm. The feed rate and cutting speed are expected to be the biggest contributors to surface roughness and delamination factor. Finally, different cutting tool geometries also influenced the fiber pull-out that affect surface roughness and delamination factor in milling kenaf fiber reinforce plastic composite materials.

Keywords: Kenaf fiber · Milling · Delamination · CCD

1 Introduction

Natural fiber is a hairy-like raw material that comes from natural resources such as animal, mineral, and plant fibers. Also, natural fiber is sustainable as well as eco-efficient; therefore, it has been used to replace glass fiber and other synthetic polymer fibers that

M. A. Zakaria et al. (Eds.): *Advances in Mechatronics, Manufacturing, and Mechanical Engineering*, LNME, pp. 85–96, 2021.
https://doi.org/10.1007/978-981-15-7309-5_9

have various kinds of applications in engineering [1]. Natural fiber is famous for its flexibility for processing because it is less delicate to health hazards and damage in machine tools during manufacturing. Furthermore, natural fiber has a lot of beneficial characteristics, for example considerably high tensile strength, satisfying aspect ratio of fiber, and low density, but due to the nature of the composite, machining of composite materials becomes a major cause of concern in the industry [1]. Natural fiber is a renewable source; hence, its cost is very low compared to man-made fibers. In addition, natural fiber is a biodegradable material. In other words, it is an environmentally friendly fiber. Therefore, it can be used to substitute non-biodegradable materials in product development.

1.1 Kenaf Fiber Composite

In a plant fiber, cellulose is the main segment. The cellulose of a plant fiber is made up from carbon, hydrogen and oxygen ($C_6H_{10}O_5$) which is a form of crystalline. Because of the hygroscopic nature of cellulose, it enables plant fibers to absorb moisture from the surrounding in relatively large amount [2]. Kenaf (*hibiscus Cannabinus*, L. family Malvacea) is one of the type of plant fibers that can be reinforced with polymers as a composite. Kenafis used as natural fiber in reinforced composite instead of others because of its large population and low in price. Kenaf fiber offers a greater characteristic compared to other natural fiber, for example, its long fiber, small diameter and a good interfacial adhesion to matrix [3]. The height of kenaf can reach more than 3 m with a 3-5 cm diameter within 3 months which indicates a rapid growth that able to deal with a great demand [2]. Wambua et al. [4] claimed that poor mechanical properties could be the result of lack of firmness between the polymers (hydrophobic) and the natural fiber (hydrophilic). Kenaf fiber offers better interfacial adhesion to matrix polymer than other natural fibers [5]. Matrix polymer can support the fibers, transfer the stresses to fibers to bear most of the load, prevent direct physical damage to fibers, and improve ductility and toughness of the whole composite [5]. Epoxy produces no toxic gases and offers high temperature resistance that enables it to be used at high temperature [6]. Hafizah et al. [7] studied the tensile behavior of kenaf fiber reinforced with several polymer composites and concluded that kenaf reinforced epoxy composite had the highest ultimate tensile strength and the Young's modulus. Various research has been conducted and it has been proven that machining parameters possess significant effects on the performance and quality of natural fiber composite [8, 9].

1.2 Milling Kenaf Fiber Composite

Ö. Erkan et al. [10] claimed that increased spindle speed worsened the damage factor and increased the plastic deformation rate on an end-milled fiber reinforced plastic composite. Davim et al. [11] claimed that feed rate was expected to affect delamination rather than cutting speed for the machining of fiber reinforced plastic. The impact of depth of cut is not as important as cutting speed and feed rate in composite machining; however, it still gives a significant effect on machining process [12]. Besides, the milled surface is getting smoother as the flute of the cutting tool increases [10]. The end-mill cutter can be made of either a high-speed steel or a carbide insert and usually has a straight shank or a tapered shank. Normally, the cutter rotates on a workpiece in a perpendicular axis but it can still be tilted in order to match the need of producing a machine-tapered or curved surface. However, milled composites always tend to have surface roughness and delamination problem that depends much on cutting parameters and the composite's characteristic [13]. Thus, an optimum setting of these parameters is needed to find out in order to control the quality of a workpiece. The advantages of a high-speed steel (HSS) cutting tool over a carbide cutting tool are higher strength to withstand cutting forces and lower cost. However, a carbide insert cutting tool is better than an HSS tool in metal cutting due to its great abrasion resistance and high temperature resistance, which allows it to be used in machining process at high speed without the need to consider overheating situation. For machining of natural fiber composites, an HSS cutting tool is sufficient to meet this research's purpose.

1.3 Delamination Factor

Delamination of a natural fiber composite can be described as the loss of adhesion between the layers that can lead to critical damage of the reinforcement layers (separation). It is one of the failure mechanisms in composite structure with reduce structural quality of material which cause poor assembly tolerance as well as has high possibility to result in long term performance regression [14]. Previous researchers concluded that low feed rate favors small damage on composites [10, 13]. The non-linear relationship between the feed rate and the delamination might be caused by opposite milling to orientation of kenaf fiber. The obtained results might be less accurate if not all the specimens are milled in direction of orientation of fiber. The damage of the composite is greater if the operation is conducted with high cutting speed and high feed rate [13]. Furthermore, the delamination factor can be worsened by increasing the feed rate and tool life can be shortened by increasing the cutting speed that causes higher torque [15]. Davim et al. (2004) studied the effects of cutting speed and feed rate on delamination of glass fiber reinforced plastic during machining with two different matrices [13].

2 Methodology

2.1 Materials Preparation

Kenaf reinforced epoxy composite materials were prepared using vacuum-assisted resin transfer molding (VARTM) method. The method was used to transfer resin to natural fiber in a vacuum condition for molding with the assistance of a vacuum pump. The kenaf fiber in mat form were cut into the dimensions of 300 mm × 300 mm × 5 mm (thickness). Two layers of kenaf fiber mat were used and mixed with epoxy resin. The resin was prepared in the ratio of 4 (800 g epoxy) to 1 (200 g hardener). The VARTM manufacturer recommended this ratio for fabricating kenaf fiber reinforce plastic composite materials. The VARTM method is gaining the interest of manufacturers due to its ease of handling and low preparation cost. Figure 1. shows the actual setup of VARTM. The composite panels were then cut into small workpiece with the approximate dimensions of 75 mm × 30 mm × (8–10 mm) for milling process.

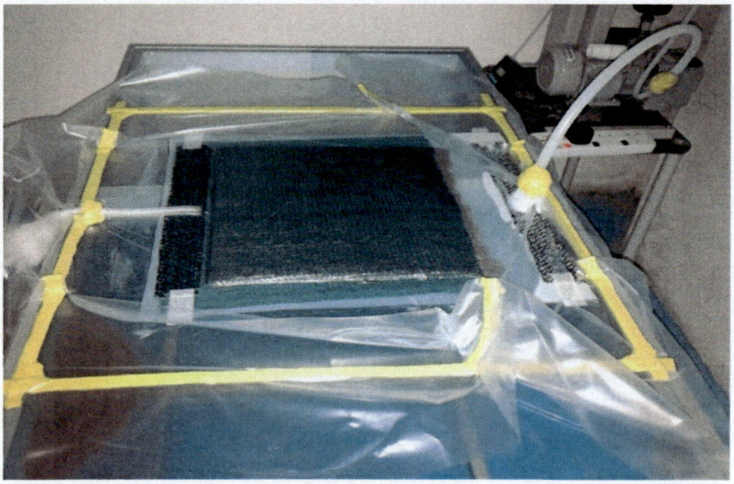

Fig. 1. The VARTM Process

2.2 Milling Parameter

Table 1 shows the parameters used in this research in which all the units were converted to suit the use of a computer numerical control (CNC) milling machine. The three parameters used in this research were spindle speed (rpm), feed rate (mm/min), and depth of cut (mm), with low and high levels for each parameter. A two-level factorial design of experiment was selected with eight experiment runs designed using Design Expert 7.0 software. Four center points were added to seek for the significance of the curvature and error testing. If the curvature of the results is insignificant, a linear modeling equation is sufficient to express the optimization. If the curvature is significant, there is a chance for applying RSM. In RSM, the CCD approach was used to obtain a quadratic modeling equation for parameter optimization. RSM was used to develop a second-order modeling equation that has better accuracy in expressing the optimization.

Table 1. Parameter setting for milling process.

Milling parameters	Low	High	Center point
Spindle speed (rpm)	500	1000	700
Feed rate (mm/min)	200	1200	750
Depth of cut (mm)	1	3	2

2.3 Measurement Delamination Factor

The results of delamination factor were obtained after milling processes. The delamination factor (F_d) can be determined using Eq. (1).

$$F_d = W_{max}/W \qquad (1)$$

Where; F_d = Factor of delamination, W_{max} = maximum width of the delamination area, W = width of original cut (Fig. 2).

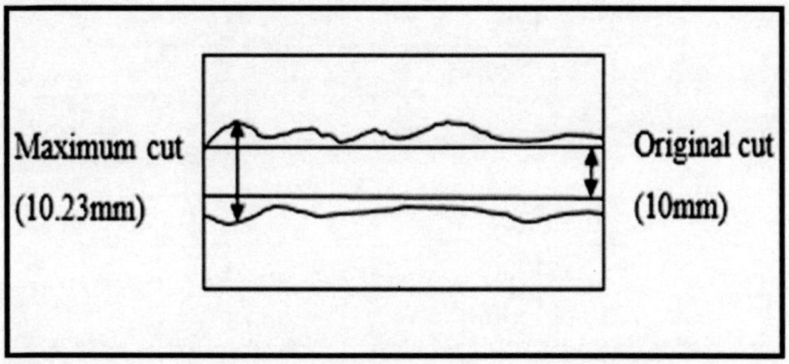

Fig. 2. Example measurement of delamination factor (F_d)

3 Result and Discussion

The delamination of milled kenaf reinforced epoxy composites were observed and measured using Scanning Electron Machine (SEM) (Fig. 3). The maximum width of cut was observed and divided it by original width of cut which was 10 mm (diameter of cutting tool) to obtain the factor of delamination using Eq. 1. The samples of delamination pictures which were taken by SEM. The larger the value of F_d indicates the severer the delamination. The relationship between the factors and the factor of delamination was investigated using ANOVA with the aid of Design Expert 7.0.

Table 2 shows the results of factor of delamination according to different combinations of milling parameters.

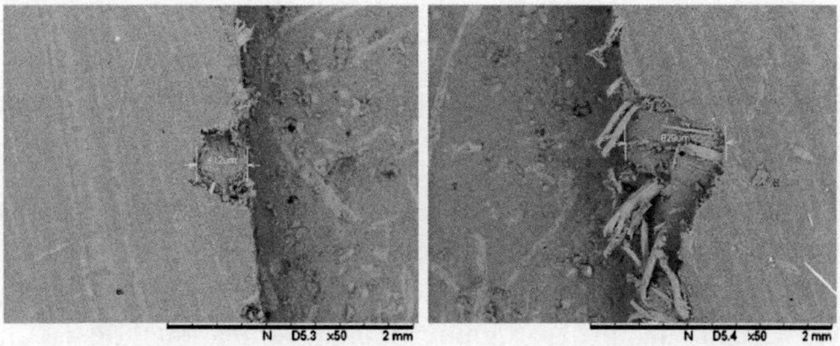

Fig. 3. Image from SEM to measure the delamination factor.

Table 2. Result for delamination factor (F_d) using screening parameter setting (2-Factorial Design).

Run	Spindle speed (rpm)	Feed rate (mm/min)	Depth of cut (mm)	F_d
1	1000	1200	1	1.113
2	500	200	1	1.039
3	500	1200	1	1.077
4	750	700	2	1.042
5	750	700	2	1.067
6	1000	1200	3	1.016
7	750	700	2	1.062
8	1000	200	3	1.134
9	1000	200	1	1.124
10	750	700	2	1.034
11	500	1200	3	1.075
12	500	200	3	1.035

3.1 Fiber Pull-Out

Refer Fig. 3, the fiber pull-out in milling Kenaf Fiber Reinforced Plastics Composite course by various combinations of machine parameter. With the help of microstructure of fracture specimen were obtain SEM, the possible failure condition of the machining process was studied to determine the nature of energy absorption as far as fiber orientation. The large quantities of fracture materials are due to the mechanical effect produced by the recoil force (recoil pressure) by tool-material interaction [16]. These pressure pulls out the fiber fragments in the surrounding area shown in the Fig. 4. The fiber pull out are due to the aforementioned difference in thermal properties between the natural fibers and the epoxy matrix. The voids over the surface of pulled-out fibers show poor interfacial bonding between fibers and matrix. From Fig. 5, the fracture of kenaf fiber is seen to show fiber pull-out and delamination. However, due to the anisotropic nature, the composite will influenced by relative direction between the beam course and the fibers [16]. Beside that fiber pull out cause the fracture in delamination due to the direction of cutting tool during machining. The occurrence of not enriched resin surrounding the fiber does not limit the sliding motion of fibers by shearing action at the interface [17].

3.2 ANOVA Screening Result for Delamination Factor (2-Factorial Design)

The collected results were then analyzed using analysis of variance (ANOVA) method and the data is displayed in Table 3. Design Expert software had been used to conduct the ANOVA and plot the interaction graphs of factors and delamination. The null hypothesis is the factor has no significant effect on delamination of milled kenaf composite. By using confidence level 95%, the factors are significant on delamination if the p-value less than 0.05 (reject null hypothesis).

Fig. 4. Pressure pull out the fiber fragments

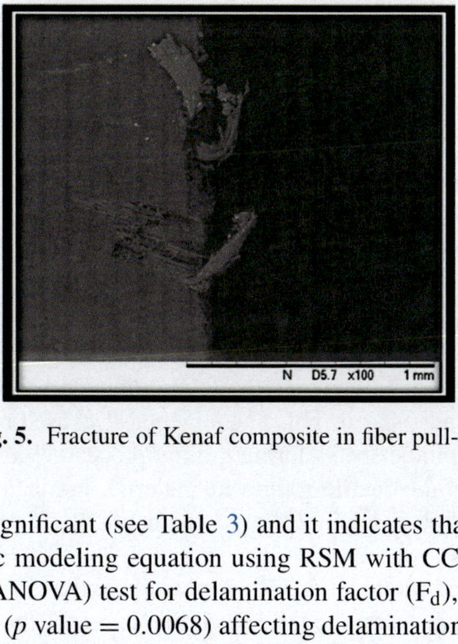

Fig. 5. Fracture of Kenaf composite in fiber pull-out

The curvature is significant (see Table 3) and it indicates that there is a chance for developing a quadratic modeling equation using RSM with CCD approach. From the analysis of variance (ANOVA) test for delamination factor (F_d), spindle speed was the most significant factor (*p* value = 0.0068) affecting delamination. The depth of cut was the second significant factor (*p* value = 0.0292) whereas the feed rate was the least factor (*p* value = 0.2237). In order to apply RSM, CCD (see Table 4) was added to further investigate the parameters for delamination by constructing a second-order modeling equation.

3.3 Results of Delamination Factor Using Central Composite Design

After obtaining the experimental result using CCD, ANOVA test was conducted again to determine the significance of the three factors and the significance of lack of fit of the quadratic modeling equation as shown in Table 5. From Table 5, the main factors significant to F_d were spindle speed with *F* value of 9.67 (0.72%), feed rate with *F*

Table 3. ANOVA result of delamination factor (F_d) using 2FI design

Source	F-value	p-value Prob > F
Model	20.75	0.0012 significant
A-spindle speed	16.27	0.0068
B-feed rate	1.84	0.2237
C-depth of cut	8.12	0.0292
AB	56.78	0.0003
Curvature	24.26	0.0026 significant
Residual	–	–
Lack of fit	0.91	0.5300 not significant
Pure error	–	–

value of 4.86 (4.4%) and the AB interaction with *F* value of 18.43 (0.06%). The AB interaction (spindle speed and feed rate) had the largest effect on kenaf composite to laminate, followed by A (spindle speed) and B (feed rate). The graphs of F_d versus A, B, and AB interaction are plotted in Fig. 6, Fig. 7, and Fig. 8, respectively.

Figure 6 shows the effect of spindle speed on F_d, where F_d increased slightly with the increase of spindle speed at feed rate of 700 mm/min and depth of cut of 2 mm. F_d decreased for lower spindle speed. This phenomenon is caused by the vibration of the cutting tool with high spindle speed during milling process. With large vibration, high F_d is produced on top of the ductile composite materials laminated. Besides, high feed rate (Fig. 7) produced high F_d. Spindle speed and depth of cut remained in the middle range.

In Fig. 8, the depth of cut maintained at 2 mm, where the red line indicates the high level of feed rate (1,200 mm/min) and the black line indicates the low level of feed rate (200 mm/min). It can be clearly observed that high feed rate (1,200 min/mm) with high spindle speed will produce the worst delamination factor. A good F_d will be produced using lower feed rate (200 mm/min) with lower spindle speed. Davim et al. shared the same viewpoint that high cutting (spindle) speed caused great damage on glass fiber reinforced plastic composites during machining [6]. The non-linear relationship between the feed rate and F_d might be caused by opposite milling to the orientation of kenaf fiber. The results obtained might be less accurate if all the specimens are not milled in the direction of orientation of fiber. In addition, previous researchers concluded that low feed rate causes small damage on composites [8, 15].

Table 4. Result for delamination factor (F_d) using Central Composite Design.

Run	Spindle speed (rpm)	Feed rate (mm/min)	Depth of cut (mm)	Fd
1	1000.00	1200.00	1.00	1.075
2	1000.00	200.00	1.00	1.134
3	500.00	1200.00	3.00	1.077
4	750.00	700.00	2.00	1.062
5	1000.00	200.00	3.00	1.113
6	500.00	200.00	1.00	1.035
7	750.00	700.00	2.00	1.034
8	500.00	1200.00	1.00	1.124
9	1000.00	1200.00	3.00	1.067
10	750.00	700.00	2.00	1.042
11	750.00	700.00	2.00	1.039
12	500.00	200.00	3.00	1.016
13	1170.45	700.00	2.00	1.083
14	750.00	−140.90	2.00	1.012
15	329.55	700.00	2.00	1.021
16	750.00	700.00	2.00	1.059
17	750.00	700.00	2.00	1.038
18	750.00	1540.90	2.00	1.087
19	750.00	700.00	0.32	1.054
20	750.00	700.00	3.68	1.046

Table 5. ANOVA result of delamination factor (F_d) using Central Composite Design

Source	F-value	p-value Prob > F
Model	10.99	0.0005 significant
A-Spindle speed	9.67	0.0072
B-Feed rate	4.86	0.0435
AB	18.43	0.0006
Residual	–	–
Lack of fit	3.21	0.1357 not significant
Pure of error	–	–
Cor Total	–	–

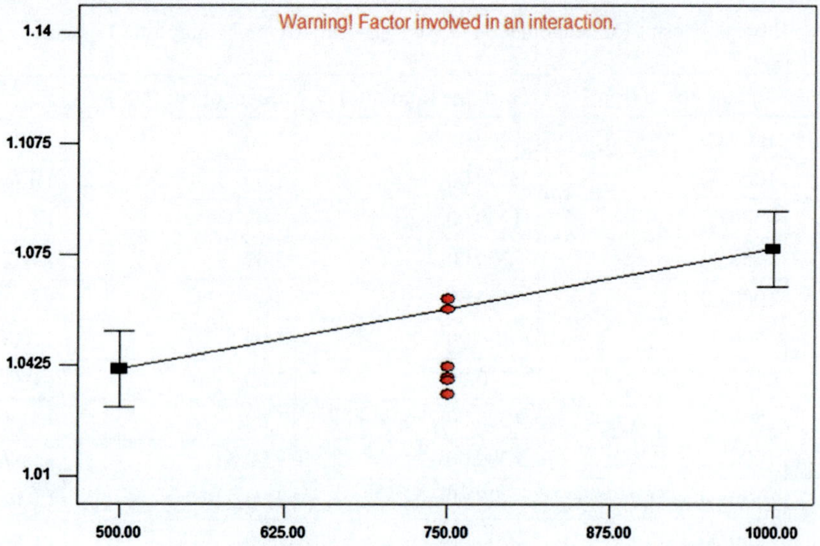

Fig. 6. Effect of A (Spindle speed)

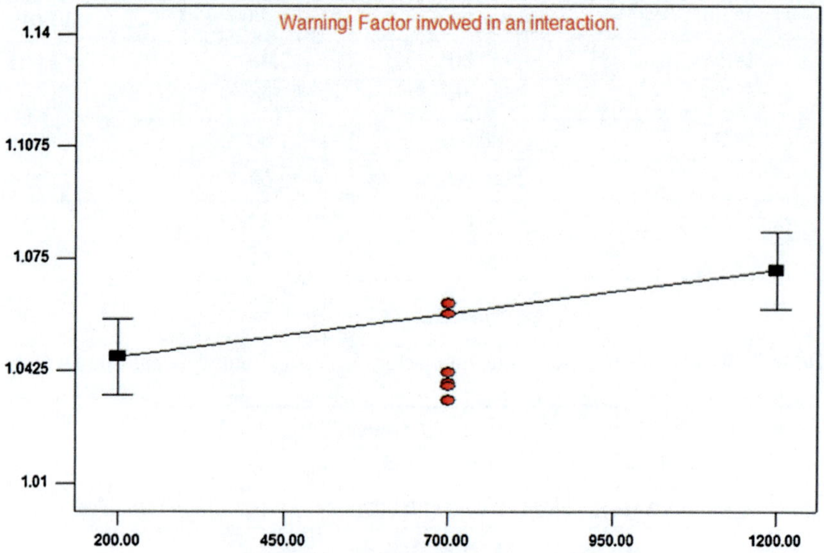

Fig. 7. Effect of B (Feed rate)

3.4 Optimum Parameter and Modeling Equation for Optimization of Delamination Factor

The second order modeling equation for optimization of milling parameters on minimizing delamination factor (F_d) is shown in Eq. (2).

$$Fd = 0.85466 + 2.492 \times 10^{-4}A + 2.163 \times 10^{-4}B - 2.55 \times 10^{-7}AB \quad (2)$$

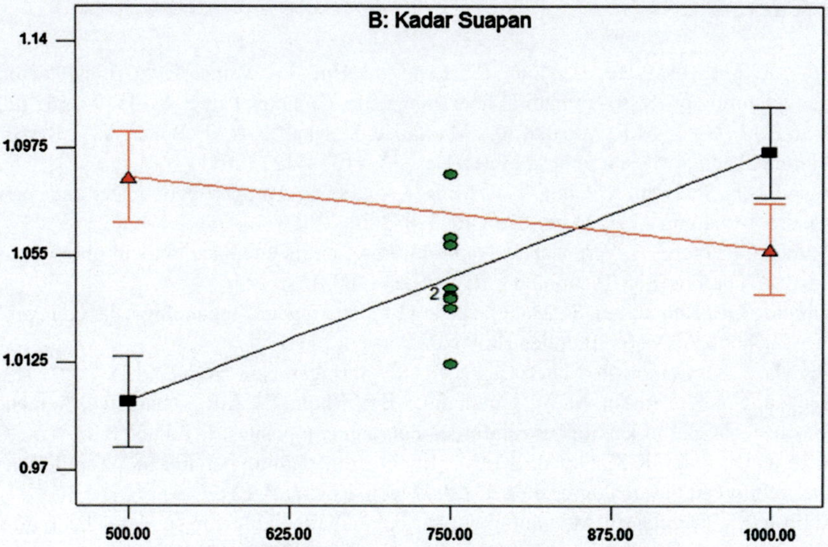

Fig. 8. Effect of AB (Spindle speed and Feed rate) interaction

Where; A = spindle speed (rpm), B = feed rate (mm/min), C = depth of cut (mm) in terms of actual value

Table 6 shows the percentage error between calculated (F_d) using the Eq. (2) and the measured F_d. The measured (F_d) was located at the prediction range (0.96–0.06) and the percentage error was only 1.93% which proved the modeling equation and the suggested optimum parameters can be accepted.

Table 6. Optimum parameter and conformation test for delamination factor (F_d)

Spindle speed (rpm)	Feed rate (mm/min)	Depth of cut (mm)	Calculated F_d	Measured F_d	Percentage error (%)
530	310	2	1.012	1.032	1.93

4 Conclusion

The suggested optimum milling parameters in this research for the minimization of F_d were 530 rpm of spindle speed, 310 mm/min of feed rate, and 2 mm of depth of cut. For the optimization of delamination, the curvature shown in ANOVA after the 2^3 factorial design of experiment was significant. This enabled the utilization of RSM with CCD approach that requires more experimental runs to collect data for constructing a second-order modeling equation. CCD approach was selected and eight additional experimental runs were added. Low spindle speed and low feed rate caused less delamination on milled kenaf composites. In addition, with deeper depth of cut, the possibility of delamination can be reduced.

References

1. Ho, M., Lee, J.H., Ho, C., Lau, K., Leng, J., Hui, D., Wang, H.: Critical factors on manufacturing processes of natural fiber composite. Compos. Part B **43**, 3549–3562 (2012)
2. Akil, H.M., Omar, M.F., Mazuki, A.A.M., Safiee, S., Ishak, Z.A.M., Bakar, A.A.: Kenaf fiber reinforced composites: a review. Mater. Des. **32**, 4107–4121 (2011)
3. Yousif, B.F., Shalwan, A., Chin, C.W., Ming, K.C.: Flexural properties of treated and untreated kenaf/epoxy composites. Mater. Des. **40**, 378–385 (2012)
4. Wambua, P., Ivens, J., Verpoest, I.: Natural fibres: can they replace glass in fibre reinforced plastics? Compos. Sci. Technol. **63**, 1259–1264 (2003)
5. Schmid, S.R., Kalpakjian, S.: Manufacturing Engineering and Technology, 6th edn., pp. 172, 216–226, 662, 662–668. Prentice Hall, New Jersey (2009)
6. Chawla, K.K.: Composite Materials, pp. 81–83. Springer, New York (2012)
7. Hafizah, N.A.K., Hussin, M.W., Jamaludin, M.Y., Bhutta, M.A.R., Ismail, M., Azman, M.: Tensile behaviour of kenaf fiber reinforced polymer composites. J. Teknol. **3**, 11–15 (2014)
8. Babu, G.D., Babu, K.S., Gowd, B.U.M.: Effect of machining parameters on milled natural fiber-reinforced plastic composites. J. Adv. Mech. Eng. **1**, 1–12 (2013)
9. Abilash, N., Sivapragash, M.: Optimizing the delamination failure in bamboo fiber reinforced polyester composite. J. King Saud Univ.-Eng. Sci. **28**, 92–102 (2013)
10. Erkan, Ö., Birhan, I., Çiçek, A., Kara, F.: Prediction of damage factor in end milling of glass fibre reinforced plastic composites using artificial neural network. Appl. Compos. Mater. **20**, 517–536 (2012)
11. Davim, J., Rubio, J., Abrao, A.M.: A novel approach based on digital image analysis to evaluate the delamination factor after drilling composite laminates. Compos. Sci. Technol. **67**(9), 1939–1945 (2007)
12. Palanikumar, K.: Modeling and analysis for surface roughness in machining glass fibre reinforced plastics using response surface methodology. Mater. Des. **28**(10), 2611–2618 (2007)
13. Davim, J.P., Reis, P., Conceição António, C.: A study on milling of glass fiber reinforced plastics manufactured by hand-lay up using statistical analysis (ANOVA). Compos. Struct. **64**, 493–500 (2004)
14. Rakesh, P.K., Singh, I., Kumar, D., Sharma, V.: Delamination in fiber reinforced plastics: a finite element approach. Sci. Res. **3**, 549–554 (2011)
15. Hocheng, H., Puw, H.Y., Yao, K.C.: Experimental aspects of drilling of some fiber-reinforced plastics. In: Proceedings of the Machining of Composites Materials Symposium, pp. 127–138 (1992)
16. Leone, C., Papa, I., Tagliaferri, F., Lopresto, V.: Investigation of CFRP laser milling using a 30 W Q-switched Yb: YAG fiber laser: effect of process parameters on removal mechanisms and HAZ formation. Compos. Part A **55**, 129–142 (2013)
17. Alavudeen, A., Rajini, N., Karthikeyan, S., Thiruchitrambalam, M., Venkateshwaren, N.: Mechanical properties of banana/kenaf fiber-reinforced hybrid polyester composites: effect of woven fabric and random orientation. J. Mater. Des. **66**, 246–257 (2014)

Implementation of Kanban-Based FIFO System to Minimize Lead Time at Automated Optical Inspection Operation - A Case Study in Semiconductor Industry

Prakit Krom[1], Rosmaini Ahmad[2](✉), Shaliza Azreen Mustafa[2], and Tan Chan Sin[2]

[1] School of Manufacturing Engineering, Universiti Malaysia Perlis, Pauh Putra Campus, 02600 Arau, Perlis, Malaysia

[2] Advanced Manufacturing System (AMS) Research Group, School of Manufacturing Engineering, Universiti Malaysia Perlis, Pauh Putra Campus, 02600 Arau, Perlis, Malaysia

rosmainiahmad@unimap.edu.my

Abstract. This paper presents an improvement project in a semiconductor industry to minimize product lead time in automated optical inspection (AOI) operation. The Lean Thinking (LT) approach is applied to drive this improvement project, which it performed based on three main stages; observation, improvement and validation. Initial observation study found a significant cause of this problem (lead time). There have improper record and control mechanisms of receiving the product lots from production department, picking the products lots for inspection and restoring the products lots after inspection. Therefore, this not only cause the high and inconsistent lead time of products lot to be inspect (due to random rack location searching during picking and restoring processes) also the right sequence of product lots from one type to another that went through the AOI operation cannot systematically managed and controlled. In improvement stage, a production control system, namely Kanban-based First-In-First-Out (FIFO) system is proposed to solve the stated problem. Generally, Kanban technique applied in this research project is to systematically guide the inspection operator for performing AOI operation based on FIFO rule. There are two keys rules behind Kanban technique application; First is to record the sequence of the product lots that been received from production department and obtained the rack location for their temporary store. Second is to provide input information for the inspection operator to perform the activities picking and restoring in the right sequence of product lot direct to the rack location that has been stored before. Another technical mechanism included in Kanban-based FIFO system is the application of supermarket buffer system to minimize the wafer lot searching activity. In validation stage, the proposed solution has been tested within the period of one month. Results show that the searching times for picking and restoring the product lots has reduced to 77.0% and overall product lead time due inspection process has reduced 54.2%. The mechanism of Kanban-based FIFO system is then proposed to the top management of the case study company to be embedded with Internet of Thing (IoT) technology to support Industrial 4.0 evolution.

M. A. Zakaria et al. (Eds.): *Advances in Mechatronics, Manufacturing, and Mechanical Engineering*, LNME, pp. 97–108, 2021.
https://doi.org/10.1007/978-981-15-7309-5_10

Keywords: Improvement project · Lean Thinking · Industry case study · Production control system · Kanban-FIFO system · Automatic optical inspection (AOI) operation

1 Introduction

Lean Manufacturing (LM) is an approach or method that is proven to give positive effects on business operations of many organizations. The role of LM is to ensure the operation is running and sustain at top efficiency and effective levels at minimum possible operation cost. Also reflects the acceptable quality of the operation output (e.g. products or services). In manufacturing related organization perspective, Papadopoulou and Ozbayrak [1] stated that LM is one of the management approaches that can be applied to achieve world class manufacturing operation through a cost reduction mechanism. The key aims of LM are to increase productivity, lead time and costs reduction and improve quality [2].

In literature, related research in LM is continuously evolves to variety of scope and they are extensively reported. Some latest scope of researches related to Lean is as follows. For example, Buer et al. [3] studied the linkage between LM and Industry 4.0. Research agenda that can benefits from this linkage was then proposed for future studies. Henao et al. [4] explored the importance needs to optimize LM and sustainability, where the focused is given in discussing the effect of LM in sustainable performance. In relation, one of the comprehensive studies related to 'Lean' is given by Stone [5], which presented four decades journey of research including theory, philosophy, principles and measurements.

The maximum benefit of LM approach application and implementation is highly depends on Lean Thinking (LT) mindset direction and usage. Womack and Jones [6] in management point of view has described LT as the "antidote" to non-value added activities/process towards "*use less, get more*" objective. In Stone [7], LT is referred as operational philosophy of an organization towards differentiates between waste and value. The author was then summarized that LT is the act of identifying and eliminating waste. In this study, we generally depict LT is the arts of observing something and make it better. Technically, it can be defined as a process of analyzing and evaluating the subject of interest via observation activities based on specific value and create/propose possible ways to maximize the benefits. The key element behind successful practicing of LT is clear understanding the concept of "*Value*" and "*Waste*". In manufacturing context, "Value" refers to any process/activities that are customer willing to pay for and "Waste" is other way around, where customers are not willing to pay. The key proactive concept of "*Value*" and "*Waste*" is that the value of any activities/process will increasing while the waste of any activities/process are eliminated or minimized. Although, in real manufacturing practice the "Waste" activities are not totally avoidable, but LT direction practice leads the company keeps improving their operations towards maximizing the "Value" via eliminating or minimizing the "Waste".

One of the key principles of LM is to practice Pull type of production system [8]. Under LM approach, Kanban technique is the main driver of Pull type production system. The main objective of Kanban technique that is conventionally highlighted in LM approach is to control the production materials flow from one process/workstation to another. In other words, the time, amount and types of products that will be produced in present workstation is based only on what are needed in the next (downstream) workstation [9]. This way of production control mechanism mainly will eliminate or minimize overproduction and bottleneck issues, thus maximize the production process value. The original application and implementation of pull system is carried out by using manual easy made cards or it called Kanban cards to transfer the information (materials order) from one workstation to another. Generally, the complete cycle of production in a workstation based on Pull-Kanban system is required three types of Kanban cards; Move Kanban, Production Kanban and Supplier Kanban [10]. Move Kanban, also known as Transportation Kanban is function to authorize the transportation of a particular container (product batches) to a downstream workstation. Meanwhile, Production Kanban is used to activate the production process at present workstation to produce the right amount and types of products needed at downstream workstation. Finally, Supplier Kanban will alert supply chain department to supply the related raw materials needed in present workstation at the right amount and types.

This study presents the application of Kanban system to minimize the lead time in inspection operation, which is out of conventional scope of Kanban system application as been discussed in previous paragraphs. The modification of conventional Kanban mechanism is proposed as a key solution mechanism to solve the stated problem (lead time). The proposed Kanban system is also ensure the inspection operation flow follows in sequence queuing system call First In First Out (FIFO). The remaining parts of this paper are organized as follows. The next section briefs the case study background, where the focus is given on problem statement description. In following section presents the methodology used in this study. The next section discusses the related results based on the three stages of the methodology. Final section ends with a conclusion and future recommendations.

2 Case Study Background

The case study presented in this paper is focus on product inspection process by using Automated Optical Inspection (AOI) machine. AOI is a high technology inspection machine to inspect the defects that may occurs on silicon wafer surfaces. There are various diameters of wafers in the market where the smallest is four inch and the largest is twelve inch of diameters, respectively. A piece of wafer may contain 100 or more than 10000 number of chips.

Due to high quality requirement, every chips that received by customer need to have zero physical defects. Therefore, silicon wafer required highly effective and efficient process of inspection and the usage of AOI machine currently fulfill the requirement of customers. Another strict customer requirement is product delivery time. The customers prefer to receive their ordered product on time with zero defect issue. Any delay or too early deliveries will incur extra cost (e.g. production lost and holding costs) for the customers. As the case study company is classified as contract-based manufacturing, these extra costs will be covered by the manufacturer. From business perspective, this operation scenario is not totally good way of practice and it will reduce the profit of the company.

In this presented study, the focus is given on product delay issue (product lead time). Top management found that the process that significantly contributes to the delay of the product is come from AOI process. It is based on recorded overall inspection process time between actual and expected inspection time. Figure 1 shows the average of overall inspection process time per day for 30 days of recorded period.

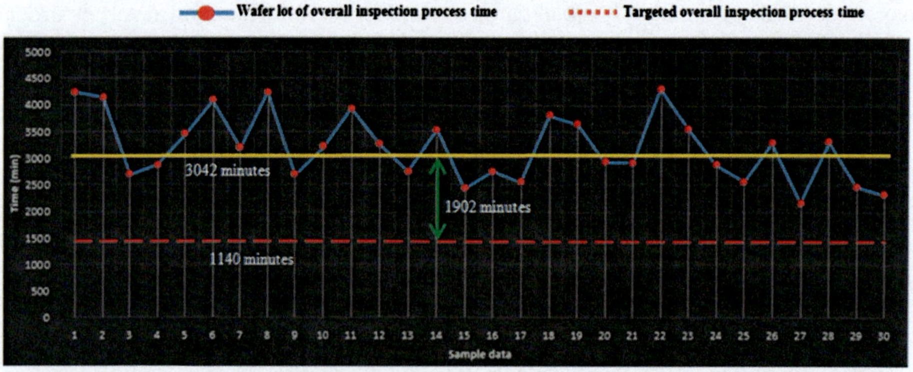

Fig. 1. Overall inspection process time pattern.

With the average number of wafer lots that went through the AOI process is 230 lots/day, Fig. 1 shows that there have no any sample that were taken followed the targeted inspection time. In average, the actual overall inspection process time is 3042 min/lot, while targeted overall inspection process time is 1440 min/lot. This data pattern reveals that currently inspection process contributes to 1902 min (31.7 h) of wafer lots delay per day. Therefore, top management has prioritized this problem as a vital improvement project to be carried out. This paper presents the detail of the improvement project to analyze, evaluate and solve the described problem.

3 Methodology

The presented improvement project was carried out based on three main stages; Stage 1: Observation, Stage 2: Improvement and Stage 3: Validation. Stage 1 focuses on further problem analysis, where the objective is to understand the mechanism of the problem and to identify the root cause(s) of the problem. In this stage, current state activities is observed, recorded and visually presented to understand the working system that currently practice in that area (AOI inspection department). The concept of Value Stream Mapping (VSM) is applied in this stage to visualize the current state activities [11]. Then, Lean Thinking (LT) approach is used to identify and evaluate the related waste activities that cause to the stated problem (lead time). Other related studies that applied LT approach is given by [12–14].

Stage 2 proceeds with solution proposal to solve the lead time issue occurs in inspection process. The integration of Kanban technique and sequence production rule of First-In-First-Out (FIFO) is proposed as the solution, it is call Kanban-based FIFO system.

This stage describes in concise and precise ways on how the proposed system works in solving the above mention problem. In this stage, again the concept of VSM is applied to visualize the future state activities flow according to the solution mechanism that is proposed.

The final stage is to validate the effectiveness of the proposed solution. The result of a pilot study is presented in this stage. This stage is follows by recommendations to sustain maximum effectiveness and efficiency of the proposed solution. Figure 2 summarizes the overall stages of the methodology used in this research project.

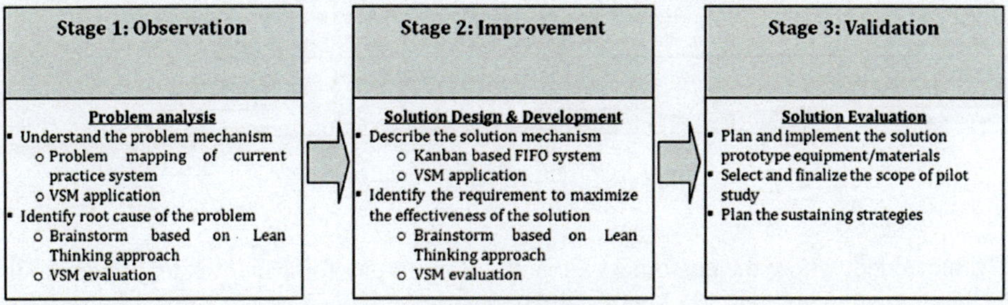

Fig. 2. Methodology stages.

4 Results and Discussion

This section presents results and discussion based on three stages of methodology described in previous section.

4.1 Stage 1 - Observation

Figure 3 visualizes the overall current work flow in AOI department. It can be divided into four activities. It started with receiving of new wafer lot from production department that was carried out by production operator (PO). The new wafer lot was placed in buffer area namely in-coming rack, which it is waiting area before the lot is proceeds to inspection. The second activity is taking the wafer lot for inspection, which it is performed by inspection operator (IO). The third activity is performing inspection using AOI machine and finally deliver inspected wafer lot to the completed inspection rack. The timeline of overall inspection process including searching (second activity) and inspection times are also presented in Fig. 3. Overall inspection process time (as initially depicted in Fig. 1) is measured once the new wafer lot was placed at the in-coming rack until the wafer lot was delivered at completed inspection rack.

Figure 3 also presents the process value classification based on three groups of values; value added, incidental and non-value added activities. In manufacturing context, value added refers to related production activities that clearly changed the value (e.g. shape, color etc) of a product from one process to another and customers willing to pay for. On the other hand, non-value added (also known as production waste) is contradicts activities

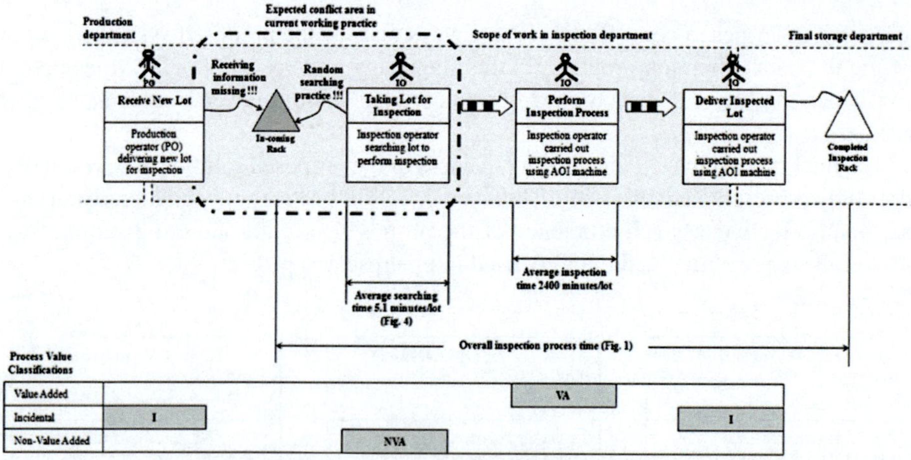

Fig. 3. Current work flow analysis.

of value added, where the customers are not willing to pay for. Meanwhile, incidental is the activities that are classified between value and non valued added activities [15, 16].

As visualizes in Fig. 3 and according to project scope area (AOI department), performing inspection is the only activity that is classified under value-added activity, while taking lot for inspection is classified as non-value added activity. Meanwhile, received new lot and delivered inspected lot are classified as incidental type activities. Under Lean Thinking approach, the general strategy that can be used to deal with non-value added and incidental activities are as follows. For non-value added activities, strategy of elimination or minimization is preferred. Meanwhile, an efficient and effective way of completing the task is the key strategy for incidental type's activities.

The analysis at this stage reveals two keys reasons that cause to high inspection process time. The first is the working system and non-value added activity that currently practices. The problem of working system that currently occurs is highlighted in Fig. 3 (bold dotted rounded triangle), where there have no proper communication (information transfer) system between receiving new wafer lot and taking new wafer lot for inspection activities. Based on LT approach, this working system eliminates the important information to be used for the next activities. The important information that was missing is the product type received and the queuing sequence of the wafer lot. Consequently, taking lot for inspection activity is time consuming and no queuing sequence of the lot was followed.

It was observed that PO randomly placed the new wafer lot to in-coming rack slot and IO also randomly taking the wafer lot for inspection from the rack slot. This current working practice justified the reason behind inconsistent and big gaps of range pattern for wafer lot searching time (taking new wafer lot for inspection activity) as shown in Fig. 4. It shows that the highest searching time was recorded for 13 min and the lowest was 3.3 min. In relation, this reason also justified the same time pattern for overall inspection process time (see Fig. 1).

Another key reason that cause to high inspection process time is existence of non-value added activity in overall inspection process, where according to this case study it

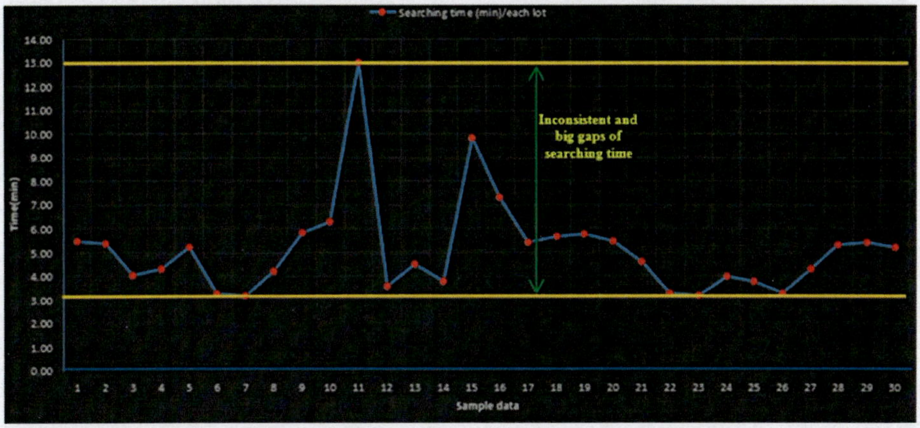

Fig. 4. New wafer lot searching time.

refers to taking lot for inspection activity. As commonly recommended in LT approach, the first priority of strategy is to eliminate the activity, however by considering the nature and needs of this activity, minimize the activity is preferred as the improvement strategy. In the following stage, the improvement solution based on recommended strategy namely, Kanban-Based FIFO system is presented. The mechanism of the proposed system is precise and concisely described.

4.2 Stage 2 – Improvement

Previous stage found that the issue of current working system and existence of non-value added activity (second activity) are the primary causes to high inspection process time. It can be concluded that the current working system between first and second activities ignored the importance of transferring and keeping the related product (wafer lot) information, thus sequence queuing flow of wafer lots for next activity does not existed and hence cause to not only high but also inconsistent waiting time for inspection process of one wafer lot to another (see Fig. 4). Meanwhile, second activity (taking lot for inspection) is evaluated as non-value added activity, where in current practice this activity is performed repetitively for every lot before inspection can be started. Thus, the implication is that the searching time for each wafer lot is included in every cycle of inspection process. Therefore, two key solution objectives to solve above described problem are formulated. First it should be able to control the right sequence queue of the wafer lot, which it starts from new wafer lot received from production department until completed the inspection process. Another solution objective is that it should be able to minimize the second activity (taking lot for inspection).

The solution called, Kanban-Based FIFO system is proposed to achieve the solution objectives presented. Figure 5 presents the mechanism of the proposed system. The proposed system basically improved the working system within first (received new lot), second (taking lot for inspection) and third (perform inspection) activities. Kanban system is introduced between first and second activities. Figure 5 shows that once PO placed a new wafer lot to in-coming rack slot, the information of the rack slot including wafer type lot are recorded in information transferring card (namely, withdraw Kanban card) and place it in waiting box order, called Heijunka Box. This type of Kanban ensures the sequence queue of the new wafer lots received follows the First-In-First-Out (FIFO) flow rule throughout the inspection activities.

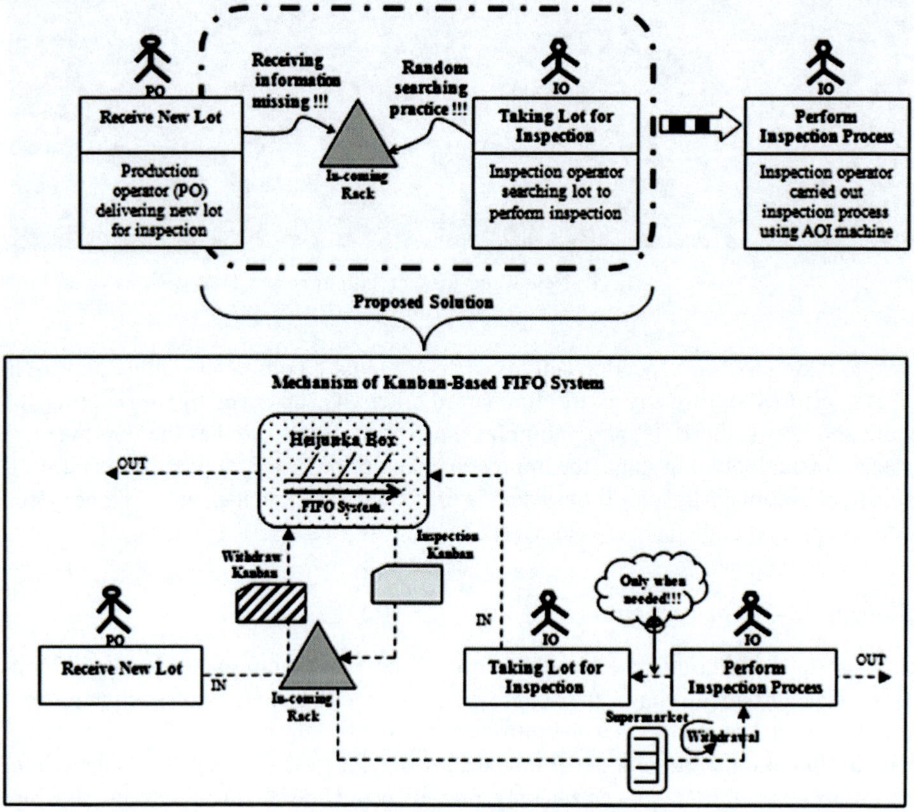

Fig. 5. Kanban-Based FIFO system

The second type of Kanban included in proposed system is called inspection Kanban. Once the second activity (taking lot for inspection) is needed, IO takes the earliest withdrawn Kanban cards that placed in Heijunka Box. The card automatically indicates the right location of rack slot that placed the right wafer lot to perform inspection and the card is return to the rack slot pocket. This Kanban mechanism is significantly shorten the searching time of second activity, which currently based on random searching of the lot. Figure 6 summaries above described working system based on proposed solution.

The second objective of solution is achieved via introduction of supermarket system (see Fig. 5). In this research project, the function of supermarket system is to ensure the wafer lot is ready for inspection and located close with the AOI machine. This supermarket mechanism directly limits the second activity to perform only when needed. Unlike previous practice, the second activity was repetitively carried out for every wafer

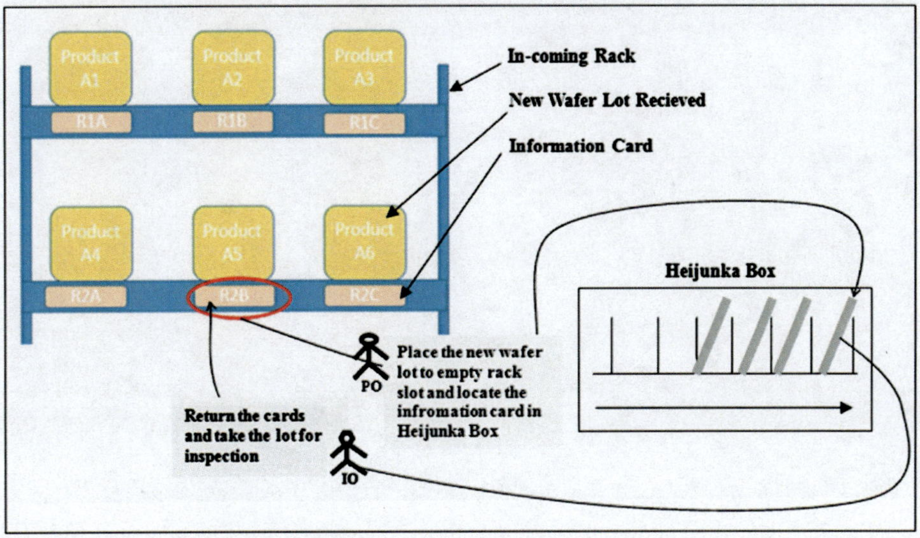

Fig. 6. Working flow of proposed solution.

lot for inspection. Therefore, the work load of IO is reduced and the IO can focus for better wafer handling tasks (loading and unloading) for inspection.

4.3 Stage 3 – Validation

The validation of the proposed solution presented in previous stage was carried out for 30 days of time period. The average of overall inspection process time of wafer lots per day was collected within the period of 30 days. Figure 7 shows the comparison pattern of overall inspection process time before and after improvement.

Majority of average value of overall inspection process time (except day 3, 23 and 24) are significantly presents the reduction, where they close to expected inspection process time. This pattern indicates the success of the proposed Kanban-Based FIFO system to solved lead time issue within inspection process. However, the average values recorded in day 3, 23 and 24 indicate the situation where the new operators training to adapt the new working system (Kanban-Based FIFO system). In average, overall inspection process time is reduced from 3042 min to 1392 min, which represent 54.2% of improvement.

In relation, random sample of wafer lots were taken per day to measure the average of searching time (second activity). Figure 8 present the patterns of searching time before and after improvement. The figure indicates two keys patterns that represent achievement of proposed solution (Kanban-Based FIFO system). First, a significant reduction of overall searching time (less than 2 min) and second is the consistent pattern of searching time. In average, searching time is reduced from 5.2 min to 1.2 min, which indicates 77% of improvement. To ensure the proposed solution is implemented in high efficient and effective way, the computerized version system that applied Kanban-Based FIFO mechanism is then implimenteded by the case study company. The computerized

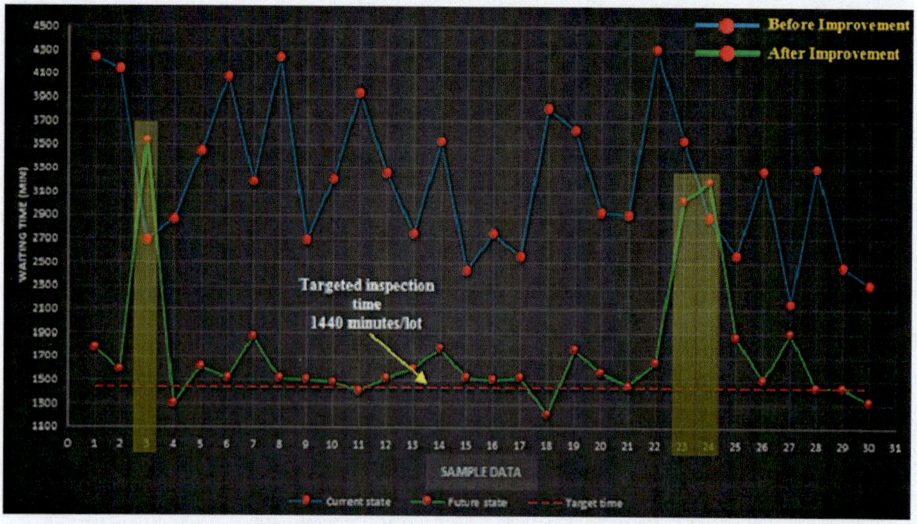

Fig. 7. Comparison of overall inspection process time before and after improvement.

version system applied Radio-Frequency Identification (RFID) and other related Internet of Things (IoT)-based technologies to ensure the working system after improvement in AOI department follows the mechanism of Kanban-Based FIFO system.

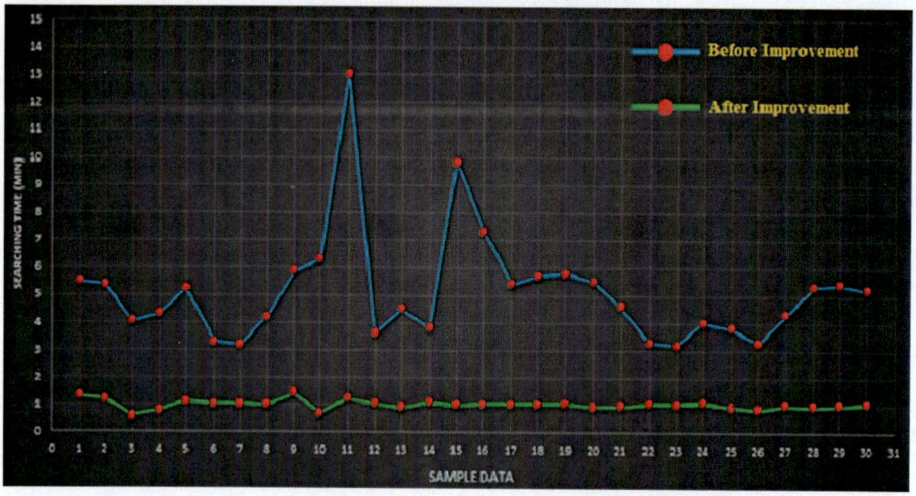

Fig. 8. Comparison of new wafer lot searching time before and after improvement.

5 Conclusion

An improvement project to minimize lead time of inspection process in a semiconductor industry is presented in this paper. The project was carried out based on three main stages; Observation, Improvement and Validation stages. The result in first stage found that the issue of current working system and existence of non-value added activity are the primary causes to high inspection process time. In second stage, a solution namely, Kanban-Based FIFO system is proposed. Kanban concept is applied between the first and second activities to improve the current working system, thus the movement of wafer lots followed the FIFO rule. Supermarket buffer system is introduced to limit the wafer lot searching activity only when needed. The result in third stage shows that the average of inspection process time of wafer lot before and after improvement is reduced from 3042 min to 1392 min. Meanwhile, Supermarket buffer system contributed not only reduction of wafer searching time from 5.2 min to 1.2 min, but also consistency pattern of searching time. The mechanism of the proposed solution presented in this article, was then converted to fully computerized version system that in-line with current company policy towards industrial revolution (IR) 4.0 implementation.

Acknowledgement. Authors wishing to acknowledge assistance and encouragement from colleagues from School of Manufacturing Engineering Universiti Malaysia Perlis, also top management and technical staffs of the case study company especially in providing data and related information of the project. Authors wish to thank for the financial support from Universiti Malaysia Perlis in completing this publication. Finally, the authors also thank the referees for their useful suggestions to improve the quality of the article.

References

1. Papadopoulou, T.C., Ozbayrak, M.: Leanness: experiences from the journey to date. J. Manuf. Technol. Manag. **16**(7), 784–807 (2005)
2. Shah, D., Patel, P.: Productivity improvement by implementing lean manufacturing tools in manufacturing industry. Int. Res. J. Eng. Technol. **5**(3), 3794–3798 (2018)
3. Buer, S.-V., Strandhagen, J.O., Chan, F.T.S.: The link between Industry 4.0 and lean manufacturing: mapping current research and establishing a research agenda. Int. J. Prod. Res. **56**(8) (2018). https://doi.org/10.1080/00207543.2018.1442945
4. Henao, R., Sarache, W., Gómez, I.: Lean manufacturing and sustainable performance: trends and future challenges. J. Clean. Prod. **208**, 99–116 (2019)
5. Stone, K.B.: Four decades of lean: a systematic literature review. Int. J. Lean Six Sigma **3**(2), 112–132 (2012)
6. Womack, J.P., Jones, D.T.: Lean thinking banish waste and create wealth in your corporation. Free Press, Simon & Schuster Inc, New York (1996)
7. Stone, K.B.: Lean transformation: organizational performance factors that influence firms' leanness. J. Enterp. Transform. **2**(4), 229–249 (2012)
8. Jasti, N.V.K., Kodali, R.: An empirical study for implementation of lean principles in Indian manufacturing industry. Benchmarking Int. J. **23**(1), 183–207 (2016). https://doi.org/10.1108/BIJ-11-2013-0101
9. Junior, M.L., Moacir, G.F.: Variations of the kanban system: literature review and classification. Int. J. Prod. Econ. **125**(1), 13–21 (2010)
10. Ramnath, B.V., Elanchezhian, C., Kesavan, R.: Application of kanban system for implementing lean manufacturing - a case study. J. Eng. Res. Stud., 138–151 (2010)
11. Rother, M., Shook, J.: Learning to see: value stream mapping to add value and eliminate muda. Lean Enterprise Institute, MIT (2003)

12. Bevilacqua, M., Ciarapica, F.E., De Sanctis, I., Mazzuto, G., Paciarotti, C.: A Changeover Time Reduction through an integration of lean practices: a case study from pharmaceutical sector. Assembly Autom. **35**(1), 22–34 (2015)
13. Blijleven, V., Koelemeijer, K., Jaspers, M.: Identifying and eliminating inefficiencies in information system usage: Alean perspective. Int. J. Med. Informatics **107**, 40–47 (2017)
14. Ahmad, R., Soberi, M.S.F.: Changeover process improvement based on modified SMED method and other process improvement tools application: an improvement project of 5-axis CNC machine operation in advanced composite manufacturing industry. Int. J. Adv. Manuf. Technol. **94**(1–4), 433–450 (2018)
15. Galankashi, M.R., Helmi, S.A.: Assessment of hybrid Lean-Agile (Leagile) supply chain strategies. J. Manuf. Technol. Manag. **27**(4), 470–482 (2016)
16. Chaple, A.P., Narkhede, B.E.: Value stream mapping in a discrete manufacturing: a case study. Int. J. Supply Chain Manag. **6**(1), 55–67 (2017)

Changeover Monitoring Tool as the Measure of Time Loss in Automotive Production

A. H. Abdul Rasib[1](✉), Z. Ebrahim[2], R. Abdullah[1], A. N. Mohd Amin[3], and Z. F. Mohamad Rafaai[4]

[1] Fakulti Teknologi Kejuruteraan Mekanikal dan Pembuatan, Universiti Teknikal Malaysia Melaka, 76100 Durian Tunggal, Melaka, Malaysia
amir.hamzah@utem.edu.my

[2] Fakulti Kejuruteraan Pembuatan, Universiti Teknikal Malaysia Melaka, 76100 Durian Tunggal, Melaka, Malaysia

[3] Pusat Pengajian Siswazah, Universiti Teknikal Malaysia Melaka, 76100 Durian Tunggal, Melaka, Malaysia

[4] Fakulti Kejuruteraan Mekanikal, Universiti Tenaga Nasional, 43000 Kajang, Selangor, Malaysia

Abstract. Time Loss is a critical issues in automotive production through non-valued added activities and effected to the overall productivity. This paper introduces the changeover monitoring tool as a function for monitoring time loss during changeover activities in automotive production. The tool for monitoring changeover can be specified known as Non-valued Changeover Time (NVCOT). The structure of NVCOT is developed through the analysis of literature study on assembly processes. The subsequent step is to do a verification of NVCOT's structure in manufacturing industries. The verification is verified by the industry's practitioners through the step of interview sessions. Finally, the structure of NVCOT is validated through case studies of five automotive manufacturing companies. In this study, the results of time loss based on non-valued added activities in automotive production is commonly affected by four basic factors; (i) Front and Rear, (ii) Right and Left, (iii) Model Variety, and (iv) Product Variety. In conclusion, overall steps of studies are based on changeover activities in automotive manufacturing. The case study of five manufacturing firms of manufacturing companies has helped to prove the robustness of NVCOT. Thus, the NVCOT can be used as changeover monitoring tool for catering changeover time reduction improvement.

Keywords: Time loss · Changeover · Non-valued added · Lead time · Productivity

1 Introduction

In the new era of producing product, product variety has been recognized as one of the foremost competitive edges for manufacturing companies to meet customers' diverse

M. A. Zakaria et al. (Eds.): *Advances in Mechatronics, Manufacturing, and Mechanical Engineering*, LNME, pp. 109–120, 2021.
https://doi.org/10.1007/978-981-15-7309-5_11

demands [1]. Thus, companies must offer a variety of products and improve their manufacturing operations performance with minimal or zero Time Loss (TL). Besides, Manufacturing enterprises are in an intensive competition in order to offer products with the best quality at reasonable costs and with a minimal lead time [2]. In this case, it is important to identify the non-value added along the manufacturing lead time taken.

Assembly processes are part of activities in the manufacturing lead time. As the number of product variety increases, the frequency of changeover time increases as well. In this case, better operational performance on flexibility in assembly process is vital in order to provide the customer with a product that matches their requested specifications and volumes during product change in operation. A flexibility is one aspect of manufacturing ability to influence a number of significant work related attitudes and behaviour that need to be given more attention to achieve an organizational competitive advantage similar to the capability in the areas of cost, quality, and delivery [3–6]. In the manufacturing industries, setup and adjustment activities consume a significant amount of time [7]. Ferradas and Salonitis [8] defined NVCOT as the time between the last generation product from previous order leaving the process and the first generation product coming out from the following process. Singh and Khanduja [9] have mentioned that setup is a sequence dependent changeover, activities which are carried out before starting the production of any product. On the other hand, McIntosh et al. [10] claimed that changeover is consisting with activities of run-down, setup, and run-up. According to Moxham and Greatbanks [11], internal setup refers to setup that are implemented while a machine is stopped while external setup relates to activities that can be managed while a machine is in. The internal set-up contributes to time loss as the machine have to be stopped. Thus, the internal set-up activities are considered as the NVCOT. In this regard, time losses by the NVCOT is the factors that contribute to the major wastes. According to McIntosh et al. [10], the use of resources for any work other than the creation of value for the end user is waste and target for TL reduction or elimination during changeover activities.

Assembly processes in automotive industry are definitely required for NVCOT as involve different parts/components for different model and different assembly position (i.e. left-right, front-rear). Singh et al. [12] stated that setup activities during changeover was the main factor which was responsible for increased downtime. Therefore, the purpose of this paper is to introduce NVCOT as the measure of time loss in assembly processes. Apart from that, this paper clarifies the effect of NVCOT to the assembly productive time in the context of left-right parts/components, front-rear parts/components, and different models in the automotive assembly line.

2 Understanding of Non-valued Changeover Time

The non-value added activity such as adjustment time for restarting the production are very significant issues because of directives related to the operation setup and trial that effected to flexibility and performance of the manufacturing operation [13–15]. According to Van Goubergen and Van Landeghen [16], product line changeovers may require minutes, hours, or even days before they are complete. Thus, setup times need to be minimized to maximize the capacity available for production. In Lean Manufacturing, the non-value added activities are related to the ineffective time that had been spent to the activities or works. Singh and Khanduja [9] claimed that to reduce set-up time, non-value adding activities should be avoided and online activities should be externalized to off line. According to Chen and Tan [17], Just In Time (JIT) implementation through the same basic elements of JIT like 5S and Multi-Skill Employee (MSE) can effectively improve operational performance especially to eliminate ineffective time during changeover activities. These are due to the absence of proper method in the operational activities especially on changeover activities [14]. The absence of proper method may cause by lack of knowledge and skill among the workers. According to Gilmore and Smith [18], the remainder of the changeover process was ineffective time, accounted for mainly by waiting time when key resources such as setters, fitters and tooling were not available. Waiting happens because of maintenance, changeover, setup, inspection activities and other quality issues [19]. Furthermore, Miller et al. [20]; Chen and Tan [17]; Olson and Villeius [21], had strongly agreed that inefficient time in operation process affected the operational performance. It is crucial to provide NVCOT knowledge and skill among the workers in order to avoid NVCOT. According to Boysen et al. [22], in a mixed-model assembly line, the application of flexible workers and machinery leads to a substantial reduction in NVCOT, so that different products can be jointly manufactured in intermixed product sequences (lot size of one) on the same line. Ferradas and Salonitis [8] had mentioned that the internal activities being the ones that can be performed only when the machine is shut down and external being those that can be conducted during the normal operation of machine, when it is still running. McIntosh et al. [10], defines the run-up stage as the time when stable state manufacture is being re-established, with productivity and quality based on requirements.

2.1 Structure of Non-valued Changeover Time

In order to develop the proposed NVCOT's structure, the element of changeover time should clarify through a literature studies in the aspects of manufacturing operations (see Fig. 1). Generally, the overall structure is came from the main topic of Changeover Time (COT). Then, the structure is divided into Valued Changeover Time (VCOT) and NVCOT. The VCOT is related to the 'External Setup'. Then the NVCOT is related to 'Internal Setup' and Run-up (RU). These internal and external setup activities involve different operations, such as preparation, after-process adjustment, checking of materials, mounting and removing tools, settings, trial runs [8]. However, the NVCOT is the focus in this study because it's contributed to the TL as had been discussed in the previous sections.

In this study, NVCOT is divided into internal setup and RU. The 'Internal Setup' is referred to setup and changeover activities that are performed while a machine is not running [10] while Mileham and Cully [23] define the RU as a time of fine adjustments and inspection that are carried out during restart of production until an acceptable condition level and output speed have been reached. In this study, the Current Model (CM) and Next Model (NM) terms were introduced to get more details and understanding in order to categorize those existing elements in 'Internal Setup'. Therefore, the CM consists the 'No. of Tooling Removing Step' and 'No. of Comp. Removing Step'. Then, the NM consists the 'No. of Tooling Contact Step', 'No. of Tooling Fastening Step' and

'No. of Setting Step'. According to Benjamin et al. [24], the actual attachment (contact and fastening) or removal of dies or tooling would be classified as an internal activity during setup time. Patel et al. [25] claimed that a further consideration with respect to set-ups is the potential for error in the setting process which is productive time lost when a machine is being set. Then, the RU is important to understand how many time the trials (adjustment and inspection) were conducted until achieving the acceptable level. McIntosh et al. [10] defines the RU period as the time when consistent condition manufacture is being re-established, with optimal productivity and quality level. Normally it includes in trial's activities such as adjustments and inspection.

In order to measure the NVCOT efficiently, all the elements should be measured in time based unit. Because, one effective measure is the adoption of time-based manufacturing [26]. Therefore, the items in the NVCOT (i.e. 'Tooling Removing Step Time', 'Component Removing Time', 'Tooling Contact Step Time', 'Tooling Fastening Time', 'Setting Step Time' and 'Trial Time') are used the time unit in seconds. In the meantime, the NVCOT's structure should be ensured through the verification. Therefore the verification of the structure has been carried out through the interview session with the manufacturing practitioners.

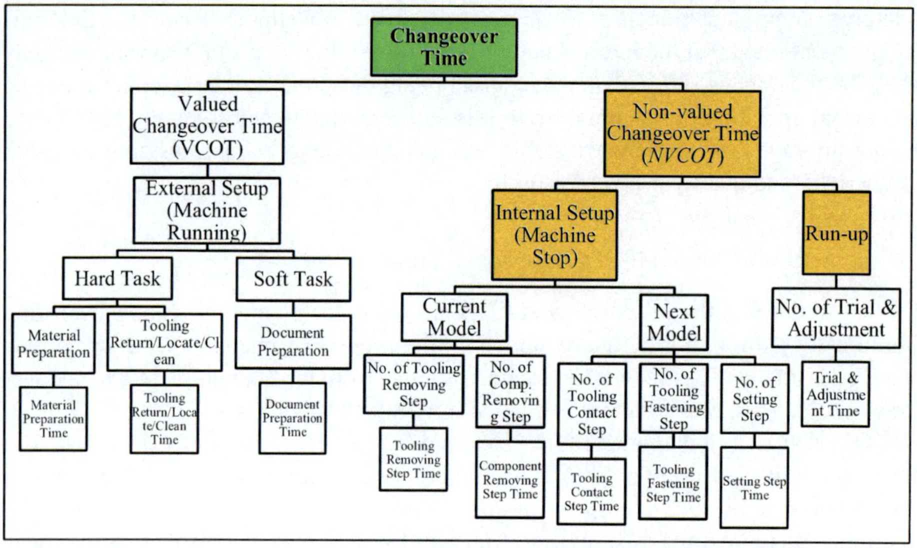

Fig. 1. Proposed non-valued changeover time's structure

2.2 Verification of Non-valued Changeover Time's Structure

The verification is carried out in order to propose more realistic NVCOT's structure. The objective is to check and confirm the suitability and realize in realistic situation. Then, the verification has been done by getting the comments from manufacturing practitioners. In order to get the more professional feedbacks, the practitioner's level also had been considered. Therefore, the engineer level and above had been chosen to get the appropriate comments. In the meantime, the verification has covered as the following items below:

- Overall changeover structure
- Setup structure
- Run-up structure

In order to perform the verification in systematic steps, the specific form had been used to conduct the task during the verification activity (see Fig. 2). The purpose of this form is to get the feedback from practitioners in manufacturing industries. In order that, the practitioners will give the comments based on the proposed NVCOT's structure (see Fig. 1). Then, the score had been evaluated by the Appropriate Level (AL). Level 1 is equal to not appropriate, Level 2 is equal to partly appropriate and Level 3 is equal to appropriate. Since received the feedback from the manufacturing practitioners, the proposed NVCOT's structure had been presented according to the summarized of practitioners' feedback as mentioned in Table 1.

Thus, the important outcome of this paper is to recognise the nonvalued changeover activities for doing better improvement on uptime in manufacturing assembly processes. In the meantime, develop the NVCOT's equation based on NVCOT's structure in order to measure the NVCOT in assembly processes.

Initial 'Framework of Time Loss Measurement Structure' Comments by Industrial Practitioner
(Please tick circle your selection in applicable)

No	**Item**	**Slide No.**	**Appropriate Level (Circle)**	**Comment**
(1)	Overall 'Initial Framework of Time Loss Measures Structure's suitability and effected to the Time Loss increasing.	10	**1 2 3**	
	(a) TL Components suitability and effectiveness.	10	**1 2 3**	
(2)	Overall Changeover Time structure suitability and effected to the Time Loss increasing.	11	**1 2 3**	
	(a) Setup structure suitability and effected to the Time Loss increasing.	11	**1 2 3**	
	(b) Run-up structure suitability and effected to the Time Loss increasing.	11	**1 2 3**	

Fig. 2. Practitioner's feedback form

3 Non-valued Changeover Time Equation

The NVCOT's equation has been developed based on the proposed NVCOT's structure as mentioned in figure (see Fig. 1). The CM is defined as a removing activities for CM's tooling when the production line just stopped. Then, the NM is defined as an installing activities just after the CM's tooling removed from production line. Then, the RU is defined as a product trial. Thus, all the activities in CM, NM and RU are considered non-valued. The non-valued activities are contributed to the TL. Therefore, the general NVCOT's equation is generated from the sum of CM, NM and RU as shown in Eq. (1). Therefore, each element in NVCOT's equation have been provided with the specific equation accordingly as stated in Eq. (2), (3) and (4).

$$NVCOT = CM + NM + RU \tag{1}$$

Where, NVCOT ≥ 0

$$CM = x_{tr}t_{tr} + x_{pr}t_{pr} \tag{2}$$

Where, CM ≥ 0

$$NM = x_{tc}t_{tc} + x_{tf}t_{tf} + x_{s}t_{s} \tag{3}$$

Where, NM ≥ 0

$$RU = x_{t}t_{t} \tag{4}$$

Where, RU ≥ 0

However, the non-valued activities are recognised as well, the important things need to be clarified is the frequency condition of the activities as mention in Table 2. In order to achieve the appropriate condition, all the possibilities during the NVCOT especially internal setup and run-up had been considered as mentioned in Table 3.

Table 1. Practitioners' feedback

Item	Sub-item	Practitioners' comments	AL
Changeover	Overall	Need to divide the value added and nonvalued activity	2.87
	Setup	1. External setup may affected by worker absenteeism or temporary worker	2.27
		2. Need to add component preparation	
		3. Setup element should have 4 M factor because affected by Time Loss	
		4. Mainly Ichikoh has used DLPI (Direct Level Productivity Indicator)	
		5. Need to follow SMED concept	
	Run-up	Need to consider the inspection in run-up	2.53

Table 2. Non-valued changeover time's elements definition

	Sub-element	Number of Frequency	Time
Internal setup	Current model	x_{tr}: Number of tooling removing step in current model- Internal setup	t_{tr}: Time of tooling removing step in current model- Internal setup
		x_{cr}: Number of component removing step in current model- Internal setup	t_{pr}: Time of component removing step in current model- Internal setup
	Next model	x_{tc}: Number of tooling contact step in next model- Internal setup	t_{tc}: Time of tooling contact step in next model- Internal setup
		x_{tf}: Number of tooling fastening step in next model-Internal setup	t_{tf}: Time of tooling fastening step in next model-Internal setup
		x_s: Number of setting step in next model- Internal setup	t_s: Time of setting step in next model- Internal setup
Run-up		x_t: Number of trial	t_t: Time of trial

Then, since the finalized NVCOT's structure has been developed, the important thing is how effective the finalized NVCOT's structure is applied on assembly processes in manufacturing. Because, certain manufacturing companies typically do monitoring the total of NVCOT rather than the specific activity in details. Therefore, the equation is created based on the activities during the changeover as mentioned in Eq. (2), (3) and (4). In this regards, the equations are considered as a tool for TL measures during changeover time. Then, regarding the RU, certain companies are ignored the RU as contributed to TL. According to McIntosh et al. [9], SMED methodology does not cover the RU period as a part of the changeover reduction strategy. Because, even the trial as a RU activity but the 1st product was considered as a production output in spite of using the 1st trial as a sample test for evaluation. Consequently, the RU is depended to the company's practice either to declare the 1st product of trial as an output or evaluation purpose.

4 Validation of NVCOT Equation

The main objective was to validate the NVCOT equation. This objectives was achieved. Thus, the results presented in this section demonstrate the potential for merging theory and practice. Consequently, this section presents the validation of NVCOT equation provided by the case studies results about measuring of TL during changeover at assembly process in five automotive companies in Malaysia. It consists of important stage as mentioned below:

- Data collection
- Data Analysis
- NVCOT's results

4.1 Data Collection

In this case study, each company has their own specific production line according to what product they are produced. The data collection is taken for five consecutive years for Company A and three consecutive months for others. In the meantime, the data collection is based on their individual production line's documents. The documents such as production changeover record and production schedule. However, only two companies has a production changeover record. Therefore, for another three companies the data had been taken by stopwatch in order to measure the changeover time as a primary data. Then, the average of ten times measured by stopwatch has been used as an assumption of changeover time for those companies. Basically, the changeover data had been collected based on the assembly position types as below mentioned.

- Right and Left: Changeover between Right hand side of car's part production and Left hand side of car's part production.
- Product Variety: Changeover for different category (i.e. CITY, ACCORD, CRV, CIVIC).
- Model Variety: Changeover for different type of series (i.e. 1.8 cc, 2.0 cc, 2.4 cc).
- Front and Rear: Changeover between front side of car's part production and rear side of car's part production.

4.2 Data Analysis

The data analysis is based on the data collection from the individual automotive company. Therefore, the Microsoft Office Excel Software is used in order to support the data arrangement with formularization. Basically, the calculation of sum of NVCOT is based on production changeover record whereas the calculation is based on the average assumption as mentioned in data collection section. Thus, the output of multiplication between the average and assembly position change frequency as become the NVCOT outcome. This section has explained the analysis of the data through the case study in five automotive companies. In the meantime, the result of the analysis will be presented in the next section towards validation of NVCOT's equation.

4.3 NVCOT's Results

The results of this study are consistent according to the individual company's data analysis. This result is important to show the NVCOT's outcomes. Basically, the results are depended to the each assembly production line in each automotive company.

In the meantime, the presented results are the sum of three consecutive months whereas for the five consecutive years has been presented by the three months average. Then, all the results of NVCOT is presented in hours in order to make more significant in monitoring. The result of Overall NVCOT according to the assembly position types (see Fig. 3). Then, the NVCOT's result for individual company (see Fig. 4). Finally, the NVCOT's results based on individual assembly production line in each automotive company (see Fig. 5).

Table 3. Non-valued changeover time condition

Internal setup	Condition	Explanation
Current Model (CM)	$x_{tr} = 0$	When using the common tooling for the next model in same production line
	$x_{tr} = 1$	When using the specific tooling or it is not designed for the other model
	$x_{tr} > 1$	When using more than one different tooling
	$x_{cr} = 0$	When using the common component for the next model in same production line
	$x_{cr} = 1$	When using the specific component or it is not designed for the other model
	$x_{cr} > 1$	When using more than one different component
Next Model (NM)	$x_{tc} = 0$	When using the common tooling for the next model in same production line
	$x_{tc} = 1$	When using the specific tooling or it is not designed for the other model
	$x_{tc} > 1$	When using more than one different tooling
	$x_{tf} = 0$	When using the common tooling for the next model in same production line
	$x_{tf} = 1$	When using the specific tooling or it is not designed for the other model
	$x_{tf} > 1$	When using more than one different tooling
	$x_s = 0$	When using the same model but have cosmetic changes
	$x_s = 1$	When model change with specific specification
	$x_s > 1$	When model increase with different specification
Run-up (RU)	$x_t = 0$	When the first produced product is considered for sale
	$x_t = 1$	When model change and upon customer requested the first piece confirmation
	$x_t > 1$	When model change and upon customer requested with minimum two product trial

5 Conclusion

In the first phase, our study contributes positively to theory by strongly proposed the NVCOT's structure in order to develop the NVCOT's equation (see Fig. 1). Then, it was achieved the NVCOT's equation as shown in Eq. (1), (2), (3) and (4). In the second phase, the validation outcome has been achieved through the case studies at five automotive companies. Then, the validation results (see Fig. 3, Fig. 4 and Fig. 5). Consequently, the results based on the assembly position types in Fig. 3 shown that the 'Right and Left' is the highest NVCOT. Then, Company A is the highest NVCOT compared to others (see

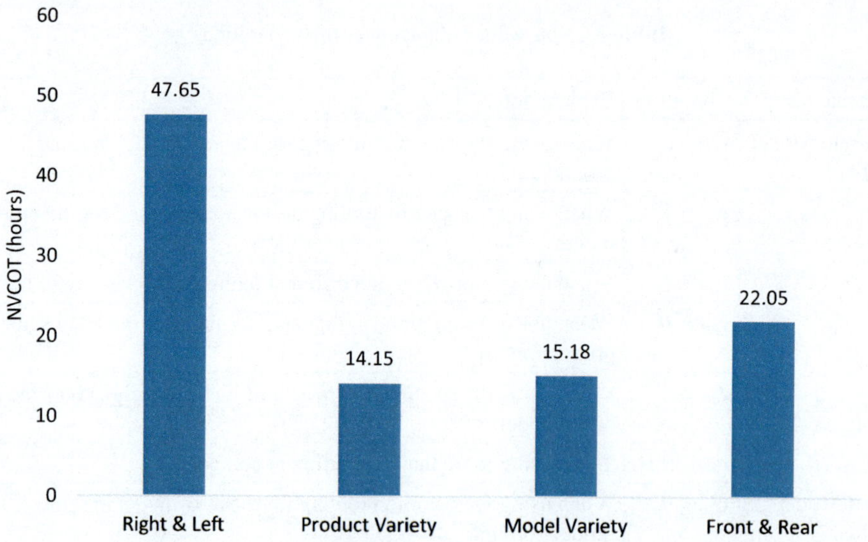

Fig. 3. Overall NVCOT's results based on assembly position types

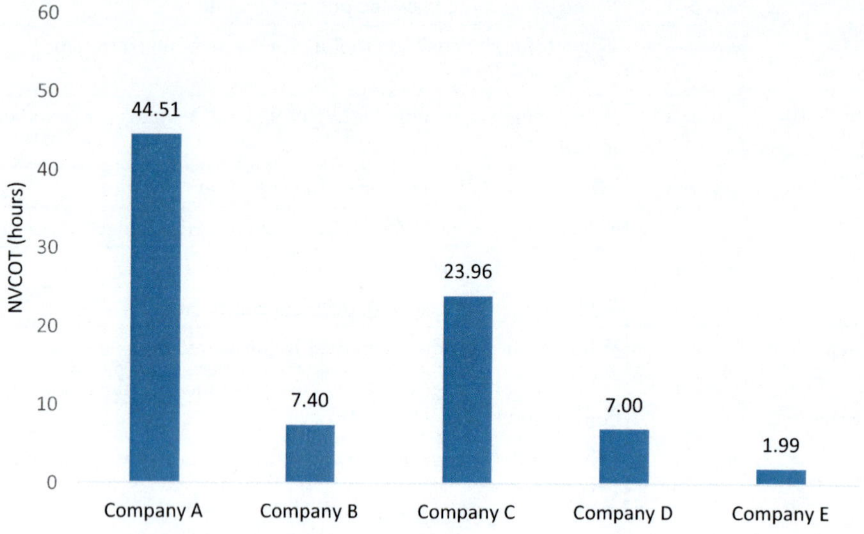

Fig. 4. NVCOT's result for individual automotive company

Fig. 4). Finally, DL Production Line is the highest NVCOT during the Front and Rear changeover (see Fig. 5). In this regards, this study is tried to identify which assembly position type, company and production line is contributed the highest NVCOT. So, the contribution of this paper is to help the practitioner in automotive manufacturing able to measure the NVCOT as changeover monitoring tool in order to improve the operation lead time and flexibility of production.

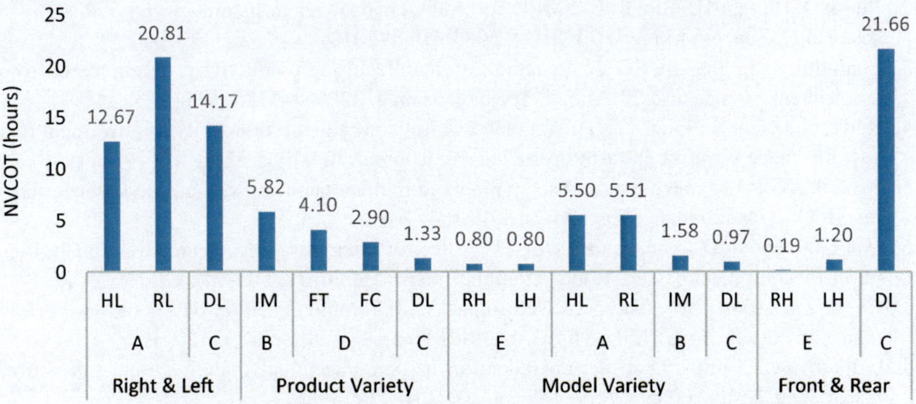

Fig. 5. NVCOT's result for individual assembly production line

Acknowledgements. The authors gratefully acknowledge funding by Pusat Pengurusan Penyelidikan & Inovasi (CRIM), Universiti Teknikal Malaysia Melaka. Also special acknowledgement to Fakulti Teknologi Kejuruteraan Mekanikal dan Pembuatan, Universiti Teknikal Malaysia Melaka and the participating automotive company for the use of the facilities and useful data in order to complete this study.

References

1. Nazarian, E., Ko, J., Wang, H.: Design of multi-product manufacturing lines with the consideration of product change dependent inter-task times, reduced changeover and machine flexibility. J. Manuf. Syst. **29**, 35–46 (2010)
2. Dombrowski, U., Mielke, T., Engel, C.: Knowledge management in lean production systems. Procedia CIRP **3**, 436–441 (2012)
3. Upton, D.M.: Flexibility as process mobility: the management of plant capabilities for quick response manufacturing. J. Oper. Manag. **12**, 205–224 (1995)
4. Ettlie, J.E.: R&D and global manufacturing performance. Manage. Sci. **44**(1), 1–11 (1998)
5. Poolton, J., Barclay, I.: New product development from past research to future applications. Ind. Mark. Manage. **27**, 197–212 (1998)
6. Baeza, M.A., Gonzalez, J.A., Wang, Y.: Job flexibility and job satisfaction among Mexican professionals: a socio-cultural explanation. Employee Relations **40**(5), 921–942 (2018)
7. James, C.D.: A review of machine efficiency in mass customization. Benchmarking Int. J. **26**(2), 638–691 (2019)
8. Ferradas, P.G., Salonitis, K.: Improving changeover time: a tailored SMED approach for welding cells. Procedia CIRP **7**, 598–603 (2013)
9. Singh, B.J., Khanduja, D.: SMED: for quick changeovers in foundry SMEs. Int. J. Prod. Perform. Manag. **59**, 98–116 (2009)
10. McIntosh, R.I., Culley, S.J., Gest, G., Mileham, A.R., Owen, G.W.: An assessment of the role of design in the improvement of changeover performance. Int. J. Oper. Prod. Manag. **16**(9), 5–22 (1996)
11. Moxham, C., Greatbanks, R.: Prerequisites for the implementation of the SMED methodology: a study in a textile processing environment. Int. J. Qual. Reliabil. Manag. **18**(4), 404–414 (2001)

12. Singh, J., Singh, H., Singh, I.: SMED for quick changeover in manufacturing industry – a case study. Benchmarking Int. J. **25**(7), 2065–2088 (2018)
13. Shahidul, M.I., Shazali, S.T.S.: Dynamics of manufacturing productivity: lesson learnt from labor intensive industries. J. Manuf. Technol. Manag. **22**, 664–678 (2011)
14. Radharamanan, R., Godoy, L.P., Watanabe, K.I.: Quality and productivity improvement in a custom- made furniture industry using kaizen. Comput. Ind. Eng. **31**(1), 471–474 (1996)
15. Kibouka, G.R.: Maintenance and setup planning in manufacturing systems under uncertainties. J. Qual. Maintenance Eng. **24**(2), 170–184 (2018)
16. Van Goubergen, D., Van Landeghem, H.: Rules for integrating fast changeover capabilities into new equipment design. Robot. Comput. Integr. Manuf. **18**, 205–214 (2002)
17. Chen, Z.X., Tan, K.H.: The perceived impact of JIT implementation on operations performance: evidence from Chinese firms. J. Adv. Manag. Res. **8**, 213–235 (2011)
18. Gilmore, M., Smith, D.J.: Set-up reduction in pharmaceutical manufacturing: an action research study. Int. J. Oper. Prod. Manag. **16**(3), 4–17 (1996)
19. Shah Satya, R.: Lean production and supply chain innovation in baked foods supplier to improve performance. Br. Food J. **119**(11), 2421–2447 (2017)
20. Miller, G.A., Felbaum, C., Tengi, R., Wakefield, P., Langone, H.: WordNet. Princeton University, Princeton (2009)
21. Olson, M., Villeius, E.: Increased usage of standardized work instructions – Development of recommendations for Autoliv Sweden AB. Department of Product and Production Development. Chalmers University of Technology, Göteborg, Sweden (2011)
22. Boysen, N., Fliedner, M., Schollet, A.: Sequencing mixed-model assembly lines: survey, classification and model critique. Eur. J. Oper. Res. **192**(2), 349–373 (2009)
23. Mileham, A.R., Cully, S.J., Owen, G.W., McIntosh, R.I.: Rapid changeover – a pre-requisite for responsive manufacture. Int. J. Oper. Prod. Manag. **19**(8), 785–796 (1999)
24. Benjamin, S.J., Murugaiah, U., Marathamuthu, M.S.: The use of SMED to eliminate small stops in a manufacturing firm. J. Manuf. Technol. Manag. **24**(5), 792–807 (2012)
25. Patel, S., Shaw, S.P., Dale, B.G.: Set-up time reduction and mistake proofing methods – a study of application in a small company. Bus. Process Manag. Jo. **7**(1), 65–75 (2001)
26. Nahm, A.Y., Vonderembse, M.A., Rao, S.S., Ragu-Nathanb, T.S.: Time-based manufacturing improves business performance - results from a survey. Int. J. Prod. Econ. **101**, 213–229 (2006)

Analysis of Non-dimensional Numbers of Fluid Flowing Inside Tubes of Flat Plate Solar Collector

K. Farhana[1,2], K. Kadirgama[1], and M. M. Noor[1(✉)]

[1] Faculty of Mechanical Engineering, Universiti Malaysia Pahang, 26600 Pekan, Pahang, Malaysia
kfarhana81@yahoo.com, muhamad@ump.edu.my

[2] Department of Apparel Manufacturing Engineering, Bangladesh University of Textiles, Dhaka 1208, Bangladesh

Abstract. The aim of this paper is to discuss the non-dimensional numbers of fluid flowing through inside the tubes of flat plate solar collectors. Empirically, to abate the cost and energy consumption or to boost up the performance and efficiency of solar collectors; computational simulation plays a vital role. In this study, CFD numerical simulation of aqueous ethylene glycol (60% water + 40%) ethylene glycol fluid flow has been done with ANSYS 15.0. Non-dimensional numbers such as surface Nusselt number, Skin friction coefficient and Prandtl number of fluids have been observed based on empirical and experimental properties. The geometry of design has been prepared using Solidworks software in accordance with the actual experimental model. The analysis revealed that the Nusselt number showed effective convection behavior, the skin friction coefficient was positive while the Prandtl number was large for both properties of aqueous ethylene glycol.

Keywords: Aqueous ethylene glycol · Nusselt number · Prandtl number · Skin friction coefficient

1 Introduction

The exhaustion of non-renewable resources and their contribution to the global warming urge to form alternative sources of energy [1]. In this context, solar energy attracted more attention as it is inexhaustible, plentiful, free and clear which does not create any pollution into the environment [2, 3]. Many solar energy harvesting devices have already been invented and flat plate solar collector is one of them [4]. In flat plate solar collectors, header and riser tubes are welded together formed heat exchanger inside which through them working fluids can flow [5]. Primitively, water used most of the cases, afterward a lot of experimental and numerical research done and still on investigating stage to improve the performance of flat plate solar collector [6–8]. It is aforementioned that empirically to abate the cost and energy consumption or to boost up the performance and efficiency of solar collectors; computational simulation plays a vital role [9]. Computational evaluation should be done by different solvers such as Computational fluid dynamics (CFD) which is one of the most important numerical solvers [10].

M. A. Zakaria et al. (Eds.): *Advances in Mechatronics, Manufacturing, and Mechanical Engineering*, LNME, pp. 121–131, 2021.
https://doi.org/10.1007/978-981-15-7309-5_12

Nomenclature	
ρ	Density (kg/m^3)
C_p	Specific heat (J/kg-K)
D	Characteristic linear dimension (m)
k	Thermal conductivity (W/m-K)
μ	Viscosity (kg/m-s)
T	Temperature (K)
v	Velocity (m/s)
τ	Stress tensor
U	Velocity vector
p	Pressure (kg/m-s^2)
g	Gravitational acceleration (m/s^2)
h	Sensible enthalpy (joule)
∇	Advection operator
EG	Ethylene glycol
R_e	Reynolds number
N_u	Nusselt number
P_r	Prandtl number
C_f	Skin friction coefficient

Ethylene glycol is a colourless, odourless, hygroscopic, non-volatile liquid with lower viscosity and a higher boiling point (198 °C) [11]. Besides, ethylene glycol is a non-flammable and low toxic substance as well as it is superior to higher volumetric energy density. Moreover, it is completely soluble with water and with other polar solvents [12, 13]. The sources of ethylene glycol are plentiful and have a wide range of application potentiality, therefore the study on composing newly applied methods concerned with energy, environment, and technology has been expanding for the last two decades [14, 15].

The non-dimensional numbers play an important role in fluid mechanics and heat transfer. These numbers can give a clear idea of whether the fluid is inertial or viscous force dominant as well as the flow is laminar or turbulent. Moreover, they define different heat transfer behaviour such as convection, momentum diffusion, heat diffusion significantly [16]. Thus, prior to the physical application of aqueous ethylene glycol in flat plate solar collectors, numerical simulation can be done repeatedly to find out its fluidic characteristics. Therefore, various non-dimensional behaviour such as Nusselt number, Prandtl number and skin friction coefficient of aqueous ethylene glycol can be evaluated by computational numerical simulation. The aim of this study is to analyses the non-dimensional numbers (Nusselt number, Prandtl number, and skin friction coefficient) of fluids flowing through inside the tube of flat plate solar collector based on empirical and experimental values. Here, the experimental design was prepared using the "SOLIDWORKS 2016" software for numerical simulation with CFD ANSYS R15.0. The ANSYS workbench with Fluent Flow Fluent analysis system used with energy and Viscous Laminar models.

2 Theoretical Expression

2.1 CFD Governing Equations

For numerical simulations of flow design and boundary conditions, the CFD method used mass conservation using the fluid flow governing equations. Here single-phase model has been fixed and steady-state condition applied. According to Versteeg and Malalasekra [17], the governing equations are the following,

Continuity Equation:

$$\rho + \nabla.(\rho U) = 0 \tag{1}$$

Momentum equation:

$$\rho U + \nabla.(\rho UU) = -\nabla \mathrm{p} + \nabla.\tau + \rho g \tag{2}$$

Energy Equation:

$$\rho h + \nabla.(\rho U C_P T) = \nabla.(k\nabla T) \tag{3}$$

2.2 Properties of Aqueous Ethylene Glycol

The empirical and experimental values of density [18], specific heat [19], thermal conductivity [20] and viscosity of aqueous ethylene glycol have been presented in Table 1. Reynolds's number and flow behaviour have been identified empirically by the Eq. (4) [21],

$$R_e \frac{\rho D v}{\mu} \tag{4}$$

Table 1. Data on the physical properties of aqueous ethylene glycol at different conditions.

Properties	Empirical values	Experimental values			
		At 30 °C temperature	At 50 °C temperature	At 70 °C temperature	At 80 °C temperature
Density (kg/m^3)	1050	1050.5	1039.0	1021.7	1011.8
Specific heat (J/kg-K)	3600.648	830	880	680	530
Thermal conductivity (W/m-K)	0.26	0.407	0.433	0.463	0.452
Viscosity (kg/m-s)	0.0022	0.00208	0.00133	0.00116	0.00121

3 Methodology

3.1 Design Modeler

The three-dimensional geometry of the designing model has been developed in the "Solidworks 2016" software. The geometry has been imported to design modeler of fluid flow fluent analysis systems in ANSYS R15.0 workbench for CFD simulation. The outer and inner diameters of the tube were 12.7 mm and 12.5 mm respectively while the total length of the tubes was 12.325 m. Figure 1 shows the schematic view of inside tubes of flat plate solar collectors.

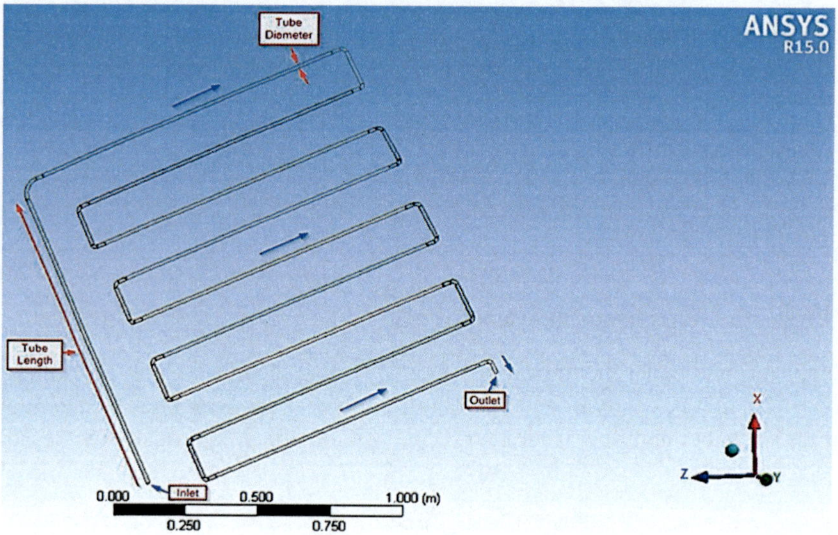

Fig. 1. The geometry of the experimental design.

3.2 Meshing

CutCell meshing has been done automatically in the three-dimensional computational fluid domain. Figure 2 presents the 3D meshed model of the design. In addition, mesh independence test studied by changing mesh sizing Relevance Center such as coarse, medium and fine which has been represented in Table 2. The minimum and maximum size of the meshing of fine, medium and coarse relevance center were different. Here, fine relevance center selected to obtain flawless solution calculation.

3.3 Setup and Solution Calculation

Pressure-based, Absolute-velocity and Time-steady solver used in this study. Energy equation and Viscous-Laminar model used [22, 23]. In Boundary Condition, inlet mass flow rate 0.0108 kg/s and temperature 311.98 K; outlet temperature 312.98 K and fluid

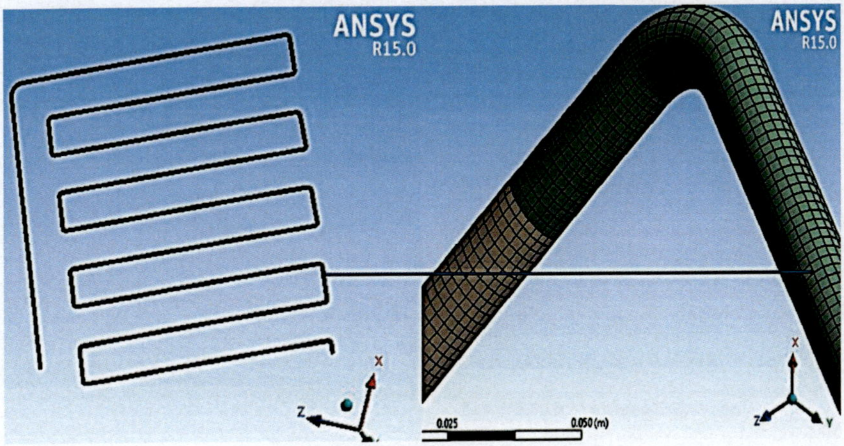

Fig. 2. The meshing of design.

Table 2. Sizing of meshing.

Size of mesh	Relevance center		
	Fine	Medium	Coarse
Min size	3.3141e−004 m	5.6576e−004 m	1.1315e−003 m
Max size	6.6281e−002 m	0.113150 m	0.22630 m

domain temperature 324.87 K were set for both types of properties [24, 25]. In solution, pressure, momentum, and energy second upwind spatial discretization fixed. Standard solution initialization selected to run the solution calculation with a hundred number of iterations and converged with the tolerance of this much.

4 Result and Discussion

The resulting numerical data of aqueous ethylene glycol showed in Table 3, as it illustrates that there is a big difference in Nusselt number between empirical and experimental properties. But Nusselt number is close among the experimental values of aqueous ethylene glycol and it was in decreasing trend with increasing temperature. Moreover, the Nusselt number in all conditions of fluids is greater than one (Nu>>1) which defines the more effective convection heat transfer of fluid layer [26]. Figure 3 and Fig. 4 show the distribution of the Nusselt number in the pipes of solar collectors. Nusselt number for both properties performed higher distribution (Fig. 3 and Fig. 4) in inlet position than the outlet position of solar collector that portrays better convection behaviour at the inlet point of solar collector [27]. Moreover, there is good relation between the Nusselt number and Reynolds number which interpret increasing Reynolds number also enhance Nusselt number of fluid [28].

Table 3. Numerical data of different non-dimensional numbers of aqueous ethylene glycol fluid.

Values of different properties	Nusselt number	Skin friction coefficient	Prandtl number
Based on empirical values			
At room temperature	287.57	0.03	30.4
Based on experimental values			
At 30 °C temperature	45.29	0.03	4.2
At 50 °C temperature	45.14	0.02	2.7
At 70 °C temperature	32.62	0.02	1.7
At 80 °C temperature	26.04	0.02	1.4

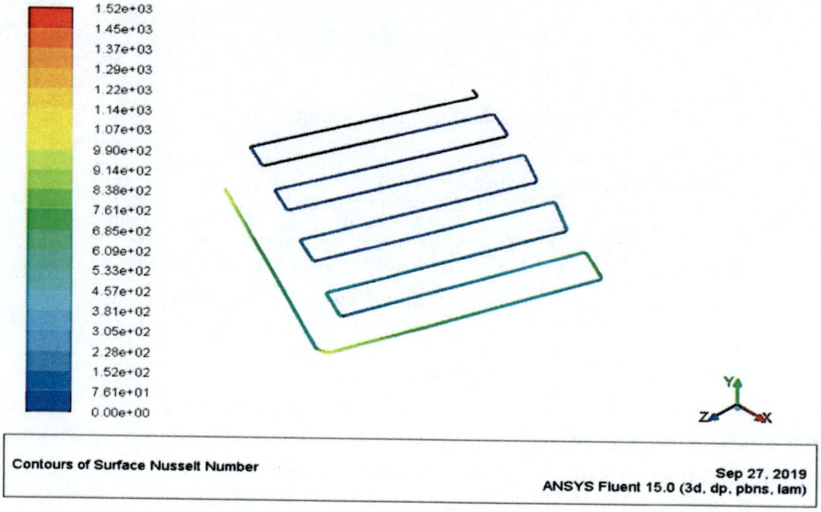

Fig. 3. Distribution of Nusselt number in the tubes for empirical properties.

In the case of the skin friction coefficient (C_f), the dragging of aqueous ethylene glycol is higher for both empirical properties and the properties at 30 °C temperature. But skin friction coefficient decreases with an increase of temperature. Besides, all properties of fluids flow show positive skin friction coefficient [29]. The skin friction coefficient can also be decreased in different approaches such as Gupta, Dwivedi [30] studied that skin friction coefficient decreased due to increases in pipe length. Figure 5 and Fig. 6 display the plot of skin friction coefficient inside the tubes of solar collectors. Aqueous ethylene glycol exhibited the development of skin friction coefficient throughout the solar collector and it appeared more in the corner area of the collector for both types of properties. This phenomenon is too similar to Moinuddin, Joubert [31] skin friction calculation of an external corner with turbulent flow study.

Finally, the Prandtl number is higher ($Pr >> 1$) than one that determines the momentum diffusivity dominates the behaviour. Besides, a large Prandtl number is responsible

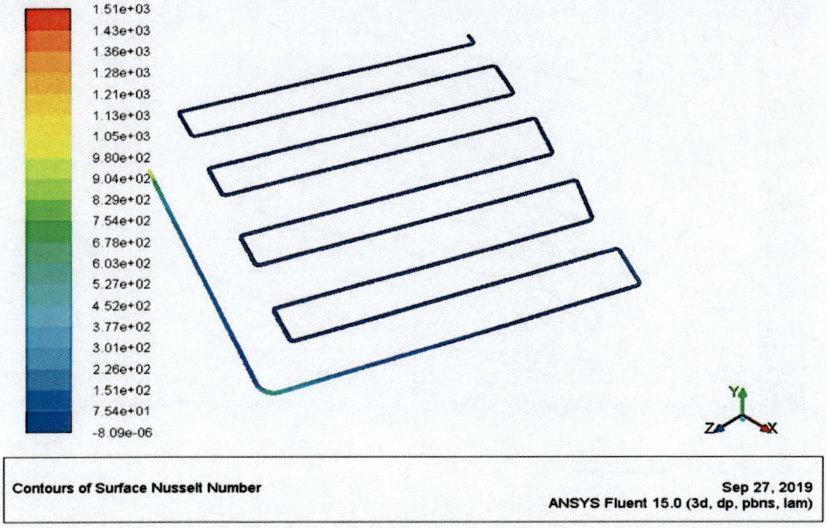

Fig. 4. Distribution of Nusselt number through the tubes for experimental properties.

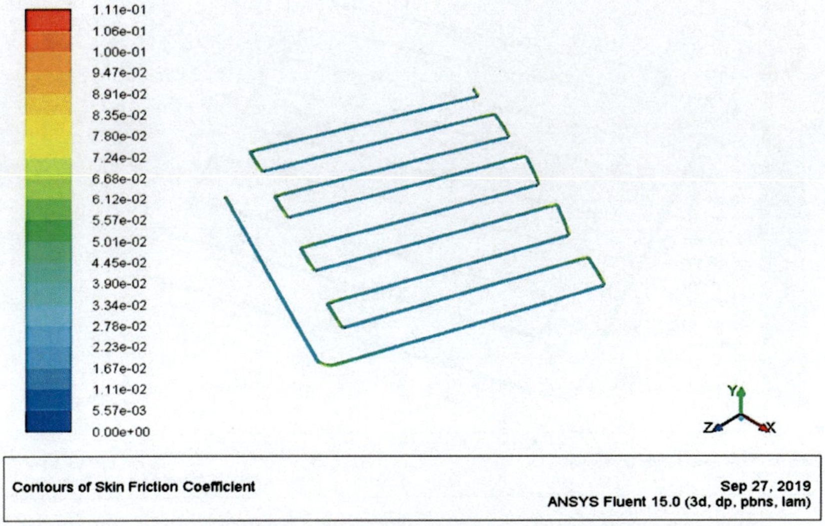

Fig. 5. Display of skin friction coefficient of tubes for empirical properties.

for quick heat diffusion in liquid metal [32]. Figure 7 and Fig. 8 presents the contour of Prandtl number for both properties accordingly. The development of Prandtl number was good throughout the collector for both types of properties of aqueous ethylene glycol. Moreover, Prandtl number depicted the temperature sensitivity character and both values showed the variation between them.

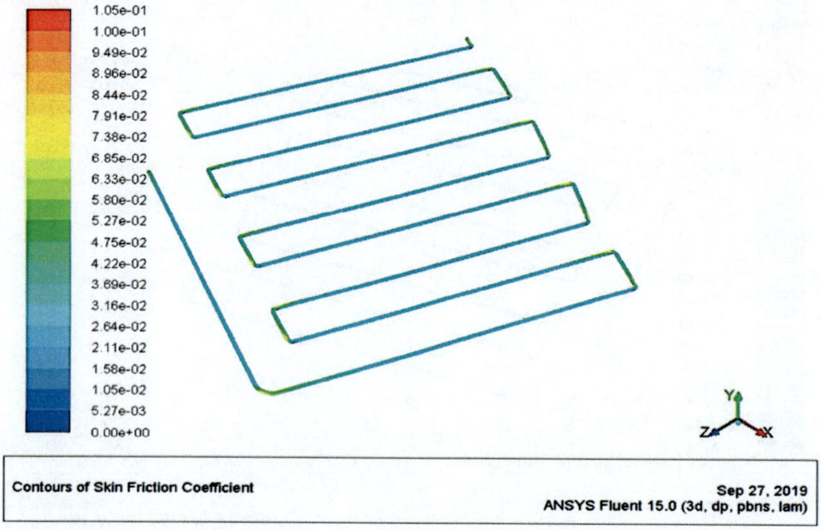

Fig. 6. Display of skin friction coefficient of tubes for experimental values.

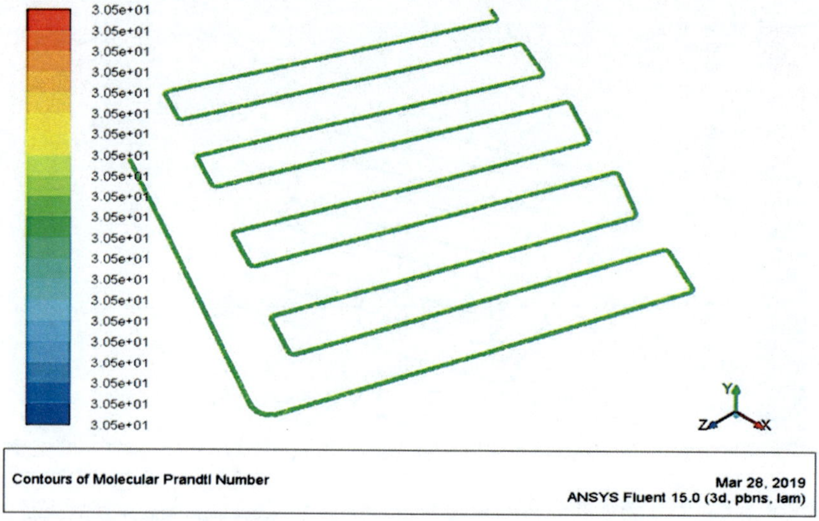

Fig. 7. The contour of the Prandtl number of empirical values.

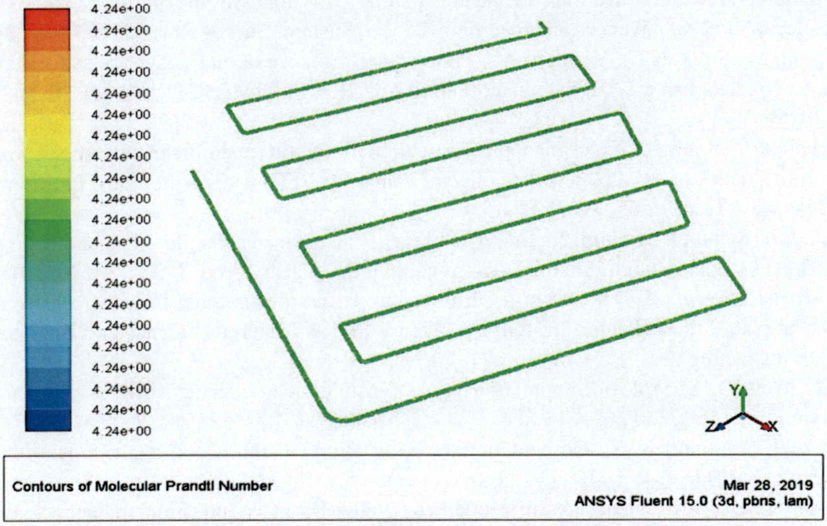

Fig. 8. The contour of the Prandtl number of experimental values.

5 Conclusion

In this computational simulation, different non-dimensional numbers of aqueous ethylene glycol have been studied with different properties (empirical and experimental) using ANSYS 15.0 workbench. The simulations revealed that the Nusselt number and Prandtl number of empirical values differ from the experimental values, but both define the similar criteria of fluid flow. However, the Nusselt number of experimental values at various temperatures has a maximum difference of about 42.5% between lowest and highest temperatures. Similarly, around 66.6% variation found in Prandtl number. On the other hand, skin friction coefficient of individual property of aqueous ethylene glycol is close among them. But there was about 33.3% difference of skin friction coefficient of aqueous ethylene glycol when it was at maximum and minimum temperature.

Acknowledgments. The authors would like to thanks to University Malaysia Pahang (UMP), Ministry of Higher Education (MOHE) of Malaysia for the Research Grants RDU 180328, 190323 and Bangabandhu Science and Technology Fellowship Trust (Bangladesh) to provide financial assistance and laboratory facilities to carry out this study.

References

1. York, R.: Do alternative energy sources displace fossil fuels? Nat. Climate Change **2**(6), 441–443 (2012)
2. Gadisa, A., et al.: Transparent polymer cathode for organic photovoltaic devices. Synth. Met. **156**(16), 1102–1107 (2006)
3. Mekhilef, S., Saidur, R., Safari, A.: A review on solar energy use in industries. Renew. Sustain. Energy Rev. **15**(4), 1777–1790 (2011)

4. Hussein, A.K.: Applications of nanotechnology to improve the performance of solar collectors–recent advances and overview. Renew. Sustain. Energy Rev. **62**, 767–792 (2016)
5. Jamil, M., Sidik, N.C., Yazid, M.M.: Thermal performance of thermosyphon evacuated tube solar collector using TiO2/water nanofluid. J. Adv. Res. Fluid Mech. Therm. Sci. **20**(1), 12–29 (2016)
6. Meibodi, S.S., et al.: Experimental investigation on the thermal efficiency and performance characteristics of a flat plate solar collector using SiO2/EG–water nanofluids. Int. Commun. Heat Mass Transfer **65**, 71–75 (2015)
7. Said, Z., Saidur, R., Rahim, N.: Energy and exergy analysis of a flat plate solar collector using different sizes of aluminium oxide based nanofluid. J. Clean. Prod. **133**, 518–530 (2016)
8. Farhana, K., et al.: CFD modelling of different properties of nanofluids in header and riser tube of flat plate solar collector. In: IOP Conference Series: Materials Science and Engineering. IOP Publishing (2019)
9. Davidson, D.: The role of computational fluid dynamics in process industries. In: Eighth Annu Symp Front Eng. (2003)
10. Koziel, S., Yang, X.-S.: Computational optimization, methods and algorithms, vol. 356. Springer (2011)
11. Yue, H., et al.: Ethylene glycol: properties, synthesis, and applications. Chem. Soc. Rev. **41**(11), 4218–4244 (2012)
12. Zhang, H., et al.: High Efficient Ethylene Glycol Electrocatalytic Oxidation Based on Bimetallic PtNi on 2D Molybdenum Disulfide/Reduced Graphene Oxide Nanosheets. J. Colloid Interface Sci. **547**, 102–110 (2019)
13. Kim, H.J., et al.: Highly active and stable PtRuSn/C catalyst for electrooxidations of ethylene glycol and glycerol. Appl. Catal. B **101**(3–4), 366–375 (2011)
14. Serov, A., Kwak, C.: Recent achievements in direct ethylene glycol fuel cells (DEGFC). Appl. Catal. B **97**(1–2), 1–12 (2010)
15. An, L., Chen, R.: Recent progress in alkaline direct ethylene glycol fuel cells for sustainable energy production. J. Power Sources **329**, 484–501 (2016)
16. Maliska, C.R.: On the Physical Significance of Some Dimensionless Numbers Used in Heat Transfer and Fluid Flow. Federal University of Santa Catarina, Florianópolis (1990)
17. Versteeg, H., Malalasekra, W.: An Introduction to Computational Fluid Dynamics: The Finite Volume Method Approach. Longman Scientific and Technical, Harlow (1995)
18. Tsierkezos, N.G., Molinou, I.E.: Thermodynamic properties of water+ ethylene glycol at 283.15, 293.15, 303.15, and 313.15 K. J. Chem. Eng. Data **43**(6), 989–993 (1998)
19. Roslan, A., et al.: The effects of ethylene glycol to ultrapure water on its specific heat capacity and freezing point. J. Appl. Environ. Biol. Sci **7**(7S), 54–60 (2017)
20. Deng, C., Zhang, K., Yang, T.: Thermal Conductivity of 1, 2-Ethanediol and 1, 2-Propanediol Binary Aqueous Solutions at Temperature from (253 to 373) K. arXiv preprint arXiv:1711.07189 (2017)
21. Reynolds, O.: An experimental investigation of the circumstances which determine whether the motion of water shall he direct or sinuous, and of the law of resistance in parallel channels. Philos. Trans. R. Soc. Lond. **174**, 935–982 (1883)
22. Hawwash, A., et al.: Numerical investigation and experimental verification of performance enhancement of flat plate solar collector using nanofluids. Appl. Therm. Eng. **130**, 363–374 (2018)
23. Noor, M.M., Wandel, A.P., Yusaf, T.: Simulation of biogas combustion in MILD burner. J. Mech. Eng. Sci. **6**, 9 (2014)
24. Ranjith, P., Karim, A.A.: A comparative study on the experimental and computational analysis of solar flat plate collector using an alternate working fluid. Procedia Technol. **24**, 546–553 (2016)

25. Said, Z., et al.: Performance enhancement of a Flat Plate Solar collector using Titanium dioxide nanofluid and Polyethylene Glycol dispersant. J. Clean. Prod. **92**, 343–353 (2015)
26. Mostafazade, A.A., Afshin, H.: Development of Nusselt number and friction factor correlations for the shell side of spiral-wound heat exchangers. Int. J. Therm. Sci. **139**, 105–117 (2019)
27. Sorokes, J.M., Hardin, J., Hutchinson, B.: A CFD primer: what do all those colors really mean? In: Proceedings of the 45th Turbomachinery Symposium. Turbomachinery Laboratories, Texas A&M Engineering Experiment Station (2016)
28. İnan, T., Başaran, T., Erek, A.: Experimental and numerical investigation of forced convection in a double skin façade. Energies **10**(9), 1364 (2017)
29. Barros, J.M., Schultz, M.P., Flack, K.A.: Measurements of skin-friction of systematically generated surface roughness. Int. J. Heat Fluid Flow **72**, 1–7 (2018)
30. Gupta, S.K., et al.: Investigation of coefficient of skin friction and axial velocity of fully developed turbulent flow through pipe. History **2**(7), 312–318 (2016)
31. Moinuddin, K., Joubert, P., Chong, M.: Skin friction CFD calculation for complex flow: turbulent flow along an external corner (2004)
32. Raju, B.H.S., Nath, D., Pati, S.: Effect of Prandtl number on thermo-fluidic transport characteristics for mixed convection past a sphere. Int. Commun. Heat Mass Transfer **98**, 191–199 (2018)

Design of a Drag and Lift Type Blade for Power Generation via Air Turbine

W. S. W. A. Najmuddin[1], M. T. Mustaffa[2(✉)], M. S. Abdul Manan[2], A. F. Annuar[2], and A. Atikah[2]

[1] School of Manufacturing Engineering, Universiti Malaysia Perlis, Kampus Pauh Putra, 02600 Arau, Perlis, Malaysia
[2] Structural Mechanics and Dynamics Research Group, School of Manufacturing Engineering, Universiti Malaysia Perlis, Kampus Pauh Putra, 02600 Arau, Perlis, Malaysia
mohammadtaufiq@unimap.edu.my

Abstract. Compressed air is a vital medium for transferring energy in industrial processes. Compressors are efficient and reliable in meeting need and demand, especially in the industry while at the same time helping to reduce carbon footprint. However, there are several constraints using an air turbine as an operational element in generating electricity from compressed air. One of the key factors involved is the design of a blade that capable to increase the performance of the turbine and electricity generation from compressed air. Thus, this paper objectives are to develop a proposed design of a drag and lift type blade design and to evaluate the power generate that able to be harnessed from the force of the air. Additionally, the power coefficient that is harnessed from the generator using these new designs of blade also evaluated. In the experiment, the model is set up using the proposed concept of Savonius and Darrieus typed in order to generate electricity for small power equipment. Results show that these blade designs able to generate of 21-v for Savonius whereas 24-v for Darrieus from 0.3 MPa input source. Based on the result of an experiment, this study reveals that both type of blade design able to generate electricity constantly however, the performance of Darrieus vertical axis blade design is better compared to Savonius vertical axis blade design.

Keywords: Compressed air · Drag force · Blade design · Air turbine

1 Introduction

Several of the renewable energy has shown huge attention among researcher to meet higher demanding of the clean energy. Natural of airflow known, as wind power is one of a renewable energy sources that shown a massive growth trend globally. Such rapid development and human population is largely lead by increasing of energy demand and at the same time the important need to reduce environmental impact. At the end of 2017, the global cumulatively installed wind capacity has reached 539,123 MW, compared to 318,697 MW in 2013, which has increased by more than 169% [1]. Harnessing wind energy is refers to the process of creating electricity using the wind, or airflows that occur naturally in the earth's atmosphere. However, the characteristics of fluctuation and

M. A. Zakaria et al. (Eds.): *Advances in Mechatronics, Manufacturing, and Mechanical Engineering*, LNME, pp. 132–138, 2021.
https://doi.org/10.1007/978-981-15-7309-5_13

intermittence wind speed create a major obstacle in delivering constant power output to the electricity demand. This condition also brings the negative effect onto the operation efficiency of wind turbine system, life expectancy and mechanical structures [2].

Currently, there are two categories of modern wind turbines based on the rotation of the axis namely horizontal axis wind turbines (HAWTs) and vertical axis wind turbines (VAWTs). Numerous researches have been carried out on the effect of blade design in performing higher efficiency for electric generation. Selection of design parameters is crucial for optimization of wind turbine performance. The effects of various design parameters such as number of blades, overlap ratio, cross sectional area of Savonius rotor on the performance of rotor were studied [3, 4], and [5]. According to the driven force, the vertical axis wind turbines can be classified into the lift and drag force. There are several study presents two designs of wind turbines consist of movable guide vanes and a shaft that were measured by drag force and drag coefficient [6]. The proposed design had increases the torque of the model on the side that rotates to wind direction, hence, increased the drag coefficient and reduced the negative torque on another side of the frame that rotates opposite to the wind.

Wind turbines are used to produce electricity from off shore wind farms to residential smaller scale wind turbines [7]. Lucas et al. present a paper related to wind turbines can be classified into two categories which are horizontal axis and vertical axis wind turbine [8]. Vertical axis wind turbine is the main concern in this experiment. Savonius is an open overlapping two half cup is used which is very helpful to wind turbine design. Simplicity, low cost construction, silence and angular velocity are the advantages of Savonius. In addition, it can overcome extreme weather conditions without any damage and also can adopt wind from different direction. On the other hand, the drag type of Savonius vertical axis wind turbine receives less wind energy by having a higher torque so it cannot turn faster than the wind speed.

Apart from that, the most representative model of a lift-type vertical axis wind turbine is the Darrieus turbine. In order to generate Darrieus turbine to move quicker than the wind flow, aerodynamic airfoil is used to create a lift force. Furthermore, blades of vertical axis wind turbine can have a constant shape along their length. These are much easier to design, fabrication and replication of the blade which can influence in a cost reduction. Jonny Hylander and Göran Sidén evaluate the power coefficient of wave power by using Betz's law [9]. This paper present a design and modelling of VAWT drag force blade using Savonius concept and lift force type using Darrieus concept. The model of air turbine is assembled and measure by using oscilloscope to determine the output voltage. Data collection is based on the output measurement of pressure gauge with seven different pressure indexes and then, the output voltage is measured. The result present the performance of the blade type of turbine related to the specified compressed air pressure.

2 Methodology

Dimensional drawing of prototype that has been proposed is designed using CAD design software. The designed part is transformed to digital model data in order to produce into a model by using 3D Printer. Figure 1 and Fig. 2 shows the vertical axis blade of Savonius and Darrieus concept respectively that has been designed.

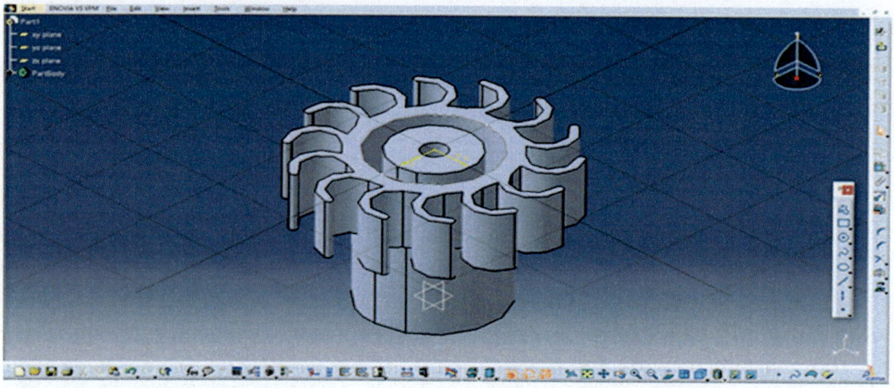

Fig. 1. Savonius vertical axis blade design model.

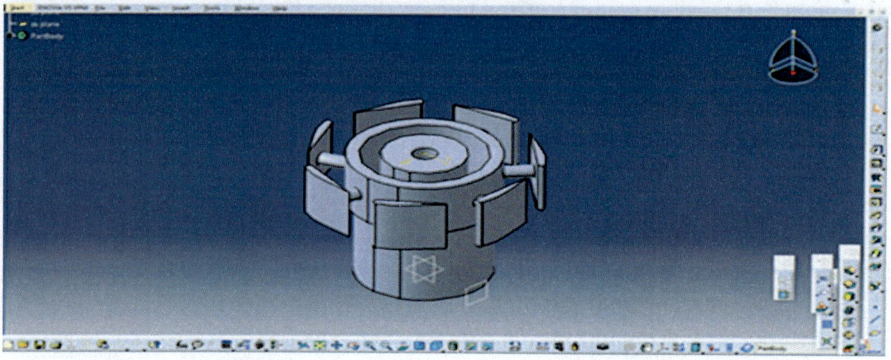

Fig. 2. Darrieus vertical axis blade design model.

The proposed type of a blade design, the Savonius vertical axis blade that consists with semicircular shapes is illustrated in Fig. 3 whereas the Darrieus concept of blade design is shown in Fig. 4. The schematic view of air turbine main parts is shown in Fig. 5.

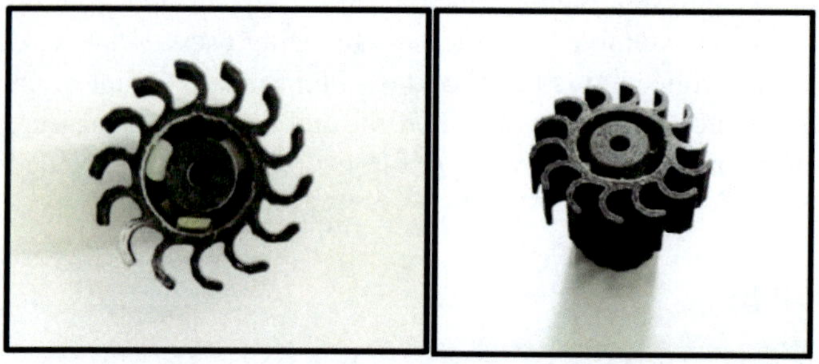

Fig. 3. Savonius vertical axis blade proposed model design.

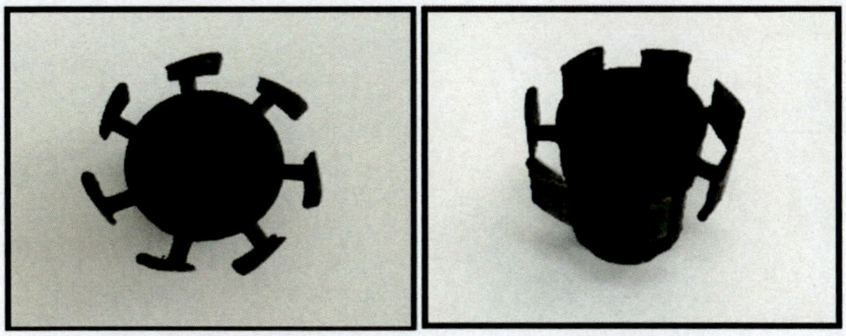

Fig. 4. Darrieus vertical axis blade proposed model.

Fig. 5. Main parts of air turbine.

In this stage, all the apparatus is prepared and set up as shown in the Fig. 6. The filter regulator lubricator is switched on to allow the compressed air flow to the pressure gauge. The particular load value of input pressure is determined by using pressure gauge. After the pressure has been adjusted to the desired input, the tap valve is switched on. Then, the compressed air flow to the air turbine and rotates the blade from the air energy. Air turbine generator that attached to the electrical wire outputs is actuated once the blades inside turbine start to rotate. The output voltage is recorded at oscilloscope. These steps are repeated by changing the pressure to particular load from 0.05 MPa, 0.08 MPa, 0.10 MPa, 0.15 MPa, 0.20 MPa, 0.25 MPa and 0.30 MPa.

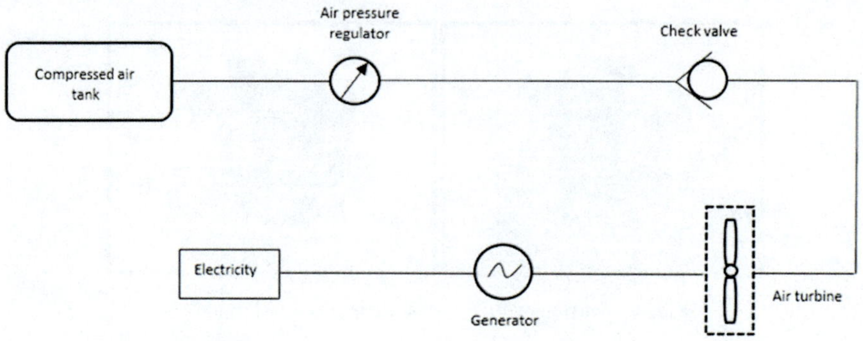

Fig. 6. The schematic of experimental set up for air turbine electric generation.

3 Result and Discussion

In this experiment, Savonius vertical axis blade design used to discover the output voltage. Table 1 shows both design of air turbine is tested by setting the value of input pressure gauge with 0.05 MPa, 0.08 MPa, 0.10 MPa, 0.15 MPa, 0.20 MPa, 0.25 MPa and 0.30 MPa.

Table 1. Data of voltage output for Savonius and Darrieus blade design.

Air pressure, MPa	Savonius blade design (Volt)	Darrieus blade design (Volt)	Comparison of Darrieus over Savonius (%)
0	0	0	0
0.05	0	0	0
0.08	0	9	900
0.10	8	14	75
0.15	9	18	100
0.20	13	19	60
0.25	16	20	25
0.30	21	24	14

Figure 7 illustrates the voltage generated when using the proposed types of blade design. The experiment is conducted with seven different pressure indexes and the output voltage is measured.

From the graph, Darrieus design generated the highest voltage of 24 V whereas Savonius design produces 21 V. Minimum pressure of 0.08 MPa is able to produce of 9 V of power using Darrieus design concept. However, for Savonius design blade, 8 V of power is generated with minimum pressure of 0.1 MPa applied. Increment value of compressed air pressure relatively will increase the power output. Thus, the blade design that equipped with lift type of Darrieus blade design generated able to generate more energy with the escalation of load pressure.

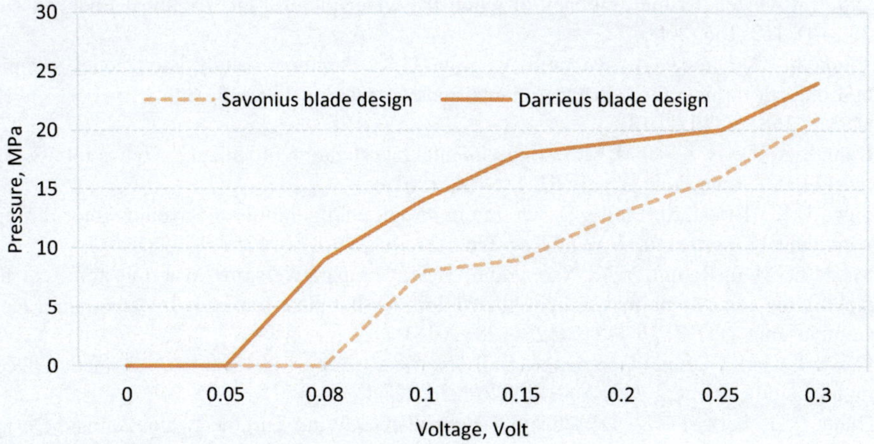

Fig. 7. Savonius and Darrieus blade design output voltage versus pressure generated.

Therefore, the Darrieus blade design produces high velocity and power coefficient taking into account the lift type of blade design that assemble in the casing compartment which use direct flow source of air compressed.

4 Conclusion

The main focus of this research is to develop air turbine using compressed air system and the effects of the blade design based on the performance of the air turbine. An air turbine is used to directly drive a generator that transforms compressed air pressure to generate electricity. The experiment has reveals that the Darrieus type generates higher voltage of 24 V compare to Savonius drag force type blade design with 21 V. Thus, Darrieus design able to generate more electricity than Savonius typed. Others, Darrieus type also able to produce power at minimum pressure of 0.08 MPa compared 0.1 MPa in Savonius type. In fact, Darrieus typed can generates large torque primarily in the high-speed range compare to Savonius typed which generates large torque in the low speed range. This experiment also shows capability of the design to produce amount of energy that can be applied in small electrical equipment. Significantly, other options that involve the number of blades and the types of blade design should be recommended for improvement in terms of increasing the efficiency of the air turbine system.

Acknowledgement. The authors would like to thank Research Management and Innovation Centre of Universiti Malaysia Perlis (UniMAP) for funding support, and Universiti Pertahanan Nasional Malaysia (UPNM) for providing the facilities during experimental work.

References

1. Global Wind Energy Council. Global statistics n.d. http://www.gwec.net/global-figures/graphs/. Accessed 13 July 19

2. Gul, T., Stenzel, T.: Intermittency of wind: the wider picture. Int. J. Global Energy Issues **25**(3–4), 163–186 (2006)
3. Talukdar, P.K., Sardar, A., Kulkarni, V., Saha, U.K.: Parametric analysis of model Savonius hydrokinetic turbines through experimental and computational investigations. Energy Convers. Manag. **158**, 36–49 (2018)
4. Damak, A., Driss, Z., Abid, M.S.: Experimental investigation of helical Savonius rotor with a twist of 180. Renewable Energy **52**, 136–142 (2013)
5. Saha, U.K., Thotla, S., Maity, D.: Optimum design configuration of Savonius rotor through wind tunnel experiments. J. Wind Eng. Ind. Aerodyn. **96**(8–9), 1359–1375 (2008)
6. Mustaffa, M.T., Rahim, Y.A., Najmuddin, W.S.: Comparative study of a drag force for flat plate frame and a combined straight bladed darrieus-flat plate frame. J. Telecommun. Electr. Comput. Eng. (JTEC) **10**(1–14), 125–128 (2018)
7. Howard, Clark National Geographic Daily News, 20 August 2012. http://news.nationalgeographic.com/news/energy/2012/08/120820-helix-wind-collapse/
8. Dunn, C.G., Sarkar, D.S., Deisadze, L.: Vertical Axis Wind Turbine Evaluation and Design (2013)
9. Sidén, G., Ambrosio, M.D., Medaglia, M.: Vertical axis wind turbines: history, technology and applications master thesis in energy engineering – May 2010 Supervisors: Jonny Hylander Authors (2010)

Effect of Dimple Diameter and Pattern on Frictional Properties of Macro-Dimpled Aluminium Surface

Rahimi Ramli[1] and Izwan Ismail[1,2](✉)

[1] Faculty of Manufacturing and Mechatronic Engineering Technology, Universiti Malaysia Pahang, 26600 Pekan, Pahang, Malaysia
izwanismail@ump.edu.my

[2] Center of Excellence for Advanced Research in Fluid Flow (CARIFF), Universiti Malaysia Pahang, Lebuh Raya Tun Abdul Razak, 26300 Pahang, Gambang, Malaysia

Abstract. This paper presents the effect of different dimple size and dimple distribution on the reduction of surface friction and to produce dimple surface texture using ball-nose milling method. The dimple is a geometrical feature created on flat steel as surface texture to improve tribological performance. The dimple surface texture produced by using ball-end-nose 3-axis CNC milling method. The dimple shape is circular with size varies between 0.5 mm to 3.0 mm. The height of dimples was set constant at 0.25 mm. Dimples were distributed with two arrangement which is linear and radial. Five samples with different dimple size produce for each dimple arrangement. The geometry of dimples then measured using optical microscopy and image analysis. The samples than tested for surface friction with and without lubricant using Universal Testing Machine (UTM) equipped with ASTM D1894 complied apparatus. The lubricated friction test conducted under the constant volume of hydraulic-oil ISO-68 with the face-to-face condition. The results show that the size and distributions of dimple have a significant impact on the friction performance of the textured surface. The friction reduction behaviour was significantly observed on samples with surface texture compared with untextured one, the specimens. When the test load increases, the effect of friction reduction from all forms of dimple size decreases. This study provided insight on improving life of moving parts in machinery and automotive components.

Keywords: Surface texture · Dimple structure · Surface friction

1 Introduction

In recent years, there has been an increase in the technology of the friction reduction system. The motivation behind it is the friction-related energy losses contributes to limits the efficiency of the mechanical system especially when two metals rubbing or sliding each other [1]. There are varieties of method to overcome this problem, one of them is by altering topographical characteristic of the surfaces of the materials [2]. This method which is also called as surface texturing can reduce friction significantly [3].

M. A. Zakaria et al. (Eds.): *Advances in Mechatronics, Manufacturing, and Mechanical Engineering*, LNME, pp. 139–146, 2021.
https://doi.org/10.1007/978-981-15-7309-5_14

There are various methods to reduce the friction of metal surfaces, such as applying lubricant, coatings, and surface texture. Having two methods at the same time will enhance the friction reduction behaviour. This idea can be achieved by producing a surface texture that consists of dimple. The dimple retaining applied lubricant and provide lift in the hydrodynamic state during tribological action [4].

During sliding of metal surfaces such as aluminium, micro debris produced during friction build micro wedges in which scratching the surface itself. This behaviour increases surface friction. Lubricant tends to squeeze away from the flat sliding surface thus losses of hydrodynamic pressure occurred. This behaviour induced sticking problems of the sliding surface. The presence of micro dimple acts as a trap for the micro debris which reduced the chance of wedge formation. Dimples also change the lubrication state by acting as an additional lubricating reservoir and affecting the distribution of hydrodynamic pressure.

With less friction for mechanical components, it can directly impact efficiency and reduce energy consumption. Reducing friction provides added benefits to reduce wear, which can extend the life of the mechanical parts. Some researchers try to characterize how the shape, diameter size, depth and surface density affect the surface friction velocity. Even sliding velocities between two surfaces have been shown to have different effects depending on dimple size on the friction coefficient. Generally, many have concluded that the presence of micro dimple can improve the reduction-of-friction.

Shape, diameters and density of dimpled surfaces bring impact on friction reduction on two sliding surfaces. Piston rings with micro-dimples structure reduce friction from 800 N to 380 N [5]. This result is translated directly into energy efficiency and fuel savings.

Dimple structure surface can be produced by several different process methods such as using micro end milling. This is one of the methods that can produce dimple-surface-structure with a different diameter. This method brings similar benefits that conventional machining processes have. However, micro-milling tends to produce a dimpled surface with higher accuracy and more flexible machining. The high accuracy of this milling method is mainly due to the cutting tool is controlled by computer. Accuracy and precision control of the pattern of dimples are very crucial because it will affect the performance of the surface when sliding with another reference surface. Feed rate, depth of cut and the spindle speed is the critical machining parameter that should consider when producing the desired dimple structure on the surface [2].

Each of the dimple geometries is highly depended on the angle of inclination of the tool, spindle speed and feed rate. Every tooth can successfully remove material if spindle speed is selected correctly. Moreover, high feed rate could make sure that the dimple structure is not lapping over when the tool cut the material [6]. Cutting tool movement with a high feed rate during the milling process could produce fine dimple-structure [7].

In this study, the effect of dimple size and dimple distribution on the reduction of surface friction was investigated. Both kinetic and static friction was measured. The dimple surface texture produced using ball-nose milling method was selected as a method to prepare the sample.

2 Methodology

The effect of different dimple size and dimple distribution on the reduction of surface friction (kinetic and static) was investigated through experimental study. One factor at a time (OFAT) experimental design used to investigate effects. Eight samples of Al 6010 with various dimple size and distribution were prepared. Prior surface texture processing, the Al 6010 were cut to size of 60 mm × 50 mm × 10 mm using a milling machine. The cutting speed and feed rate were kept constant to produce surface roughness of (~Ra = 1.5 μm ± 0.05 1.5 μm). The surface roughness was measured using Universal Testing Machine (UTM).

The process of dimple machining was done by using the eyes of a ball end nose tool mounted on a grinding CNC machine. Depth of cut is set at a depth of 0.5 mm. The rotation speed is set at 1000 RPM and the feed rate is set at 950 mm/min. The dimple depth is set to the same value for all diameters, the size of the error and the distance of the puncture of the specimen differ according to the size of the dimple.

In this study, there are 4 different dimple sizes namely 1 mm, 3 mm 5 mm and 6 mm. The change in friction to the difference in dimple size can be clearly seen when the interval is greater than 20%. Duplicate arrays in 2 polar arrays namely in-line and radial.

Samples that do not have any dimple are used as a reference for friction effect experiments compared to dimple samples. All tests are using Universal Testing Machine.

CNC ball-end milling process was used to produced four different dimple sizes at a constant depth of 0.5 mm depth. The size of all these samples is 60 mm × 50 mm × 10 mm. The tool size that used to make the dimple are different for every size. The size of cutting tools of 1 mm, 5 mm, 12 mm and 18 mm were selected to produce dimple diameter of 1 mm, 3 mm, 5 mm and 6 mm respectively. This difference is happened because of the constant depth that makes a shallow penetration.

There are two types of arrangements for the dimple are produced. It is in-line and radial as shown in Fig. 1a and Fig. 1b. Although there are two types of arrangement, the amount of dimple produced from the machining process is different for each sample. Gap has been set for each dimple is 2.5 mm. The amount of dimple produced for the smallest size of dimple 1 mm is 295 dimples in radial and 256 dimples in in-line. Whereas, the largest of the dimple size for 6 mm is 39 dimples in radial and 36 dimples in in-line.

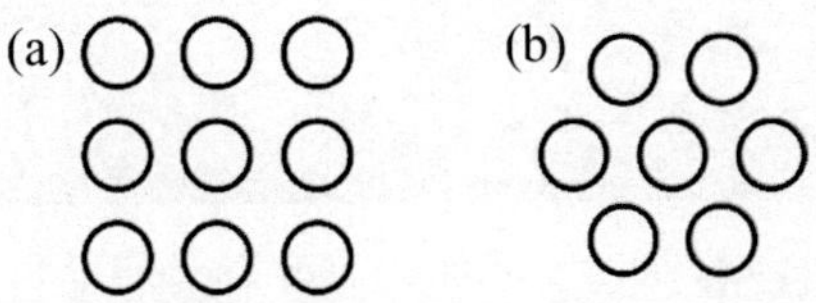

Fig. 1. Arrangement of dimple (a) in-line and (b) radial.

In the friction test, the samples are tested to obtain friction against the 8 samples that have been provided. Each sample has a different dimple size, and this enables different friction results in each test. Testing is done by using Universal Testing Machine (UTM).

UTM testing Machine is one of the recommended machines to perform this test. ASTM D1894 is a testing standard that covers the determination of the friction coefficient of starting and sliding. Testing apparatus and methodology such as sample size, sliding plate materials, pulling string, reference plane, pulley materials and force measuring device (UTM) are comply with this standard. UTM was fitted with a specially designed apparatus. It is then fastened to a rope at the connection between the sample and clamping on the UTM 50KN Load Cell machine and across the bearing aimed at pulling the sample onto the sliding plate material.

3 Result and Discussion

Dimples produced on Al 6010 samples have circular in shape from the top view as shown in Fig. 2. Figure 2(a) and (d) are samples with radial pattern while figure (b) is sample with in-line pattern. Consistent dimple size produces in each sample.

There are 8 different patents per sample. The sizes of each dimple are 1 mm, 2 mm 5 mm and 6 mm and are arranged in 2 order shapes that are in-line and radial. The distance for each dimple is fixed at 2 mm and the smaller the dimple the larger the number will be. Figure 2(a) and (e) displays a 1 mm size, dimple (b) and (f) shows a 3 mm size dimple, (c) and (g) show displays a 5 mm size dimple and last (d) and (h) displays are 6 mm size dimple.

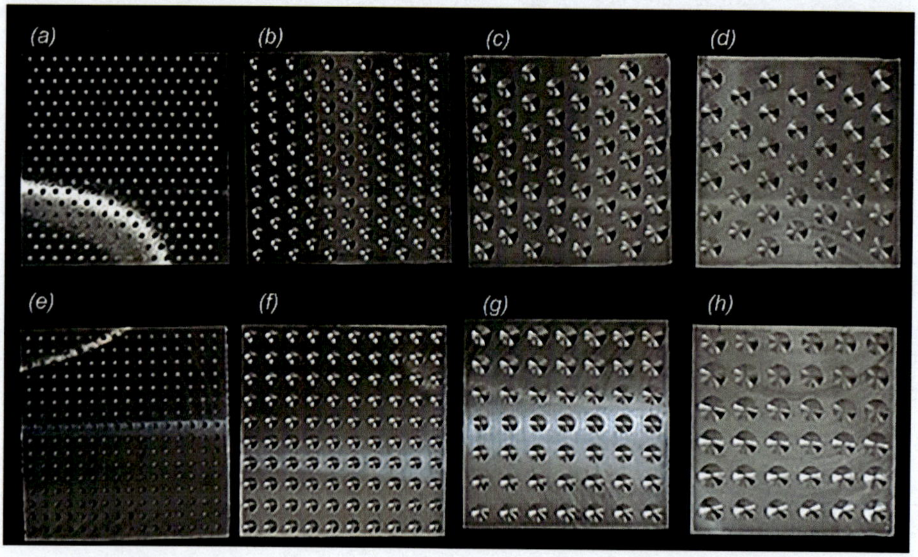

Fig. 2. Photograph of sample surface for in-line and radial arrangement

Figure 3 shows the overall comparison of static friction coefficient among all the sample. Among all the samples that had tested, samples with radial dimple arrangement under lubricated were the lowest static friction coefficient. Sample with in-line dimple arrangement under non-lubricated surface condition obtained the highest value of static friction coefficient followed by radial dimple arrangement with non-lubricated surface and in-line dimple arrangement under lubricated condition under lubricated surface condition. These results showed that the lubricated surface had more significant effect on reducing static friction coefficient compare to dimple arrangement. Radial dimple arrangement was better dimple arrangement compared to in-line dimple in term of reducing static friction. However, with the help of lubricant, dimple with in-line arrangement would reduce more static friction compared to radial arrangement under the non-lubricated surface. Data for Fig. 3 are presented in Table 1.

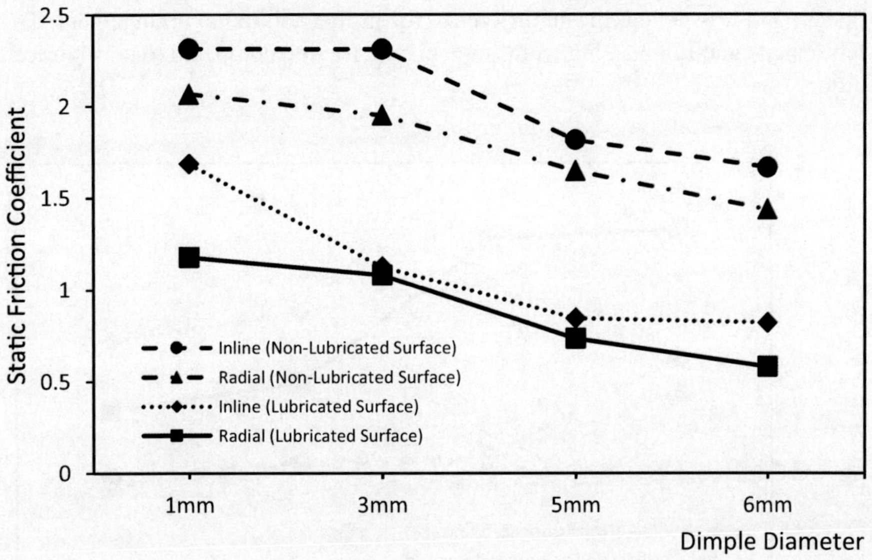

Fig. 3. Overall comparison of static friction coefficient

Table 1 shows measured static friction of samples with different dimple diameter size and arrangement with and without lubricant.

Table 1. Overall comparison of static friction coefficient

Dimple diameter	In-line (Non-lubricated)	Radial (Non-lubricated)	In-line (Lubricated)	Radial (Lubricated)
1 mm	2.313	2.067	1.686	1.178
3 mm	2.312	1.953	1.126	1.081
5 mm	1.817	1.648	0.845	0.735
6 mm	1.667	1.438	0.82	0.581

A larger dimple diameter enhanced the performance of reducing static friction. Lubricated samples with radial dimple arrangement perform the best performance in reducing static friction, larger dimple would make the reducing effect more effective. In Fig. 3,

6 mm shows lower static friction on every different arrangement and lubrication condition while 1 mm had the highest followed by 3 mm and 5 mm. However, 3 mm under in-line arrangement and radial arrangement with both lubricated surfaces had very small differences of static friction coefficient. It is due to the boundary of dimple diameter which obtains almost the same hydrodynamic pressure with in-line and radial arrangement.

Results on kinetic friction are shown in Fig. 4 where the overall comparison of average kinetic friction coefficient of all the samples presented. Compared to the static friction coefficient, average kinetic friction coefficient had more complicated results. Radial dimple arrangement under non-lubricated had the least average kinetic friction coefficient and it is no significant difference for in-line and radial arrangement. Then, the lubricated condition had higher average kinetic friction compared to non-lubricated condition.

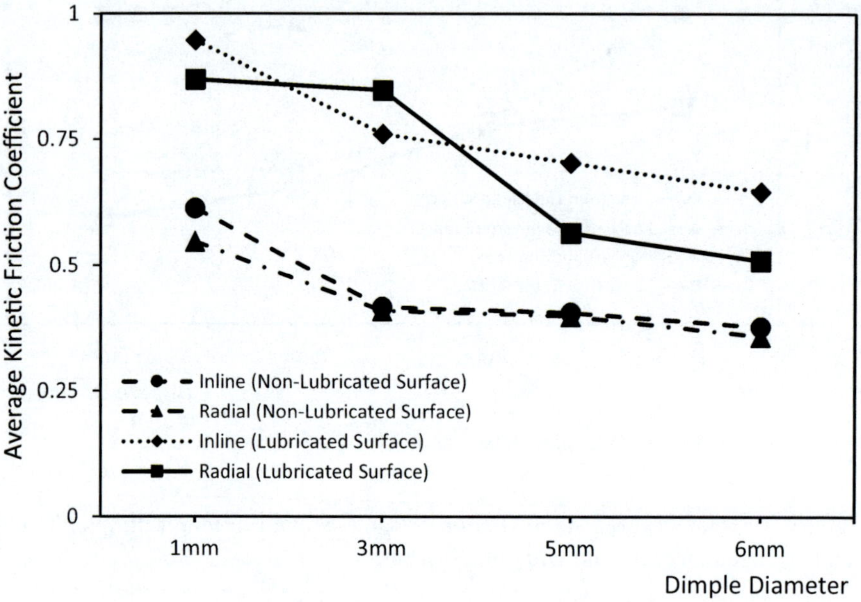

Fig. 4. Overall comparison of average kinetic friction coefficient

Table 2 shows measured kinetic friction of samples with different dimple diameter size and arrangement with and without lubricant.

At lubricated surface condition, radial dimple arrangement was better in reducing average kinetic friction. For 3 mm dimple diameter, in-line dimple arrangement which had 0.761 was better compared to radial arrangement which had 0.848. In term of dimple diameter, larger dimple diameter still performing better in reducing friction compare to another smaller dimple diameter. Thus, surface larger dimple diameter and radial dimple arrangement still would obtain smaller average friction coefficient. However, the reaction of lubricant and aluminium surface was the reason that affecting the performance of dimple-structure in reducing the average friction coefficient. Therefore, non-lubricated

Table 2. Overall comparison of average kinetic friction coefficient

Diameter	In-line (Non-Lubricated)	Radial (Non-Lubricated)	In-line (Lubricated)	Radial (Lubricated)
1 mm	0.613	0.547	0.947	0.869
3 mm	0.416	0.409	0.761	0.848
5 mm	0.405	0.398	0.703	0.564
6 mm	0.377	0.358	0.646	0.509

surface condition was the choice of lubrication condition for reducing aluminium surface kinetic friction.

4 Conclusions

Findings of this study conclude that dimple holes were proven to reduce friction. We see in static condition for friction that small dimple with non-Lubricated surface and in-line become high friction. In average dry (non-lubricated) condition, the in-line with non-lubricated become higher friction at dimple size of 1 mm. Then the friction reduced at larger the dimples size. This finding is similar in the lubricated situation.

Moreover, in kinetic conditions, large dimple areas are found to have a low friction coefficient and reduce friction better than small dimple. This is because every dimple on the surface will have the ability to produce hydrodynamic pressure and of course have greater hydrodynamic pressure contributing to the reduction of coefficient of friction. On the other hand, in lubrication, larger dimple diameters and higher densities will reduce surface friction.

Acknowledgement. The authors would like to acknowledge the Ministry of Education Malaysia (MOE) for the funding of this research under the Fundamental Research Grant Scheme (FRGS/1/2016/TK03/UMP/02/12) RDU160132.

References

1. Hutchings, I., Shipway, P.: Tribology: friction and wear of engineering materials. Butterworth-Heinemann (2017)
2. Graham, E., Park, C.I., Park, S.S.: Fabrication of micro-dimpled surfaces through micro ball end milling. Int. J. Precision Eng. Manuf. **14**, 1637–1646 (2013)
3. Yu, H., Deng, H., Huang, W., et al.: The effect of dimple shapes on friction of parallel surfaces. Proc. Inst. Mech. Eng. Part J J. Eng. Tribol. **225**, 693–703 (2011)
4. Ji, J.-H., Guan, C.-W., Fu, Y.-H.: Effect of micro-dimples on hydrodynamic lubrication of textured sinusoidal roughness surfaces. Chinese J. Mech. Eng. **31**, 67 (2018)
5. Denkena, B., Grove, T., Schmidt, C.: Machining of micro dimples for friction reduction in cylinder liners. Procedia CIRP **78**, 318–322 (2018)

6. Matsumura, T., Takahashi, S.: Micro dimple milling on cylinder surfaces. J. Manuf. Process. **14**, 135–140 (2012)
7. Hassan, S., Ghani, J.A., Haron, C.H.C., et al.: A review of the milling process to fabricate a dimple structure. J. Mech. Eng. **1**, 175–192 (2017)

Feasibility Study of Wave Energy Converter Using Compressed Air to Generate Electricity

W. S. W. A. Najmuddin[1], M. T. Mustaffa[2](✉), M. S. Abdul Manan[2], A. F. Annuar[2], A. Atikah[2], and M. N. Azzeri[3]

[1] School of Manufacturing Engineering, Universiti Malaysia Perlis, Kampus Pauh Putra, 02600 Arau, Perlis, Malaysia

[2] Structural Mechanics and Dynamics Research Group, School of Manufacturing Engineering, Universiti Malaysia Perlis, Kampus Pauh Putra, 02600 Arau, Perlis, Malaysia

mohammadtaufiq@unimap.edu.my

[3] Faculty of Science and Defence Technology, Universiti Pertahanan Nasional Malaysia, Kem Sg. Besi, 57000 Kuala Lumpur, Malaysia

Abstract. Wave energy is one of the promising resources for harnessing a vast amount of renewable energy. Wave energy converter is being defined by a particular design and the power take-off system which signify the mechanism of absorbing the wave energy and convert it into electricity. Addressing the significant potential of wave energy, the wave converter using a pneumatic system and compressed air as energy storage is designed and fabricated. The concept is presented and tested in the wave maker to analyze the performance of the prototype. This project aims to study the capabilities of small-scale, low cost and portable wave energy converter using compressed air to generate power. In this project, an experiment is conducted using the force of wave to generate linear motion of double acting cylinder that specifically creates compressed air in a small-scale tank. From the experiment, it is observed that the wave energy able to compressed air at pressure up to 0.5 bar. Then, the compressed air which is stored in the storage tank is released and it flows through water turbine that works as a generator to produce a power output of 18 V. The feasibility study reveals that the experiment able to harness electricity using low equipment tool. Further improvement of the design can increase the efficiency of the pneumatic generated system and able to generate an electric power.

Keywords: Wave energy · Energy conversion · Pneumatics

1 Introduction

In recent years, the ocean energy research has attracted industry and researchers in this field since it has many potentials to generate electricity. As part of renewable energy, ocean energy can be divided into six types of different origins and characteristics: ocean wave, tidal range, tidal current, ocean current, ocean thermal energy, and salinity gradient [1–3]. The globally distributed resources associated with most ocean energy sources able to provide renewable energy with the potential contribution of energy supply and mitigate climate change in future [2]. With the exception of tidal range energy, the

M. A. Zakaria et al. (Eds.): *Advances in Mechatronics, Manufacturing, and Mechanical Engineering*, LNME, pp. 147–153, 2021.
https://doi.org/10.1007/978-981-15-7309-5_15

energy derived from the ocean involved the technologies that have reached conceptual up to demonstration but still at early stage of development.

Although, ocean energy technologies are at an initial stage of development, there is an extensive effort on undergoing research and development and encouraging participation between government, learning institutions as well as industrial sector. Due to its nascent stage of development, ocean energy potential significant contribution lies in the constant of energy supply and reduction of carbon emission in overcoming climate change. Thus, this study aims to develop a device for electric generation using the wave energy resource with zero carbon emission. The concept of the proposed prototype differs from the current concept as it involves a mechanism system of pneumatic and compressed air to capture wave energy. The prototype is adaptable to be placed either onshore or offshore. The buoyancy is placed on water surface level which connected with a pneumatic cylinder and the tank on the platform. Compressed air obtains from the wave will be stored in the miniature portable tank as energy storage. The energy source of compressed air then can generate electricity.

There are various types of compressors used in the pneumatic system which generates a high compressed air pressure as energy storage. Today, a huge variety of different CAES concepts exist at different levels of development and applications substantiated by a broad historical background on how compressed air energy storage (CAES) has evolved over time [4]. Portable and hand-carrying size of a miniature compressed air storage system have many advantages in terms of its sustainability, clean, small size, easy access and low maintenance. The capability of the miniature compressed air storage system during the reciprocating process by using a piston type of double-acting cylinder as a compressor has been investigated and analyzed [5].

Several technologies are developed to exploit wave energy with different operating principles for wave energy conversion. The energy is equally separated between the potential energy component, where the water is forced against gravity from the wave trough and crests and the kinetic energy component, that is, the water oscillating velocity [6]. Many patents have been published, with each project being defined via particular design and power take-off system includes hydraulic, air, electrical, and mechanical which represent the mechanism capturing the wave energy and convert it into electricity [7, 8]. Waves are categorized into two namely wind seas, the waves generated locally and swell, the waves generated by distant winds. In specific, the energy density of swell wave is more consistent and it is important for the wave energy converter industry [9].

2 Methodology

2.1 Fabrication Process

In this project, the experiment about wave energy converter and compressed air system for small-scale application is conducted using the wave maker. Wave energy is an input and a double acting round cylinder is used to compress air into the miniature tank in order to produce the compressed air as power output. In order to ensure the air is compressed from the wave energy to generate electric power, a wave energy converter is designed with three main parts, namely float, connectors and frame that hold a pneumatic cylinder. The float is made up of layers of Styrofoam and the front side of the float is designed in curved shape to make sure more force of wave is impacted on that portion as shown in Fig. 1. Layers of styrofoam are coated with fiberglass to strengthening the structure of the float. Additionally, the frame is made up to withstand forces of float from the wave formed which creates a linear motion of double acting pneumatic cylinder. An iron plate is attached to the body of the frame to hold the upper part of the cylinder.

Fig. 1. The float made with layers of styrofoam.

2.2 Experiment

The experiment is carried out in Senibina Kapal laboratory that is equipped with flume wave maker at Faculty of Science and Defense Technology, Universiti Pertahanan Nasional Malaysia (UPNM). The dimensions of the wavemaker is 9.8 m × 2.0 m × 1.4 m (length × width × height) with rectangular section as shown in Fig. 2. At one end is fitted with wave generator whereby wave absorbing beach at the other end to receive the wave energy impact.

Fig. 2. The wave maker laboratory at National Defence University of Malaysia (UPNM).

The connecting bar provides the mechanical linkage between float and a pneumatic cylinder that attached to the frame. The float is fixed at the end of connecting bars whereas another end is attached to the lower part of the pneumatic cylinder which also acts as a lever with a bearing pivoted to its center to allow the linear movement of the piston to produce compressed air and stored in the miniature tank. All the parts of the wave energy converter which has been made are placed on the standing platform as shown in Fig. 3.

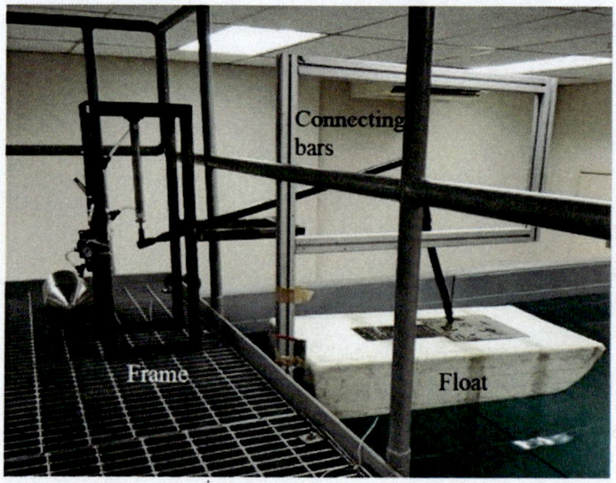

Fig. 3. Installation of wave energy converter in to the wave maker.

2.3 Experimental Process

The float is put on the water while the frame is located on the standing platform. Both of them are linked using a connecting bar to transfer the energy force created from the wave. The energy from the wave then moves the pneumatic cylinder and produce compressed air that kept into the portable tank. Figure 4 shows the schematic diagram of air compression that is developed for the experiment.

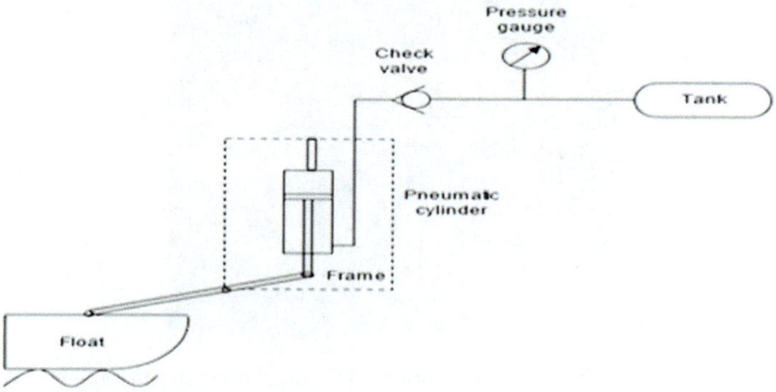

Fig. 4. Schematic diagram of wave energy air compression.

The first phase of the compressed air system is completed which allows the pressurized air flows through the storage tanks following the extraction and retraction of the piston during the compression process due to the constant wave. The compressed air stored in the portable tank able to generate electricity via a small rotating turbine which gives the power output displayed on the oscilloscope.

This second phase involved the data collection of power output from the compressed air inside the tank which presents the capability of wave energy to generate electricity.

3 Result and Discussion

In this experiment, the wave input is obtained from the wave maker and the data of actuator is displayed which varies with time. The actuator load is the force applied by the actuator in the wave maker to create a wave similar to the ocean waves. For each second, there would be a different actuator load value which shows the variation of the forces to generate a wave. Figure 5 shows the graph of the actuator load of wave maker against time when the wave parameter is set accordingly as shown in Table 1. The actuator is driven by a 13.8 MPa pressurized oil supply acting on a 9.5 cm diameter double acting piston with the capacity to drive the ram at a maximum velocity of 1 m/s.

Table 1. The setting of wave parameters

Sea state type (wave condition)	Regular sea wave (monochromatic)
Wave period	1 (s)
Wave height	0.09 m

The wave energy provides the force to move the float and consequently produces an extension and retraction of a double acting pneumatic cylinder that attaches at frame via a connecting rod. The compressed air created via pneumatic mechanism system from constant wave energy is then stored in the air storage tank. Besides, the pressure of compressed air is identified by using the pressure gauge. The linear motion of the piston rod in a pneumatic cylinder is extended when the float lifted and retracted when the float back to the normal level. Figure 6 shows the extension and retraction of the piston rod in a pneumatic cylinder through the wave produced.

The compressed air which has been stored in the tank released and converted the mechanical motion in a small turbine that connected in the model system to generate power output and recorded in terms of a voltage signal. At air pressure of 0.5 bar, a small turbine generates electricity and the power output is displayed on the oscilloscope as illustrates in Fig. 7.

The reading in oscilloscope shows the voltage peaked generates a current of 18 V. Based on the feasibility study, the experiment and data collected reveal that the prototype capable to generate electricity. Besides, with further improvement of this prototype, it can be able to produce high pressured air.

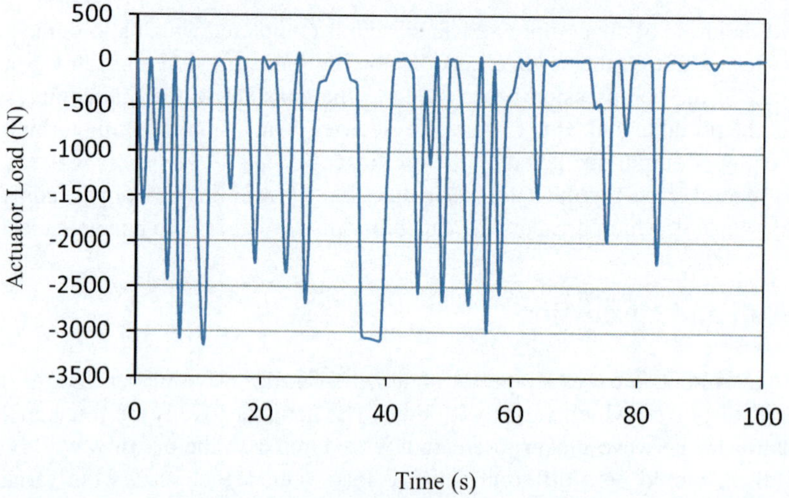

Fig. 5. Actuator load versus time.

Fig. 6. Extension and retraction of piston rod in pneumatic cylinder due to motion of float.

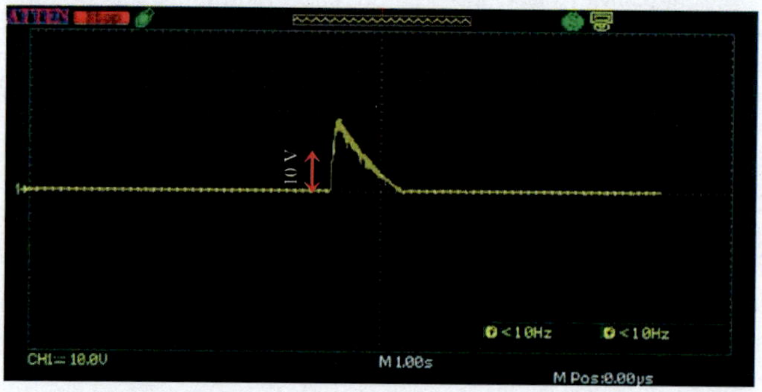

Fig. 7. The power output generate from wave maker by pneumatic system.

4 Conclusion

In conclusion, a small-scale compressed air system is applied in this project as a wave energy converter. Compressed air energy storage is known for its environmental friendly, zero carbon emission and requires lower cost compare to other energy sources such as natural gas and fossil fuels. The performance of the wave energy converter is tested using a wave maker with selected parameters. The wave energy has been successfully converted into compressed air and stored in the small-scale tank, followed by generation of power output in terms of a voltage signal. In the future, it is recommended to make several improvements of the design and reduce lost energy due to mechanical linkages in order to capture more wave energy that will deliver cost-effective and better efficiency of the prototype. By using a large compressed air tank, the pressure can be increased as well as electricity generating time. Hence, the experimental study is feasible for further research in terms of generates electricity for local demand and low carbon emission using high potential of renewable energy.

Acknowledgement. The authors would like to thank Research Management and Innovation Centre of Universiti Malaysia Perlis (UniMAP) for funding support, and Universiti Pertahanan Nasional Malaysia (UPNM) for providing the facilities during experimental work.

References

1. Huckerby, J., Jeffrey, H., Jay, B., Executive, O.: An international vision for ocean energy. Ocean energy systems implementing agreement. Lisbon, Portugal (2011)
2. Mitigation, Climate Change: IPCC Special Report on Renewable Energy Sources and Climate Change Mitigation. Cambridge University Press, Cambridge (2011)
3. Brito, A., Villate, J.L.: Annual report. Implementing agreement on ocean energy systems. IEA-OES, The Executive Committee of Ocean Energy Systems. Lisbon, Portugal (2014)
4. Budt, M., Wolf, D., Span, R., Yan, J.: A review on compressed air energy storage: basic principles, past milestones and recent developments. Appl. Energy **170**, 250–268 (2016)
5. Najmuddin, W.S.W.A., Mustaffa, M.T., Awang, A.H., Hussain, S.: Development of miniature compressed air storage system using solenoid valves for dynamic pneumatic actuator. J. Telecommun. Electron. Comput. Eng. (JTEC) **10**(1–14), 83–86 (2018)
6. Falnes, J., Kurniawan, A.: Fundamental formulae for wave-energy conversion. Roy. Soc. Open Sci. **2**(3), 140305 (2015)
7. Titah-Benbouzid, H., Benbouzid, M.: An up-to-date technologies review and evaluation of wave energy converters (2015)
8. Têtu, A.: Power take-off systems for WECs. In: Handbook of Ocean Wave Energy, pp. 203–220. Springer, Cham (2017)
9. Rusu, E., Onea, F.: A review of the technologies for wave energy extraction. Clean Energy **2**(1), 10–19 (2018)

Manufacturing Transformational Change Through Asset Orchestration

R. Abdullah[1](✉), R. H. Weston[2], H. O. Mansoor[3], P. M. Jackson[4], and S. King[4]

[1] Faculty of Mechanical and Manufacturing Engineering Technology, Universiti Teknikal Malaysia Melaka, Malacca, Malaysia
rohana_abdullah@utem.edu.my
[2] Manufacturing Modelling Ltd., Loughborough, UK
[3] College of Administration and Economics, University of Al Fallujah, Al Fallujah, Iraq
[4] University of Leicester, Leicester, UK

Abstract. By documenting a case of transformational change in a Multi-National Company this paper brings together conceptual thinking from management and systems engineering schools to propose a new systematic approach to the application of emergent asset orchestration theory. When so doing the paper illustrates how collective decision making amongst multiple management teams, responsible for large scale organisational change, can be systematically integrated and enabled at multiple-levels of abstraction.

Keywords: Asset orchestration · Transformational change · Dynamic systems modelling · Resource based view

1 Introduction

Firms must be flexible to allow them to adapt strategically and respond rapidly and effectively to market change [1]. When so doing, typically they must adjust their operational processes to manage inherent complexities within the many systems they deploy [2]. Literature from strategic management schools reports upon the so called 'resource based view' (RBV) of firms [3]. Much of this literature explains that strategy realisation should focus on matching co-specialized resources and combinative capabilities to existing and new opportunities in external markets. Barney [4] states that for any firm to sustain a competitive advantage it will require rare, valuable, inimitable and non-substitutable resources. Yet the RBV of firms has been criticised as it does not on its own adequately explain how firms compete and maintain competitive advantage. Adner and Helfat [5] posit that previous studies of firms' performances have neglected the relationship between management decisions and a firm's outcomes [6]. They also define the 'Dynamic Managerial Capabilities' (DMC) of any firm as being the capabilities with which managers incorporate, build and reconfigure organisational tools and competencies. Adner and Helfat [5] further note that DMC is based on three prime attributes, namely (i) managerial human capital (ii) managerial social capital and (iii) managerial cognition capabilities.

M. A. Zakaria et al. (Eds.): *Advances in Mechatronics, Manufacturing, and Mechanical Engineering*, LNME, pp. 154–160, 2021.
https://doi.org/10.1007/978-981-15-7309-5_16

Sirmon and Hitt [7] and Sirmon et al. [8] describe a model of resource management to enable firms to cope with external change. This model is based on three concepts; (a) structuring of the company's asset portfolio (b) bundling and deploying capacity building resources and (c) leveraging the capabilities of companies.

Helfat et al. [9] and Teece [10] state that the key function of DMC is to realise asset orchestration (AO). They go on to define AO as comprising: managerial 'search and selection' typically followed by managerial 'configuration and coordination' of resources and capabilities. Sirmon et al. [8] also describe AO as a systematic method of structuring, building and exploiting a company's assets to deliver value to its customers and competitive advantage to the company. They and other authors further emphasise that AO should be realised by managers to sustain competitive behaviours in firms, thereby highlighting the importance of a plethora of managerial abilities that sustain coordinated deployments of specialized and co–specialized assets.

Taylor and Helfat [11] consider characteristics of critically important roles played by top level managers in firms. Whereas, these authors posit that middle managers also have a critical role in linking upper and lower management levels; and when so doing they should organise the use of complementary assets resulting in their downstream configuration and/or reconfiguration. These authors also deliberate the integrating functions of middle managers to be a critical aspect of AO.

Chadwick et al. [12] explain that in order for top executive teams to achieve congruence between the firm's assets and changing market settings, typically decisions taken by top level managers must be supported by information, expertise and actions provided by middle-level managers. Teece [10] argued that at times of significant change the redeployment of co-specialised assets invariably has a key role to play in firms. Whereas Taylor and Helfat (2009) add that having complementary assets is crucial but not generally enough alone to achieve new transitions. Taylor and Helfat [11] further highlight the importance of dynamic capabilities possessed by middle management, placing particular emphasis of key middle management roles which act as connectors in aligning and linking the operationally distinctive organizational units that possess complementary assets.

According to Sirmon et al. [8], the relationship between resource orchestration mechanisms and various levels of any asset hierarchy have yet to be clearly clarified. Nonetheless, Sirmon points to this as a literature gap which indicates that detailed explanation is needed to describe the typical relationships that exist between asset orchestration mechanisms and three levels of management commonly observed an organizational hierarchy.

Based upon these reported findings, this paper then goes on to propose the use of an AO road map which is underpinned via simple multi-level of abstraction cognitive models in order to systemise the critical management decision-making aspects of AO within firms. Further, since the exploratory nature of this study in a contemporary setting and given the goal of achieving an in-depth close up look at the phenomena of AO mechanisms, this research study adopts building theory from case study as a research methodology [3, 9]. Accordingly, semi-structured interviews were chosen as the main method for the data collection within the case study company. The interviews conducted within three level of the structural hierarchy of the firm "manager in the top, middle and low level".

2 Proposed Reference Model of Asset Orchestration

To begin to address the observed gap in understanding about 'how AO mechanisms can be mapped onto common organising structures used by firms' this paper: (a) describes a new, visual reference model of asset orchestration; (b) gathers case study evidence about AO processes that have been used successfully to realise an actual scenarios of significant and transformational change in a world class firm.

The present authors have observed a gap in the current asset orchestration (AO) literature which will limit its systematic and practical application within different firms. Through thorough review of these literatures, the authors conceived a new reference framework for the implementation of AO processes.

This reference model is shown in Fig. 1 and has been developed with functionality in mind to provide a ready-made visual guide on how companies 'resource based view' and the 'dynamic capabilities view' could be broadly and consistently applied in a variety of companies. The reference model is in a schematic form to demonstrate the multi-level mapping of key AO processes (of 'search', 'selection', 'configuration' and 'deployment') on a company's primary resources (especially its strategic, co-specialised and make or buy assets) so that responses can be made to a more flexible, sustained and competitive market changes.

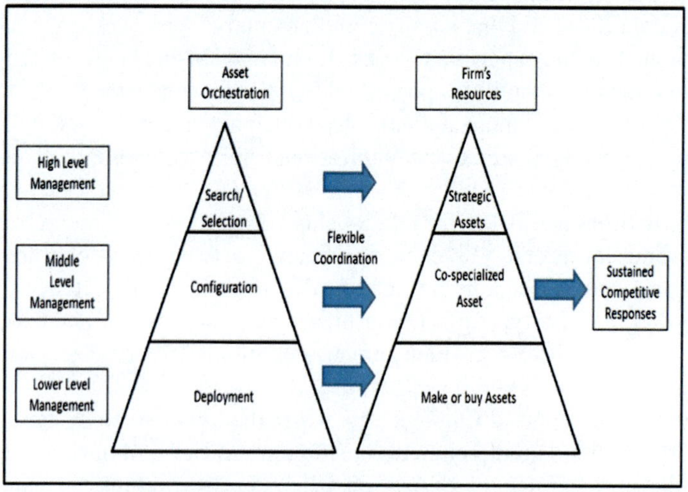

Fig. 1. Reference model of application of AO in a firm

In Fig. 1, reference is made to the 'depth' and 'breadth' found in the traditional decision making of a company. Here 'depth' refers to the different levels and roles of the organizational management within the firm, at which managerial decisions and actions occur, whilst 'breath' deals with the influence and the impact of any managerial decision on the firm.

Figure 1 also shows at each primary hierarchical tier, a clear description of a typical AO processes (thus, decision makings) that are carried out and also the common actions or outcomes that affect the key assets are listed. Nonetheless, AO processing across the hierarchical tiers will need to be consistent, requiring team effort amongst the managers. Importantly the range of detail (depth) and control (breadth) of the AO processing will vary depending on inherent characteristics of each AO process; the unique properties of each firm; and the particular environmental dynamics of the company.

Importantly, in order to maintain the success in this competitive market environments, at least some degree AO analysis must continue leading to new searches, market selections, subsequent acquisitions of resources, configurations and installations over the life of the company and thus, over the life span of its current and prospective products and/or the existing and new services provided by the company.

In summary, although not taking into account the extremely complex nature of the companies, and their evolving markets environments and the requisite AO processes, Fig. 1 provides a clear visual reference model of the present AO thinking. The following section discussed the case study work performed and evidence of the practical application of this reference model.

3 Case Study Results at GMS

GMS is a leading and global producer of semiconductor components with prime end-customers in a range of industrial sectors covering electronics, automotive, and medical. In this paper we retrospectively report on a substantial case of AO processing in GMS which followed the design and prototyping of a new electronic product. The AO processing described led to the distributed manufacture and global supply of that new product. Based upon the following reasoning, the magnitude and nature of the reported change led the present authors to characterise this as being a case of 'transformational change'.

A semi-structured interviews were held with the managers with individual responsibilities at each tier of the AO management and decision making within GMS. The interviews were partially in a free form but Fig. 1 and Fig. 2 were shown to the managers during the interviews to align their responses to recently published notions about AO. Figure 2 is an alternative version of Fig. 1 and was developed by the present authors to work alongside Fig. 1 as a 'visual interviewing' method to simplify, guide and assist to organise the questions and interview outcomes.

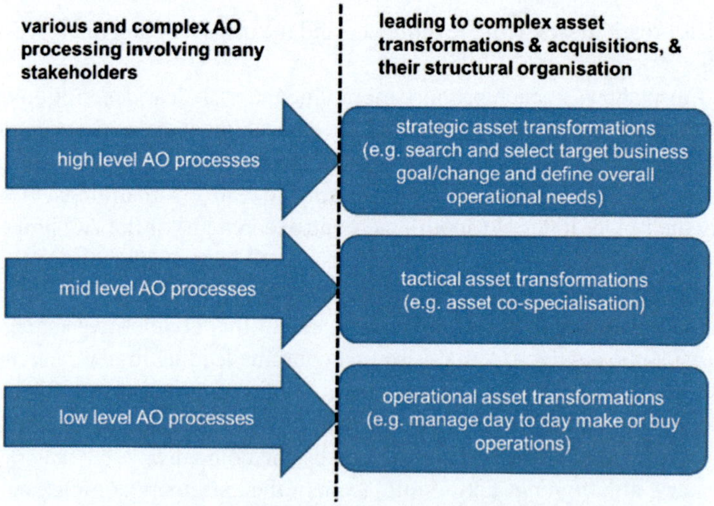

Fig. 2. Visual aid used to semi-structure the questioning of managers

4 Discussions

In our literature review, we found that real associations between asset orchestration mechanisms and different tiers of any asset hierarchy such as those proposed by McGee [13] and Sirmon et al. [8] have yet to clearly explained. Further we observed in our case study research that there is a significant gap in delivered understanding about how complex, multi-levels of AO management decision making should be decomposed, delivered and reintegrated into a coherent and effective whole. Such a gap in the delivery of holistic management decision making within firms needing a sustained ability to compete in dynamic markets had also been widely discussed within the general systems engineering literature.

In seeking to address these outstanding concerns, the present authors characterised and then used a simplification of emergent AO literature into a proposed 'reference model of AO processing'. Initially this model was formulated into a simple to use, visual interviewing tool for characterising and classifying procedural aspects of large scale change projects. This new reference model does not cover all aspects of AO in the organisations to maintain its ease of use. Never the less it was found to usefully structure and advise questioning during retrospective interviewing of mangers whom had contributed to the development and deployment of many collective decisions within a specific-case complex (project and organisational) decisional hierarchy; which had recently underpinned a major programme of change in a multi-national business. Further a direct outcome of the structured interviewing was a case study example of emergent AO thinking in action; which in itself usefully extends the base of emergent AO literature, by providing concrete examples of mappings between needed AO processes, well-structured AO decisions and actions, leading to consequent asset transformations.

The simple visual reference model was also found during interviewing to usefully structure and document the positioning of multi-level descriptions of AO processes and resources descriptions that had to be realised as part of this particular transformational change project. Indeed, despite significant differences in their company position and parent discipline the reference model (and its embedded AO thinking) was rapidly understood by those managers interviewed and was accepted as a ready means of communicating change project ideas across role, disciplinary and organisational boundaries.

5 Conclusion

In summary the proposed new 'systematic approach to AO application' was conceived and designed by the present authors to help 'underpin, visualise and integrate' critical aspects of specific case change projects needed within firms to 'sustain competitive behaviours within uncertain environments'. The approach is systematic but simplistic. It does not itself direct specific decision making; it only describes the primary decision making phases that should be actioned and example uses of underpinning cognitive maps which must have shared ownership and essentially invoke common understandings by the various members of transient teams of people who will action phases of the change project.

Further research includes mapping the AO processes used in the case study company to the reference model to demonstrate how the reference model with case information and cognitive models could direct the senior management teams' thinking as they formulate and evaluate alternative strategic futures.

Acknowledgements. Sincere thanks are offered to: Professor Paul Foley, MD of Tech4i2 Ltd, UK for his practical insights into Asset Orchestration and Universiti Teknikal Malaysia Melaka, University of Leicester and University of Al- Fallujah for funding this study.

References

1. Porter, M.E.: What is strategy? Harvard Bus. Rev. **74**(6), 61–78 (1996)
2. Efthymiou, K., Pagoropoulos, A., Papakostas, N., Mourtzis, D., Chryssolouris, G.: Manufacturing systems complexity: an assessment of manufacturing performance indicators unpredictability. CIRP J. Manufact. Sci. Technol. **7**(4), 324–334 (2014)
3. Eisenhardt, K.M., Martin, J.A.: Dynamic capabilities: what are they?. Strateg. Manage. J. **21**(10–11), 1105–1121 (2000)
4. Barney, J.B.: Firm resources and sustained competitive advantage. J. Manage. **17**(1), 99–120 (1991)
5. Adner, R., Helfat, C.E.: Corporate effects and dynamic managerial capabilities. Strateg. Manage. J. **24**(10), 1011–1025 (2003)
6. Sirmon, D.G., Hitt, M.A.: Contingencies within dynamic managerial capabilities: interdependent effects of resource investment and deployment on firm performance. Strateg. Manag. J. **30**(13), 1375–1394 (2009)
7. Sirmon, D.G., Hitt, M.A.: Managing resources: linking unique resources, management, and wealth creation in family firms. Entrep. Theory Pract. **27**(4), 339–358 (2003)

8. Sirmon, D.G., Hitt, M.A., Ireland, R.D., Gilbert, B.A.: Resource orchestration to create competitive advantage: breadth, depth, and life cycle effects. J. Manage. **37**(5), 1390–1412 (2011)
9. Helfat, C.E., Finkelstein, S., Mitchell, W., Peteraf, M., Singh, H., Teece, D., Winter, S.G.: Dynamic Capabilities: Understanding Strategic Change in Organizations. John Wiley & Sons, Hoboken (2007)
10. Teece, D.J. Dynamic capabilities: Routines versus entrepreneurial action. J. Manage. Stud. **49**(8), 1395–1401 (2012)
11. Taylor, A., Helfat, C.E.: Organizational linkages for surviving technological change: complementary assets, middle management, and ambidexterity. Organ. Sci. **20**(6), 718–739 (2009)
12. Chadwick, C., Super, J.F., Kwon, K.: Resource orchestration in practice: CEO emphasis on SHRM, commitment-based HR systems, and firm performance. Strateg. Manag. J. **36**(3), 360–376 (2015)

Check for updates

Sign Language Translation System Using Convolutional Neural Networks Approach

Vinothini Kasinathan[1], Aida Mustapha[2](✉), Hui Shan Hew[1], and Vazeerudeen Abdul Hamed[1]

[1] Faculty of Computing, Engineering and Technology, Asia Pacific University of Technology and Innovation, Technology Park Malaysia, Bukit Jalil, 57000 Kuala Lumpur, Malaysia
vinothini@apu.edu.my

[2] Faculty of Computer Science and Information Technology, Universiti Tun Hussein Onn Malaysia, 86400 Parit Raja, Batu Pahat, Johor, Malaysia
aidam@uthm.edu.my

Abstract. Sign language is one of the ways to communicate by not making sound, as it is communicated by using hands and expressions. Sign language does not get much attention from the public because it is used by the minority of the society, hence the limited resources available for learning them. This paper proposes a sign translation system called ReadMe that uses deep learning approach, specifically the Convolutional Neural Networks (CNNs) to train the recognition model. However, training ReadMe using the CNNs revealed a low accuracy of 39.8% due to small size of training dataset, hence testing is aggravating. In order to increase the recognition accuracy in the future, CNNs algorithm in ReadMe will be trained using a larger dataset from both American Sign Language (ASL) and Malaysian Sign Language (BIM). In addition, the system is also hoped to enable users to train new gestures. Only by means of crowdsourcing that the system able to expand its vocabulary without facing the knowledge engineering bottleneck.

Keywords: Sign language · Convolutional Neural Networks · Deep learning

1 Introduction

According to the National Institute on Deafness and Other Communication Disorders (https://www.nidcd.nih.gov/health/hearing-aids), in the United States alone, from 90 to 95% of deaf children are born from hearing parents. That means that when a deaf child is born the parents of this child have to learn a sign language in order to help their child able to acquire a language to communicate. Parents will always choose hearing aids for their deaf child over sign language if it is an option. According to [1], very few parents learn basic sign language before their children get their implants, and there were problems came out for example like if the child is not wearing implant, the parents have no other way to communicate with their children. If the parents and the children did not learn sign language, both will have problem in communicating. In addition, child is also frustrated because the parents do not know what the child wanted and the parents do not know how to respond to a specific reaction.

M. A. Zakaria et al. (Eds.): *Advances in Mechatronics, Manufacturing, and Mechanical Engineering*, LNME, pp. 161–171, 2021.
https://doi.org/10.1007/978-981-15-7309-5_17

Deaf children of deaf parents have better reading skills as compared to deaf children of hearing parents [2] as the deaf parents has already acquired a language to communicate which is sign language and it can help their children to acquire a language faster and more efficient way to process language and reading words. This can be supported by the book Linguistic Interdependence Hypothesis by [3] that suggested the knowledge of first language can help when learning second language. This means that if the deaf person already has knowledge in using ASL, they might have better chance in getting better when learning to read English, but as stated above only 90% of deaf people were born to deaf parents, those with hearing parents might be slow in getting the process in learning a language.

American Sign Language (ASL) has its own vocabulary and syntax, it is heavily depending on visual like the gesture and the facial expression of the person who is signing. According to [4], if a deaf person learned and used ASL, it does not mean that they would be better in reading. A test has been carried out to test the deaf students between words that are coded in phonological way and words that are coded in visual similar way. But the result shows that the students had more problem remembering words that are coded similar in phonological way. This shows that people who are using visual coded language like ASL does not necessarily be better in remembering words in printed form that would require them to read it visually [5].

ReadMe is proposed to reach more audience within the minority of society in effort to remove the communication barrier between deaf people with hearing people that has no knowledge in sign language. The over-simplistic idea is that users can sign in front of camera and the camera will capture the sign motion and translates the sign. The remainder of this paper is structured as follows. Section 2 presents the related works, Sect. 3 presents the proposed sign translation system called ReadMe. Section 4 presents evaluation of ReadMe, and finally Sect. 5 concludes with plans for future work.

2 Related Work

There are limited sources of commercial sign language translator. Kinect Sign Language Translator, for example, is developed by Microsoft but has not been released in commercial package [6]. Kinect Sign Language Translator uses the Kinect, a motion sensing input device to track the hand gesture, motion and posture of the person and translate the sign language. This system can translate American Sign Language (ASL) and Chinese Sign Language to any spoken language. This system has two modes, the translator mode and communication mode as shown in Fig. 1.

Augmented Reality Sign Language (ARSL) is a sign language translation mobile application prototype that is developed by a group of students at New York University [7]. This application has speech recognition that can translate speech to sign language with using Augmented Reality (AR) technology. ARSL also can track the motion and gesture of a person and translating sign language in real-time as shown in Fig. 2. Another example is SignAll, which is an automatic sign language translation system [8]. It uses camera to translate ASL into written English and can recognize speech and translate into written form. This system can detect body movement, facial expressions and finger shapes. The technologies this system uses are computer vision, machine learning and natural language processing algorithms as shown in Fig. 3.

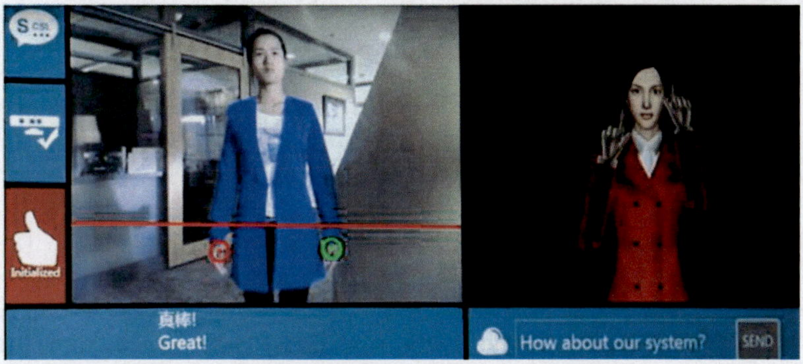

Fig. 1. Kinect Sign Language (KSL) translator [6]. The research blog can be assessed from https://www.microsoft.com/en-us/research/blog/kinect-sign-language-translator-part-1/.

Fig. 2. Augmented Reality Sign Language (ARSL) [7]. Refer the complete presentation at https://www.youtube.com/watch?v=NfIeM2VkRbw.

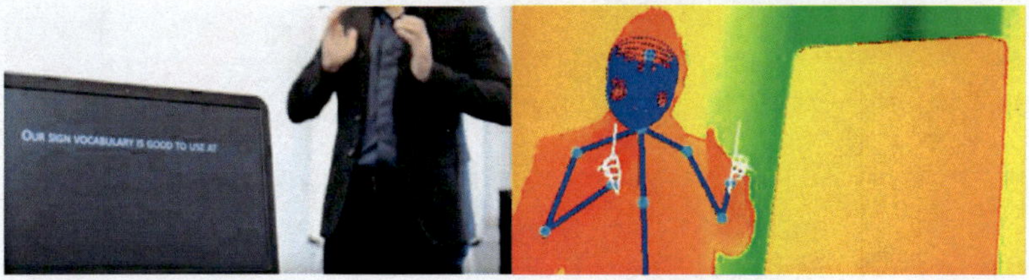

Fig. 3. SignAll [8]. The presentation in BusinessWire 2018 is available at https://www.youtube.com/watch?v=vEx6JQR66Mg.

3 ReadMe: Sign Language Translation System

This paper presents a sign language translation system called ReadMe that uses deep learning approach to train the prediction model to recognize sign languages. In specific, this research uses the Convolutional Neural Networks (CNNs) to recognize image due to its multilayer properties as compared to classical neural networks.

ReadMe system was developed using Python as the programming language. Python is commonly and widely used in machine learning due to its ability and the libraries it offers for various tasks in machine learning. Meanwhile, the Integrated Development Environment (IDE) used was PyCharm by JetBrains. PyCharm provides intelligent code completion, error checking and quick-fixes (https://www.jetbrains.com/pycharm/). PyCharm also provides a good debugger with clean and easy to use interface.

The libraries chosen for this project included the OpenCV, Keras, and TensorFlow. Open Source Computer Vision Library also known as OpenCV was designed for real-time computer vision for image processing, hence crucial for real-time language translation. Keras is a high-level neural networks API that is built to enable fast experimentation. Keras is chosen is because this project will be using CNNs and Keras support CNNs. TensorFlow is an open source machine learning framework that supports Keras.

3.1 System Design

ReadMe is designed to capture user hand movement when signing and translate the signs into text. It is trained with CNNs to translate alphabet in static motion. The main feature of Read for Me is using the camera to capture video and doing real time translating. The deaf can sign in front of the camera and the hearing and read the translated text produced by the system and both parties can communicate without any problem. The use-case diagram for ReadMe is shown in Fig. 4 and the described in Table 1.

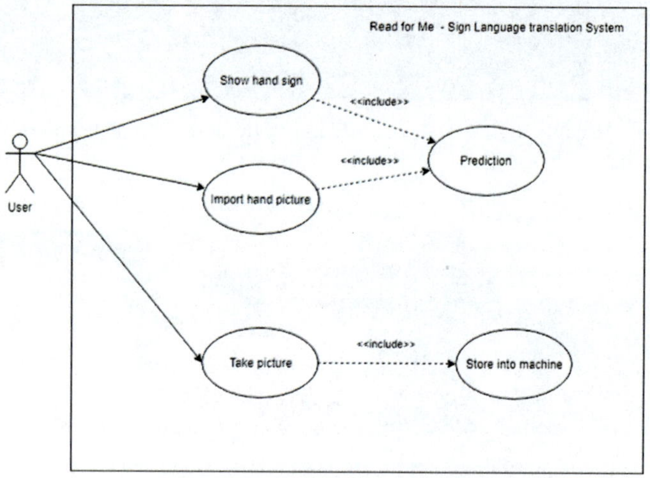

Fig. 4. ReadMe use-case diagram.

3.2 User Interface

ReadMe has two main windows. The first window is the camera capturing the video, named "Capture" window. This capture window will show real-time image of what the camera captured. Within this window, there is a green box for the users to put their hand

Table 1. Use-case descriptions.

Use cases	Use case 1	Use case 2	Use case 3
Name	Show hand sign	Import hand picture	Take picture
Description	Users show the sign language in front of the camera	Users upload a picture containing the hand sign into the system to translate	Users take picture of their hand with system
Actor(s)	User	User	User
Triggers	User	User	User
Precondition	The users run the system and the camera window is displayed	The users run the system and enter the file path to the picture	The users run the system and the camera window is displayed to allow the users to take picture of their hand
Post-condition	A prediction of the translation will be showed on screen	A prediction of the translation will be printed on screen	The picture will be stored into the machine of the users

and instructions to tell the users to put their hand in the green box or to press s to exit the system. The second window is the region of interest window, named "ROI" window. This window will crop the image of user's hand and focus, where next the images will be captured and processed before passed to the system for prediction. Below the green box is where the translated text will be displayed. Figure 5 shows the translation window.

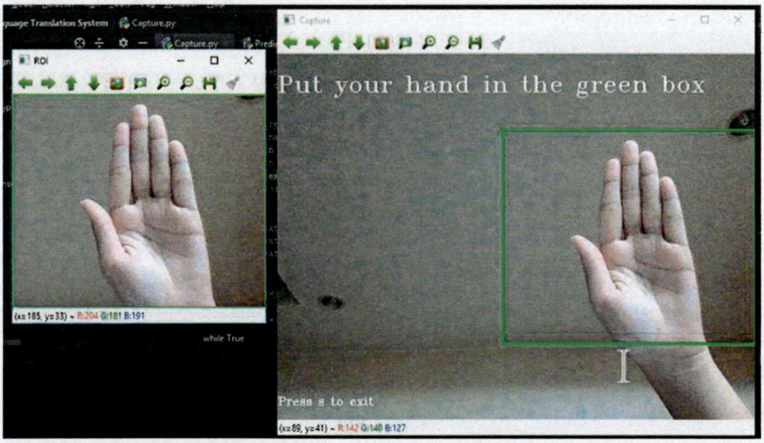

Fig. 5. User interface for translation window.

When users want to take picture of a hand, the users can run this function and two windows will be displayed. The "Take Picture" allows users to put their hand and press

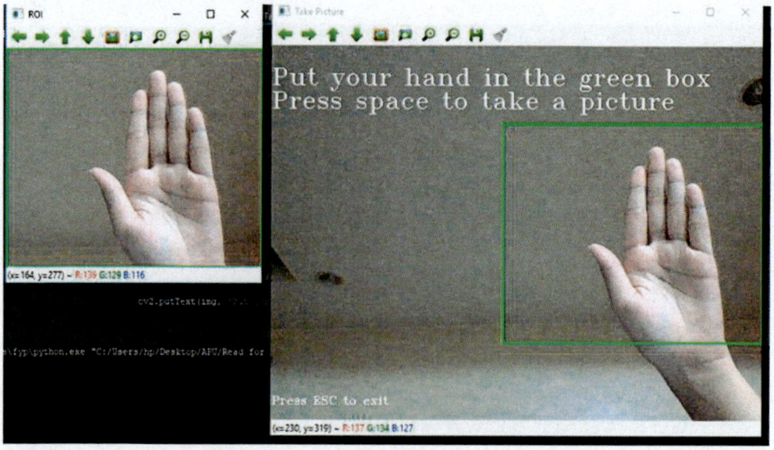

Fig. 6. User interface for image capture window.

space to take the picture. The second ROI window will display only the image of hand by cropping the image from camera. Figure 6 shows the image capture window.

Next, the stored image will be processed and converted into grayscale color image. This image then stored in a 64 by 64 array because the model trained is in 64 by 64 and the continued with prediction.

3.3 Convolutional Neural Networks (CNNs)

Convolutional Neural Networks (CNNs) are a type of deep learning algorithms first known as Artificial Neural Networks (ANNs) inspired by the functionality of the brain [9–12]. CNNs are designed to work with grid-structured inputs and have strong spatial dependencies in local regions of the grid [13]. When an image is being passed to CNNs, the image will be divided into grids, and the adjacent grids are dependent on each other as they will have similar color values. Each spatial will store grids with similar color values, then another dimension will capture a different color to create a 3-D input as volume. A convolution in CNNs is an operation of dot-product operation between a grid-structured set of weights and similar grid-structured inputs drawn from different spatial in volume.

CNNs arrange their neurons in a 3-dimensional way according to the spatial grid structure, whereby each layer will inherit from the layer before it because it is based on the small local spatial region in the previous layer. The weights are the same and shared in all neurons in the hidden layer. This makes image recognition much powerful as CNNs can identify the object and does not care which position the object is in the image [14]. CNNs is different from a typical network as shown in Fig. 7 where CNNs is 3-D with height, width and depth while traditional is 2-D.

In Phyton library, the CNNs is made up of 5 layers, input layer, convolution layer, RELU (rectified liner unit) layer, pool later and FC (fully-connected) layer. The input layer is the raw pixel values of an image. The convolution layer is the layer where CNNs will capture a grid of the image and pass it to another neuron in the next layer, the

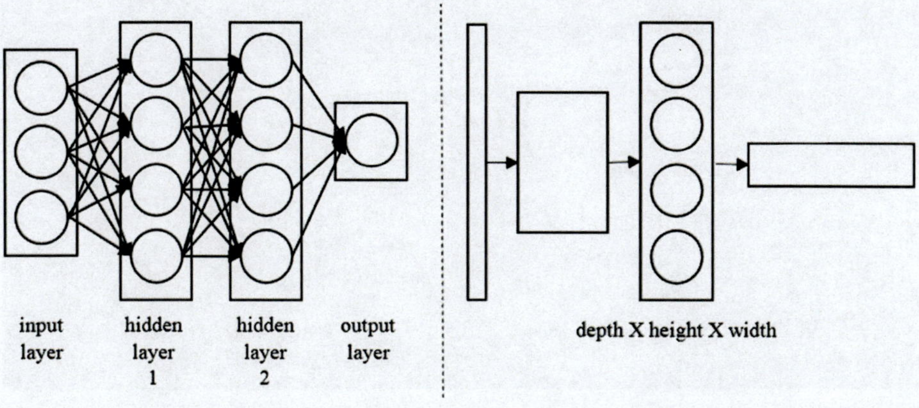

Fig. 7. Difference between typical neural networks and CNN.

neurons in the convolution layer is not fully connected to the next layer, only some are connected. RELU layer is where activation happens, the output of a neuron will map to the highest positive value, which means if the value is negative, it will be mapped to 0. In pooling layer, the neurons will be condensed into a single output this can be seen in Fig. 8.

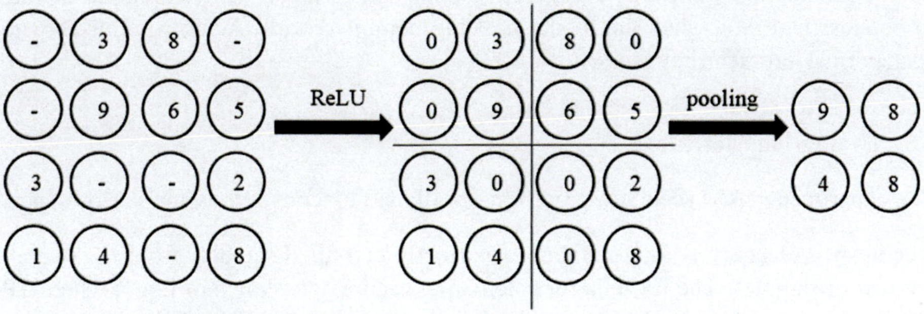

Fig. 8. CNNs transform from RELU layer to pooling layer.

In the FC layer, all the neurons will be connected to the neurons in the previous layer, this part of CNNs is the same as the typical neural networks.

3.4 CNNs Prediction

ReadMe used the Convolutional Neural Networks (CNNs) from the Python library for training the sign images. Training is performed using 10-fold cross validation method Fig. 9 shows the source codes to train the CNNs with 12 layers of sequential models.

The first layer is a convolutional layer which has the filters of 64, and kernel size of 4. It uses the "RELU" activation function with the input shape of (64, 64, 3) representing width, height, and dimension. The second layer is also a convolutional layer and have the same parameters as the first layer. The third layer is a dropout layer which purpose

```
model = Sequential()

model.add(Conv2D(64, kernel_size=4, strides=1, activation='relu', input_shape=TARGET_DIMS))
model.add(Conv2D(64, kernel_size=4, strides=2, activation='relu'))
model.add(Dropout(0.5))
model.add(Conv2D(128, kernel_size=4, strides=1, activation='relu'))
model.add(Conv2D(128, kernel_size=4, strides=2, activation='relu'))
model.add(Dropout(0.5))
model.add(Conv2D(256, kernel_size=4, strides=1, activation='relu'))
model.add(Conv2D(256, kernel_size=4, strides=2, activation='relu'))
model.add(Flatten())
model.add(Dropout(0.5))
model.add(Dense(512, activation='relu'))
model.add(Dense(N_CLASSES, activation='softmax'))

model.compile(optimizer='adam', loss='categorical_crossentropy', metrics=['accuracy'])
```

Fig. 9. Source code for training the CNNs.

is to avoid over-fitting by dropping out some of the neurons. The fourth layer is another convolutional layer that has the filters of 128. The fifth layer is the same as the fourth layer. The sixth layer is another dropout layer to prevent over-fitting again. The seventh and eighth layer are convolutional layer with 256 of filters. The ninth layer is flattening the layer into a 1D connected layer. Then the tenth layer is another dropout layer. The eleventh layer is a dense layer with 512 dot. The last layer is another dense layer which will dense the model into a 29 classes of output as this project has 29 classes of picture to be classified into. Then the model is being compiled with 'Adam' optimizer with "categorical_crossentropy" loss.

3.5 Evaluation Metrics

The experiments used accuracy, precision, recall, and F-score to measure performance.

Accuracy. Accuracy is total number of correctly classified samples from the overall number of samples. The formula for calculating accuracy is shown in Eq. 1, where TP is True Positive, TN is True Negative, FP is False Positive, and FN is False Negative.

$$\text{Accuracy} = \frac{(\text{TP} + \text{TN})}{(\text{TP} + \text{TN} + \text{FP} + \text{FN})} \tag{1}$$

Precision. Precision the number of samples classified as positive divided by total samples are classified as positive samples. The formula for calculating precision is shown in Eq. 2.

$$\text{Precision} = \frac{\text{TP}}{(\text{TP} + \text{FP})} \tag{2}$$

Recall. Recall is the number of samples is classified as positive divided by the total sample in the testing set positive category. The formula for calculating recall is shown in Eq. 3.

$$\text{Recall} = \frac{\text{TP}}{(\text{TP} + \text{FN})} \tag{3}$$

F-Score. F-score is the weighted average of Precision and Recall, whereby the score takes both false positives and false negatives into account. The formula for F-score is shown in Eq. 4.

$$\text{F - Score} = \frac{2 * (\text{Recall} * \text{Precision})}{(\text{Recall} + \text{Precision})} \tag{4}$$

4 Evaluation

Table presents shows the accuracy results for training and testing Convolutional Neural Networks (CNNs). The results showed that the training model was able to achieve 83.95% accuracy as compared to only 39.80% from the testing model. This shows that the CNNs model did not produce accurate prediction when predicting in real time (Table 2).

Table 2. Accuracy results.

	Training	Testing
Accuracy	0.8395	0.3980

To address this issue, a model find-tuning was carried out by using different type of metrics or optimizer available in the Keras library. Training with more epochs might also help to improve the accuracy of the model, but after training the model with 20 epochs, the accuracy dropped and remain the same. Next, Table 3 shows the results for precision, recall, and F-score.

Table 3. Precision, recall and F-score results.

Alphabet	Precision		Recall		F-Score	
	Training	Testing	Training	Testing	Training	Testing
A	0.79	0.05	0.80	0.03	0.80	0.04
B	0.75	0.24	0.96	0.17	0.84	0.20
C	1.00	0.50	0.96	0.13	0.98	0.21
D	1.00	0.58	0.96	0.37	0.98	0.45
E	0.63	0.04	0.83	0.03	0.71	0.03
F	1.00	0.29	1.00	0.93	1.00	0.44
G	0.91	0.60	0.97	0.60	0.94	0.60
H	1.00	0.91	0.93	0.67	0.96	0.77
I	1.00	0.21	0.59	0.70	1.00	0.32
J	0.98	0.90	0.93	0.60	0.77	0.72
K	0.96	0.30	0.92	0.27	0.71	0.28
L	1.00	0.53	1.00	0.67	0.88	0.59

Table 3. Precision, recall and F-score results (con't).

Alphabet	Precision		Recall		F-Score	
	Training	Testing	Training	Testing	Training	Testing
M	0.71	0.26	0.84	0.37	0.92	0.30
N	0.80	0.20	0.64	1.00	0.96	0.13
O	0.82	0.34	0.95	0.70	0.72	0.46
P	0.91	0.65	0.94	0.73	0.56	0.69
Q	0.94	0.61	0.97	0.67	0.69	0.63
R	0.58	0.48	0.94	0.40	0.51	0.44
S	0.61	0.33	0.51	0.07	0.68	0.11
T	0.69	0.10	0.69	0.13	0.87	0.11
U	0.56	0.83	0.47	0.17	0.52	0.28
V	0.88	0.67	0.56	0.07	0.97	0.12
W	0.79	0.43	0.96	0.53	0.86	0.48
X	0.76	0.33	0.43	0.20	0.94	0.25
Y	0.96	0.52	0.97	0.57	0.97	0.54
Z	0.77	0.44	0.97	0.80	0.92	0.57
Average	0.84	0.44	0.83	0.45	0.83	0.38

5 Conclusions

The main purpose of ReadMe is to offer a sign language system that is easy to use by both adults and children. However, this project translated the American Sign Language (ASL) instead of Malaysian Sign Language (BIM) because there are more resources such as dataset of ASL as compared to BIM used in Malaysia. The results from training ReadMe using the Convolutional Neural Networks (CNNs) revealed a rather low accuracy due to small size of training dataset, hence testing is aggravating.

In order to increase the prediction accuracy in the future, CNNs algorithm in ReadMe will be tested using a larger dataset containing a minimum of 1000 pictures per class. In addition, the system is also hoped to enable users to train new gestures. Only by means of crowdsourcing that the system able to expand its vocabulary without facing the knowledge engineering bottleneck.

Acknowledgement. This project is sponsored by Universiti Tun Hussein Onn Malaysia.

References

1. Weaver, K.A., Starner T.: We need to communicate! helping hearing parents of deaf children learn American sign language. In: 13th International ACM SIGACCESS Conference on Computers and Accessibility (ACM), pp. 91–98 (2011)
2. Azbel, L.: How do the deaf read? The paradox of performing a phonemic task without sound. Intel Science Talent Search (2004)
3. Cummins, J., Swain, M.: Bilingualism in Education: Aspects of Theory, Research and Practice. Routledge, Abingdon (2014)
4. Musselman, C.: How do children who can't hear learn to read an alphabetic script? A review of the literature on reading and deafness. J. Deaf Stud. Deaf Educ. **5**(1), 9–31 (2000)
5. Conrad, R.: The Deaf Schoolchild: Language and Cognitive Function. HarperCollins Publishers, New York (1979)
6. Microsoft Blog Editor, Kinect Sign Language Translator - Part 1. https://www.microsoft.com/en-us/research/blog/kinect-sign-language-translator-part-1. Accessed 10 Apr 2019
7. Polunina, T.: Students Create App to Translate Sign Language. https://wp.nyu.edu/connect/2018/05/29/sign-language-app/. Assessed 10 Apr 2019
8. BusinessWire, SignAll tracking the body movement, facial expression and finger shapes of a person. https://nyunews.com/2018/04/09/04-10-news-ar/. Assessed 10 Apr 2019
9. Le, Q.V., Smola, A.J., Vishwanathan, S.: Bundle methods for machine learning. In: Advances in Neural Information Processing systems, pp. 1377–1384 (2008)
10. LeCun, Y., Bengio, Y., Hinton, G.: Deep learning. Nature **521**(7553), 436 (2015)
11. Schmidhuber, J.: Deep learning in neural networks: an overview. Neural Networks **61**, 85–117 (2015)
12. Goodfellow, I., Bengio, Y., Courville, A.: Deep learning. MIT Press, Cambridge (2016)
13. Aggarwal, C.C.: Neural Networks and deep Learning. Springer, Cham (2018)
14. Bejiga, M., Zeggada, A., Nouffidj, A., Melgani, F.: A convolutional neural network approach for assisting avalanche search and rescue operations with UAV imagery. Remote Sens. **9**(2), 100 (2017)

Assessment of Piling Machine Operation Performance Using Overall Equipment Effectiveness (OEE) During Piling Construction at Universiti Teknikal Malaysia Melaka

Mohd Rayme Bin Anang Masuri(✉) and Mohammad Hafifi Bin Tajry

Faculty of Manufacturing Engineering, Universiti Teknikal Malaysia Melaka, Hang Tuah Jaya, Durian Tunggal, 76100 Melaka, Malaysia

rayme@utem.edu.my

Abstract. Unlike the manufacturing industry which is synonymous with a lean concept, the construction industry is perceived to deal with difficulties in managing resources, such as manpower and financial, caused pollution to the environment as well as non-value added tasks. These situations motivate the construction industry to apply the lean concept. This paper aims to examine the performance of piling machine operation performance during pile driving activities at the Faculty of Manufacturing Engineering in Universiti Teknikal Malaysia Melaka using Overall Equipment Effectiveness (OEE). OEE is a tool to evaluate the effectiveness of the equipment as well as the efficiency of the organization in utilizing other resources such as manpower and materials. Stopwatch Method was applied to record the time required to complete the pile construction. From the analysis, the value of calculated OEE before the implementation of the site logistic plan was 65.7%. However, the implementation of site logistic plan has successfully minimized non-value added tasks. This was proven when the second cycle of the assessment recorded that the new value of OEE is 78.0% indicating a significant improvement in the performance of piling machine operation performance. It is proven that OEE could enhance the performance of the works and the quality of the end product.

Keywords: Overall equipment effectiveness · Lean concept · Lean construction · Stopwatch method

1 Introduction

The construction industry is a challenging trade as its activities are carried out in an uncontrolled environment. It is perceived that the construction industry has put significant efforts in achieving its key performance indicators; within budget, within time, acceptable quality, environmental performance, and safety [1]. Besides, the construction industry is synonymous with overspending, an extension of time, improper management of manpower, poor maintenance of machinery, illegal wastes disposal and pollution. This situation has inspired professionals and academics in the construction industry to enhance

M. A. Zakaria et al. (Eds.): *Advances in Mechatronics, Manufacturing, and Mechanical Engineering*, LNME, pp. 172–182, 2021.
https://doi.org/10.1007/978-981-15-7309-5_18

construction performance by implementing a lean concept. In the manufacturing industry, it is proven that the lean concept has successfully minimized overproduction and defects, eliminates inappropriate processing and unnecessary motion as well as reducing excessive transportation. As a result, implementing the lean concept in the construction industry would improve the performance of construction activities [2]. Work-time wastage study, for instance, is to measure waste in each flow of construction activities [3]. Successful implementation of lean construction will encourage the company to reduce work-cost wastage in the activities which leads to the improvement of the company's turnover [4].

The building is usually divided into two segments namely superstructures and substructures. Structure components above ground level such as walls, columns, and beam are classified as superstructures. Structure components below ground level such as pile foundation (deep foundation), mat foundation, pad foundation (shallow foundation) and ground beam are classified as the substructure. Unlike the construction of the shallow foundation, construction of the deep foundation is complicated as the process will involve mobilization and utilization of piling machines, skilled manpower, and material stockpile. Although the data of soil characteristics are obtained from soil investigation, yet there are uncertainties during pile driving activities. The previous study recorded that piling activities contribute 10% to 12% of the total construction costs of the building [5]. The value is relatively small compared to building works which are 73% of the total construction cost. In another study, pile wastage contributes 17% of the total weight of construction wastes, placed second after stone tables which the wastage is recorded at 29% [6]. From a financial perspective, pile wastage represents 13% of the total waste costs including waste management, purchasing, and transport costs. This amount is considered significant to the project owners. Therefore, pile driving activities need to be carried out as effective as possible to minimize wastage by eliminating speed loss (idling and minor stoppage) and downtime loss (equipment setup and equipment failure).

According to Young and Don [7], Overall Equipment Effectiveness (OEE) is broadly applied in the manufacturing industry to evaluate the availability, performance, and quality of single equipment or multiple interrelated equipment. In the field of construction domain, OEE is applied for research purposes particularly in measuring waste and workflow in construction of superstructure [8–11]. The previous study also shows that OEE was applied to assess the performance of equipment in road construction in India [12]. The application of OEE during piling activities is an initiative to enhance the performance of building construction at the Universiti Teknikal Malaysia Melaka.

2 Lean Concept in Construction Industry

The effort to implement a lean concept in the construction industry was traced back in 1988 [13]. Since then, the awareness and research to implement the lean concept in the construction industry have grown considerably [14]. Sarhan and Fox [15] opine that the nature of the construction industry is different from the manufacturing industry. Construction is prevalently performed in an uncontrolled and complicated environment which makes it difficult to hold a lean concept. Aziz and Hafez [16] however viewed that the construction industry has significant similarities with the manufacturing industry particularly in waste elimination, cycle time reduction, reduction of variability, continuous improvement, pull production control and continuous flow. According to Egan, repetitive activities and processes are other common characteristics between the construction and manufacturing industries [17]. Egan's view probably correct for this project as pile driving activities needs to be repeated for 606 pile points. In this study, pile driving activities were carried out between February 2018 and May 2018.

2.1 Benefits of Applying Lean Concept

The main purpose of implementing the lean concept in the construction industry is to improve the ability to manage construction activities through the elimination of wastes and add value to the users [18]. At the planning and design stage, the lean concept will enhance the effectiveness of documentation management. This was echoed by Garrett and Lee [19] who claimed that the lean concept could improve the effectiveness of documentation submission and efficiency of the design review process. At the construction stage, Pasqualini and Zawislak [20] reiterate that lean tools such as Value Stream Mapping (VSM) can be used to identify waste resources. The tools would benefit the project manager to anticipate the delay in construction activities and provide a solution to it. The lean tools would also help the project manager to analyze the workflow [21]. From a financial management viewpoint, the lean concept could control the initial expenses in building construction [22].

In building and infrastructure construction project, the main objective is to increase productive time and reducing construction waste produced from processes and non-value added tasks [23]. A. comparison study done by Aziz and Hafez [16] found that productive time wastage in the manufacturing industry is recorded at 12% only compared to the construction industry with 57%. Meanwhile, another study done by Kalsaas [24] states that the design stage and construction stage make up 37% of productive time wastage mainly on safety and health issues and poor documentation systems. He also highlighted that the productive time of general workers achieve more than 60%. These situations contribute to the overall construction waste which is recorded at 55% to 65% of construction costs (Fig. 1).

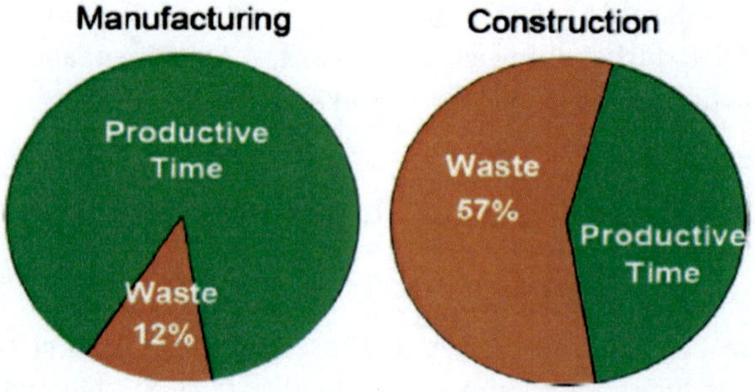

Fig. 1. Comparison of productive time and waste between manufacturing and construction.

2.2 Pile Driving Activities

In the construction of the Faculty of Manufacturing Engineering, the foundation works need to be completed before the installation of other superstructure components. Square concrete pile sizes 150 mm × 150 mm × 6 m length was installed to 606 pile points as a pile footing to receive structural loading. To ensure the quality of the piling works, the installer must observe the procedure of pile driving activities as tabulated below [25] (Table 1).

Table 1. Procedure of concrete pile installation.

Step	Activities
1 Pre-driving	– Pre-position the pile to the pile point – Removal of debris
2 Pile handling	– Lifting the pile from horizontal to vertical positio
3 Fit in pile head to hammer	– Use cable attached to the piling machine to tie the pile head – Fit in the pile head to the hammer – Use skilled workers and machine operator
4 Positioning pile in pre-driving point	– Positioning the pile shoe on the pre-driving point – Let the pile shoe sunk in the ground – Ensure the verticality of the pile
5 Pile driving	– Launch driving – Observe when the pile reached the required depth

Operating machinery, handling manpower and construction materials in the construction industry is expensive. To ensure all the resources are well managed, the lean concept can be exploited to improve machine availability, performance, and work quality.

3 Overall Equipment Efficiency (OEE)

Overall Equipment Efficiency (OEE) is widely applied in the manufacturing industry to evaluate the effectiveness of the equipment as well as the efficiency of the organization in utilizing other resources such as manpower and materials to achieve the specified quality [26]. OEE has become part of the manufacturing process to eliminate six big losses namely equipment breakdown, adjustment, and setup, idling and minor, reduced speed, start-up reject and reduced yield [27]. In the construction industry, OEE is not part of the construction process. OEE is applied for research purposes particularly in measuring waste and workflow in construction of superstructure. The previous study shows that OEE was applied to assess the performance of equipment in road construction in India. In this study, the OEE will be applied in piling construction. However, the following aspects need to be considered:

i. Planned working hours

ii. Effective working hours
iii. Work shortage
iv. Adverse weather condition
v. Maintenance
vi. Fuel consumption

The value of OEE is established by multiplying three measuring metrics namely availability, performance and quality.

$$OEE = Availability \times Performance \times Quality \quad (1)$$

where,

$$Availability = \frac{(Operating\ time - down\ time)}{Total\ operating\ time} \quad (2)$$

$$Performance = \frac{Total\ Output}{Potential\ Output} \quad (3)$$

$$Quality = \frac{Good\ Output}{Total\ Output} \quad (4)$$

3.1 Apply OEE Using Stopwatch Method

In this study, the Stopwatch Method was applied. According to Patel [28], it is the most suitable method due to its ability to adapt variation of human performance. In addition, its flexible characteristic is appropriate to record repetitive activities [29]. The procedure to apply the Stopwatch Method is demonstrated in Fig. 2.

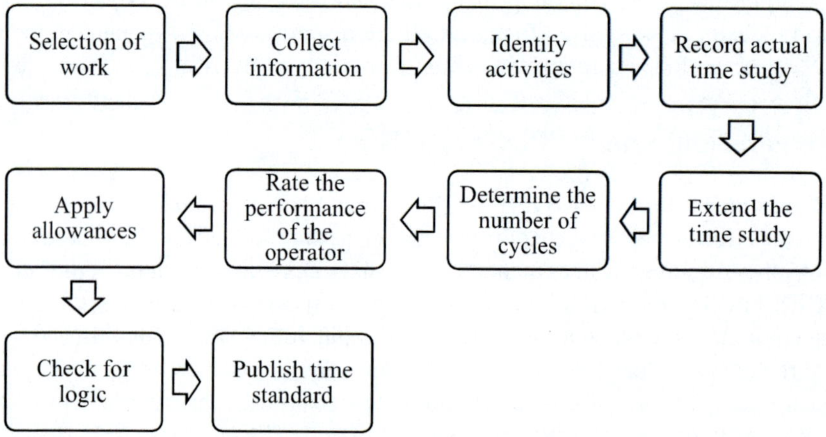

Fig. 2. Procedure to apply Stopwatch Method.

Prior to implementing OEE using Stopwatch Method, it is important to select the type of construction work and to collect all the relevant information. Identifying the related activities is crucial as this will determine the method of time recording and the number of recording cycles. The analysis will be conducted to assess the machine operation performance prior to establishing the time standard.

Selection of Work and Time Recording. In this study, the construction of building foundation work has been selected. Pile driving activity using piling machine PM01 is identified. The details of the activity are:

i. Refer drawing, identify pile point and locate piling machine.
ii. Lift and position the pile head to hammer.
iii. Pile driving until desired depth.
iv. Pile extension and welding.
v. Pile driving record.

Fig. 3. Pile extension involving multiple activities.

Figure 3 above shows the pile extension involving lifting, fitting pile head to hammer, and welding activities. These activities need to be monitored so that non-added value tasks could be minimized.

At the beginning of the study, the downtime, setup, defected pile and balance time were observed for five days as shown in Table 2.

Table 2. Time recorded for downtime, setup, defect and balance time of piling works.

	Day 1	Day 2	Day 3	Day 4	Day 5	Average
Downtime (minutes/day)	20	21	22	18	19	20
Setup time (minutes/day)	15	16	14	13	17	15
Defect (point(s))	0	2	0	2	1	1
Balance time (minutes/day)	10	8	9	12	11	10

Table 2 shows that the average operation downtime is 20 min/day while setup time is 15 min/day. From the observation, there are factors contributed to the machine downtime such as gas filling to the piling machine by the operator, movement and maneuvering of piling machine, operator fulfilling human biological needs from time to time and examining construction drawings by the operator. Preparation by the operator at the beginning of each pile driving activity contributes to setup time while pile damage causes rework. It is also recorded that at least one pile point is broken every day while the balance time is 10 min/day. Balance time is defined as the remaining time for the operator to end their tasks before 5.00 pm of the day.

In this construction project, the piling machine is operating for 9 h per day, 5 days per week. It is recorded that the break duration is 75 min/day while the target demand is 22 points/day. Therefore, the takt time can be calculated as follows:

Total working hours per day = 9 h/day × 60 min/h = 540 min/day

Available operating time = Total working hours – (Break duration + Setup time + Downtime + Balance time = 540 – (75 + 15 + 20 + 10) = 420 min/day.

Ideal cycle time = Available operating time/target demand = 420 min/22 points = 19.09 min/point/day.

Takt time = (Total working hours – breaks)/target demand = (540 – 75) minutes/day/22 points = 21.14 min/points/day.

OEE Calculation Before the Implementation of Site Logistic Plan. From the current performance at site, the average time to complete one pile point/day is 25.81 min compared to 19.09 min as targeted (Refer Table 3). Meanwhile, the actual completed pile is 17 points/day compared to 22 points/day as targeted.

Table 3. Operation cycle time for each activity before implementation of site logistic plan.

Activities	Pile point no. (min/point)										Ave.
	1	2	3	4	5	6	7	8	9	10	
1. Locate piling machine	3.4	–	–	–	–	–	–	–	–	–	0.34
2. Lift and position the pile	2.8	2.7	2.6	2.8	2.7	2.7	2.9	2.9	2.8	2.7	2.76
3. Pile driving (cycle no. 1)	4.6	4.5	4.7	4.5	4.7	4.7	4.6	4.8	4.5	4.5	4.61
4. Pile extension (cycle no. 1)	4.1	4.0	4.2	4.3	4.4	3.9	4.2	4.0	4.2	4.1	4.14
5. Pile driving (cycle no. 2)	4.8	4.9	4.8	4.9	4.7	5.0	4.8	4.9	4.5	4.7	4.80
6. Pile extension (cycle no. 2)	4.0	3.9	3.8	3.7	4.1	4.0	3.9	4.1	4.3	4.2	4.00
7. Pile driving (cycle no. 3)	4.6	4.7	4.8	4.5	4.9	4.7	4.7	4.5	4.7	5.0	4.71
8. Record	0.4	0.6	0.5	0.5	0.4	0.4	0.5	0.4	0.4	0.4	0.45
Total	28.7	25.3	25.4	25.2	25.9	25.4	25.6	25.6	25.4	25.6	25.81

The time recorded before the implementation of the site logistic plan at site, the following information is obtained:

Planned production time = Down time + Operating time + Setup time + Balance time = 465 min

Good produced part on the machine = 16 pile points

Calculation of OEE measuring metrics is established as follows:

Availability = Operating time/Planned production time = 420/465 = 0.903 => 90.3%

Performance = (Total produced part × Ideal cycle time)/Operating Time = (17 × 19.09)/420 = 0.773 => 77.3%

Quality = Good produced parts/Total produced parts = 16/17 = 0.941 => 94.1%

Therefore, the current OEE = Availability × Performance × Quality = 0.903 × 0.773 × 0.941 = 0.657 => 65.7%

Implementation of Site Logistic Plan and Improvement of Piling Works. Based on the observation at the site, poor housekeeping is identified as a factor in time wastage. Therefore, site logistic plan such as Five S's (5S – Sort, Set in order, Shine, Standardize and Sustain) need to be implemented at the site. The purpose of implementing 5S is to organize work, maximize working efficiency and to create a convenient working environment. The following 5S initiatives are implemented at the site to enhance the performance of piling works.

Sort - The concrete pile was assembled near to the piling machine based on the priority of which pile point to be installed. This initiative reduced time consumption during pile handling.

Set in order - The piling machine is strategically positioned to how the work will be performed. The operator is advised to study the construction drawing in advance and make necessary preparation such as fuel filling and greasing machine parts 10 min before going home.

Shine - The operator is responsible to keep the work area clean and free from dirt. This initiative could avoid minor faulty that cause piling machine breakdown. This would be the first step in the Standard Operating Procedure (SOP).

Standardize - SOP for piling works was established. The operator and the workers are advised to perform their activities based on the provided SOP.

Sustain - It is the responsibility of the management to monitor the compliance of the SOP is continuously adhered to. The workers are advised to keep up with the industry's best practices (Fig. 4).

OEE Calculation After the Implementation of Site Logistic Plan. After the implementation of 5S, it is recorded that the average operation cycle is 22 min/point/day (refer Table 4) compared to previous data recorded at 25.81 min/point/day. The achievement of the completed pile point has improved from 17 points/day to 20 points/day. However, the record has yet to achieve targeted 22 points/day.

The OEE calculation showed a significant improvement as below:

Operating time = 445 min (previous achievement: 420 min)

Downtime = 5 min (previous achievement: 20 min)

Fig. 4. Arrangement of piles and machine before 5S (left) and after 5S (right).

Table 4. Operation cycle time for each activity after implementation of site logistic plan.

Activities	Pile point no. (min/point)										Ave.
	1	2	3	4	5	6	7	8	9	10	
1. Locate piling machine	1.5	–	–	–	–	–	–	–	–	–	0.15
2. Lift and position the pile	1.4	1.5	1.5	1.7	1.3	1.6	1.5	1.2	1.4	1.5	1.36
3. Pile driving (cycle no. 1)	4.5	4.4	4.6	4.4	4.6	4.6	4.5	4.7	4.4	4.4	4.51
4. Pile extension (cycle no. 1)	3.0	2.9	3.1	3.2	3.3	2.9	3.1	3.0	3.3	3.0	3.08
5. Pile driving (cycle no. 2)	4.7	4.8	4.7	4.8	4.6	4.9	4.7	4.8	4.4	4.6	4.7
6. Pile extension (cycle no. 2)	3.1	3.0	2.9	3.2	3.2	3.0	2.9	3.3	3.4	3.2	3.12
7. Pile driving (cycle no. 3)	4.5	4.6	4.7	4.4	4.8	4.6	4.6	4.4	4.6	4.9	4.61
8. Record	0.4	0.5	0.6	0.4	0.5	0.3	0.5	0.6	0.3	0.6	0.47
Total	23.1	22.7	22.1	22.1	22.3	21.9	21.8	22	21.8	22.2	22.0

Total produced product = 20 pile points (previous achievement: 17 pile points)

Good produced part on the machine = 19 pile points (previous achievement: 16 pile points)

Calculation of OEE measuring metrics are as follows:

Availability = Operating time/Planned production time = 445/465 = 0.957 => 95.7%

Performance = (Total produced part × Ideal cycle time)/Operating time = (20 × 19.09)/445 = 0.858 => 85.8%

Quality = Good produced parts/Total produced parts = 19/20 = 0.95 => 95.0%

Therefore, the improved OEE = Availability × Performance × Quality = 0.957 × 0.858 × 0.95 = 0.78 => 78.0%.

Table 5. OEE value before and after the implementation of 5S

	Current data (%)	After improvement (%)
Availability	90.3	95.7
Performance	77.3	85.8
Quality	94.1	95.0
OEE	65.7	78.0

4 Conclusion

The study shows that the OEE calculations have improved from 65.7% to 78.0% after the implementation of 5S at the site (refer Table 5). The average time taken to complete one pile point has reduced from 25.81 min/point to 22 min/point. Despite a significant reduction of time, it is yet to achieve the ideal cycle time, 19.09 min/point.

There are factors contributed to the constraint for the work to achieve ideal cycle time and targeted pile point/day. Among the factors are mechanical errors which cause inconsistency of hammer drop and adjustment of pile verticality. Soil condition affected the movement of the piling machine while uncertain soil structure influenced the depth of the pile (pile set).

The study has proven that OEE can used as a tool to assess the piling machine operation performance. In the future, OEE should be consistently applied and embedded in the system of construction industry to enhance the performance of construction activities and to improve the quality of the end product.

References

1. Chan, A.P.V., Chan, A.P.L.: Key performance indicators for measuring construction success. Benchmarking Int. J. **11**(2), 203–221 (2004)
2. Koskela, L.: Application of the new production philosophy to construction. Technical report #72, Centre for Integrated Facility Engineering, Stanford University (1992)
3. Kalsaas, B.T.: Work-time waste in construction. In: 18th Annual Conference, International Group for Lean Construction, Haifa, Israel, pp. 1–13 (2013)
4. Marhani, M.A., Jaapar, A., Bari, N.A.A.: Lean construction: towards enhancing sustainable construction in Malaysia. Procedia Soc. Behav. Sci. **68**, 87–98 (2012)
5. Azman, M.A., Abdul-Samad, Z., Ismail, S.: The accuracy of preliminary cost estimates in Public Works Department (PWD) of Peninsular Malaysia. Int. J. Project Manage. **31**(7), 994–1005 (2013)
6. Bossink, B.A.G., Brouwers, H.J.H.: Construction waste: quantification and source evaluation. J. Constr. Eng. Manage. **122**(1), 55–60 (2002)
7. Young, K.J., Don, P.T.: Operational efficiency and effectiveness measurement. Int. J. Oper. Prod. Manage. **21**(11), 1404–1416 (2001)
8. Kalsaas, B.T.: Measuring waste and workflow in construction. In: 21th Annual Conference of the IGLC. International Group on Lean Construction, Fortaleza, Brazil, pp. 31–2 (2013)
9. Kalsaas, B.T., Bolviken, T.: The flow of work in construction: a conceptual discussion. In: Proceedings IGLC18. Technion-Israel Institute of Technology, Haifa, Israel (2010)

10. Kalsaas, B.T.: Further work on measuring workflow in construction site product. In: Proceedings IGLC20, San Diego State University, San Diego, US (2012)
11. Kalsaas, B.T., Gundersen, M., Berge, T.O., Koskela, L., Saurin, T.A.: To measure workflow and waste. A concept for continuous improvement. In: 22nd Annual Conference of the IGLC. IGLC, Oslo, Norway, pp. 835–846 (2014)
12. Ayane, M.N., Gudadhe, M.: Review study on improvement of overall equipment effectiveness in construction equipments. Amravati Univ. Badnera Int. J. Eng. Dev. Res. **3**(2), 487–490 (2015)
13. Wan Muhammad, W.M.N., Ismail, Z., Hashim, A.E.: Exploring lean construction components for Malaysian construction industry. In: IEEE Business Engineering and Industrial Applications Colloquium. BEIAC, Langkawi, pp. 1–6 (2013)
14. Koskela, L.: An Exploration Towards a Production Theory and Its Application to Construction. VTT Publications, Espoo (2000)
15. Sarhan, S., Fox, A.: Barriers to implementing lean construction in the UK construction industry. Built Hum. Environ. Rev. **6**, 1–17 (2013)
16. Aziz, R.F., Hafez, S.M.: Applying lean thinking in construction and performance improvement. Alexandria Eng. J. **52**(4), 679–695 (2013)
17. Murray, M.: Rethinking Construction: The Egan Report (1998). Blackwell Science, Oxford (2003)
18. Ansah, R.H., Sorooshian, S., Mustafa, S.B.: Lean construction - an effective approach for project management. ARPN J. Eng. Appl. Sci. **11**(3), 1607–1612 (2016)
19. Garrett, D.F., Lee, J.: Lean construction submittal process - a case study. Qual. Eng. **23**(1), 84–93 (2010)
20. Pasqualini, F., Zawislak, P.A.: Value stream mapping in construction - a case study in a Brazilian construction company. In: 13th International Group for Lean Construction Conference. International Group on Lean Construction, Sydney, Australia, pp. 117–125 (2005)
21. Yu, H., Tweed, T., Al-Hussein, M., Nasseri, R.: Development of lean model for house construction using value stream mapping. J. Constr. Eng. Manage. **135**(8), 782–790 (2009)
22. Lapinski, A.R., Horman, M.J., Riley, D.R.: Lean processes for sustainable project delivery. J. Constr. Eng. Manage. **132**(10), 1083–1091 (2006)
23. Nikakhtar, A., Hosseini, A.A., Wong, K.Y., Zavichi, A.: Application of lean construction principles to reduce construction process waste using computer simulation: a case study. Int. J. Serv. Oper. Manage. **20**(4), 461–480 (2015)
24. Kalsaas, B.T.: Work - time waste in construction. In: 18th Annual Conference of the IGLC. International Group for Lean Construction, Haifa, Israel, pp. 1–13 (2010)
25. Kaplan, H., Elburg, A., Tommelein, I.D.: Analysis of variability in precasting and installation of pile foundations. In: Proceedings Construction Research Congress: Broadening Perspectives 2005. ASCE, San Diego, California, pp. 1–10 (2005)
26. Dal, B., Tugwell, P., Greatbanks, R.: Overall equipment effectiveness as a measure of operational improvement - a practical analysis. Int. J. Oper. Prod. Manage. **20**(12), 1488–1502 (2000)
27. Muchiri, P., Pintelon, L.: Performance measurement using overall equipment effectiveness (OEE): literature review and practical application discussion. Int. J. Prod. Res. **46**(13), 3517–3535 (2008)
28. Patel, N.: Reduction in product cycle time in bearing manufacturing company. Int. J. Eng. Res. General Sci. **3**(3), 466–471 (2005)
29. Suryoputro, M.R., Wildani, K., Sari, A.D.: Analysis of manual material handling activity to increase work productivity (case study: manufacturing company). In: MATEC Web of Conferences. EDP Sciences, Les Ulis, vol. 154, pp. 01085 (2018)

Interactions of Lamb Waves with Defects in a Thin Metallic Plate Using the Finite Element Method

N. Ismail(✉), Z. M. Hafizi, C. K. E. Nizwan, and S. Ali

Advances Structural Integrity and Vibration Research (ASIVR), Faculty of Mechanical and Automotive Engineering Technology, Universiti Malaysia Pahang (UMP), Pekan, Pahang, Malaysia
nurazima@ump.edu.my

Abstract. A Lamb waves-based technique for damage detection is one of the promising diagnostic tools due to their ease of excitation and detection along with their ability to travel long distances. However, their dispersive and multimodal nature causing complicated wave signals and data interpretation in the damage detection process. The finite element method was performed in this study to understand their behaviour when interacting with defects for a successful implementation of this technique in Structural Health Monitoring system. FEA analysis was implemented to simulate the propagation of Lamb waves in the healthy and defective aluminium plate. The sensors are placed based on the configuration of the pulse-echo and pitch-catch method. It is noticed that there is an additional wave packet for the detected signals for defective plate based on the pulse-echo method whereas there is a significant delay in signal arrival for the pitch-catch method. This simulation study shows a significant feature extraction for the interactions of Lamb waves with defects. It is helpful for a good understanding before applying this technique for a real implementation.

Keywords: Lamb waves · Damage detection · Finite element method · Structural health monitoring

1 Introduction

Lamb waves-based technique is one of the acousto-ultrasonic approaches which are based on stress waves introduced into an analysed structure [1]. These guided ultrasonic waves can exist in plate-like structure with parallel free boundaries [2]. They can interrogate the entire thickness of the structure, affording the possibility of detecting internal and surface damage [3]. This type of characteristic makes the Lamb waves as one of the most widely used technique for structural health monitoring (SHM) system. Damage in the structure is identified by a change in the Lamb waves propagation. The propagated waveform inherently records the propagation history of Lamb waves travelling in a structure, thereby providing the information about the condition of the structure. Other than that, Lamb waves can travel over a long distance even in materials with a

M. A. Zakaria et al. (Eds.): *Advances in Mechatronics, Manufacturing, and Mechanical Engineering*, LNME, pp. 183–193, 2021.
https://doi.org/10.1007/978-981-15-7309-5_19

high attenuation ratio due to their high susceptibility to interference on a propagation path. Thus, a broad area can be quickly examined.

However, the multimodal and dispersive nature of the Lamb waves lead to the complexity of the response signals which are difficult to analyse and interpret [4]. In normal practice, damage detection is based on pattern recognition using various methods of signal processing and it showed clear advantages utilising this approach [5–8]. Nevertheless, the major concern for the Lamb waves-based SHM is the correct interpretation of the complex propagated waves. Therefore, it is necessary to use finite element method (FEM) for a good understanding of the physical principles behind waves propagation and their interaction with defects. This could significantly help in the interpretation process of damage identification results, which lead to the accuracy of the monitoring process.

Nienwenhui et al. [9] performed the simulation for the generation and detection of Lamb waves using piezoelectric wafer active sensors (PWAS). The simulation was successfully showed the expected two propagating modes, which is a symmetric and anti-symmetric mode. This result was achieved by considering the effect of finite pulse width, pulse dispersion and detailed interaction between the piezoelectric element and the transmitting structure. The finding provided more accurate predictions of the mode selectivity for the Lamb waves generation which is very crucial for the minimising the dispersive and multi-modal effect. However, the simulation works do not extend the analysis for damage detection. Liu et al. [10] investigated the use of FEM to simulate various SHM methods using PWAS. One of the methods is using wave propagation method for crack detection on the beam structure. The fundamental symmetric and anti-symmetrical modes were also successfully simulated at low frequency. Silva et al. [11] utilised the FEM to understand the general idea of the wave response in the presence of damage. The defect was simulated by clamping two sets of nodes across the thickness for having neither rotation nor displacement. The authors just applied a simple approach of discontinuity. Therefore, this paper is aim at investigating the interactions of the Lamb waves with defects which are simulated approximately as the real applications.

Time-domain analysis including the time of arrival (TOA) and signal attenuation was performed in order to study in details the effect of defects to the Lamb waves propagation. The changes of these parameters may be used as an indicator for the existence of damage. Finally, a short-time Fourier transformation (STFT) is applied to observe any changes in the frequency content. All the extracted features can be used as the defects signature for the SHM diagnostic system.

2 Theory of Lamb Waves

Lamb waves are guided waves which are refer to propagation of elastic waves in a solid plate with free boundaries. The displacements occur both in the direction of wave propagation and perpendicularly to the plane of the plate [12]. They were formed from a superposition of longitudinal and transverse waves. This superposition phenomenon leads to the generation of symmetric and anti-symmetric modes, which exists simultaneously and propagates independently of each other. S_i and A_i is the denotation for symmetric and anti-symmetric modes, respectively, where i is the order or number of the mode ($i = 0, 1, 2, 3, \ldots$). The i value depends on the excitation frequency and thickness (*fd*) product. The motion of the symmetric mode and anti-symmetric mode is illustrated in Fig. 1(a) and (b), respectively.

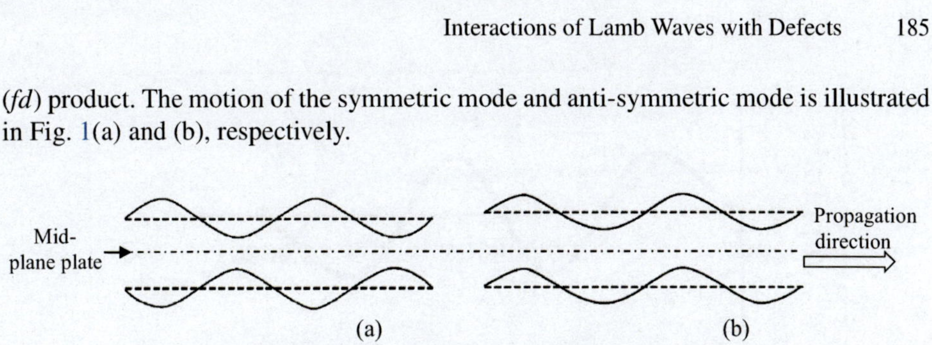

Fig. 1. Motion of the (a) symmetric and (b) anti-symmetric mode.

Before performing any research concerning the propagation of Lamb waves, the excitation frequency must be determined properly to minimize the number of modes. It is a common practice to apply low excitation frequencies for signals simplification, so that only fundamental modes (S_0 and A_0) exist. The appropriate excitation frequency can be determined by referring to the theoretical dispersion curve which is generated using Eq. 1 and 2 [12]. The curve was called as the phase velocity, c_p dispersion curves which are for symmetric and anti-symmetric, respectively. The other type of dispersion curve is group velocity dispersion curves which can be derived from the phase velocity curves using Eq. 3.

The dispersion curve dedicated for this present study is shown in Fig. 2. PACshare Dispersion Curves software was utilised to generate this curve. It shows that the increment of the number of modes is depending on the increment of the excitation frequency. Thus, this leads to presence of an infinite number of symmetric and anti-symmetric modes. Referring to that curve, the complexity of the recorded signals can be avoided if the excitation frequency is less than 800 kHz. The excitation frequency of 200 kHz was chosen for this study and it is illustrated in Fig. 3. Thus, only two modes exist during the wave actuation that is S_0 and A_0.

$$\frac{\tan qh}{\tan ph} + \frac{4k^2pq}{\left(q^2 - k^2\right)^2} = 0 \quad (1)$$

$$\frac{\tan qh}{\tan ph} + \frac{\left(q^2 - k^2\right)^2}{4k^2pq} = 0 \quad (2)$$

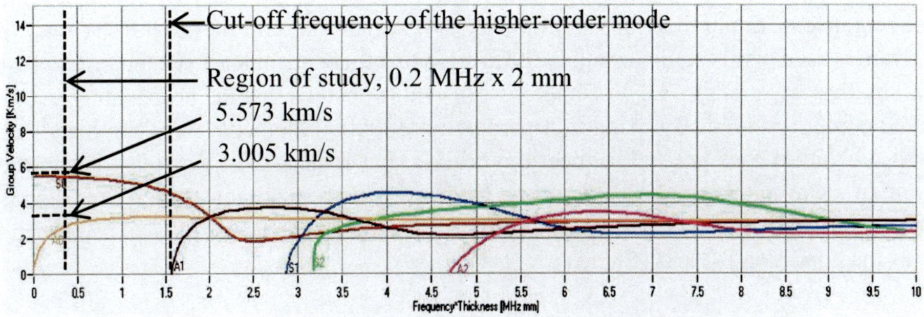

Fig. 2. Group velocity curve for aluminium plate with a thickness of 2 mm.

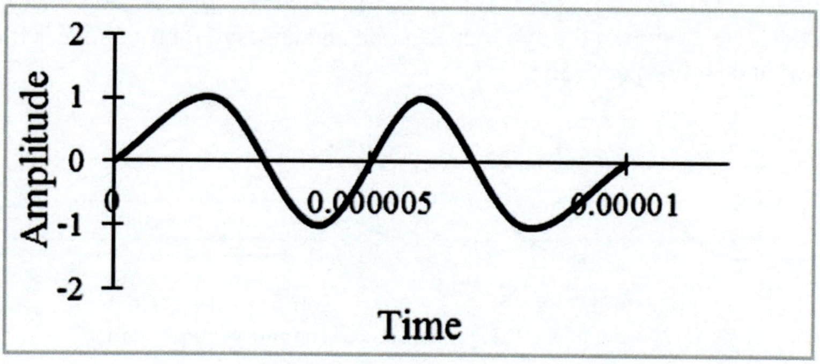

Fig. 3. A two-cycle sinusoidal impulse for excitation input.

where the parameters p and q are defined as follows:

$$p^2 = \frac{\omega^2}{c_L^2} - k^2_, \; q^2 = \frac{\omega^2}{c_T^2} - k^2_, \; h = \frac{d}{2} \text{ and } k = \frac{\omega}{c_P},$$

where ω is circular wave frequency, k is wavenumber, c_p is phase velocity, c_L and c_T are the propagation velocities of bulk waves.

$$c_g = \frac{d(kc_p)}{dk} = c_p + k\frac{dc_p}{dk} \tag{3}$$

The c_p is propagating velocity of the wave with a single frequency and c_g is the propagation velocity of the wave packet with adjacent multi frequencies.

3 Wave Propagation Modeling

A 3D FE method was used to simulate the interaction of the Lamb waves with defects in an aluminium plate specimen. The properties of the aluminium are: density = 2580 kg/m^3, Young's Modulus = 72.7 GPa and Poisson's ratio = 0.33. A commercial ABAQUS finite element software package was used to model the geometry, generate the meshing and perform the dynamics simulation of the Lamb waves propagation. Attention should be paid to two main parameters for dynamic analysis of Lamb waves propagation, i.e. element edge length, L_{min} and time step, must be chosen in order to obtain a good convergence of the numerical solution [13, 14]. A quadrilateral mesh with the L_{min} of 0.8 mm is used in this study to satisfy the criteria that there are at least 10 elements across the smallest wavelength, λ_{min} of interest. This value ensures that the propagating waves are spatially resolved. By using that parameter, 3800000 elements were produced. Δt with a value of 5 *ns* is used to meet the criteria that at least 20-time steps during the cycle of a wave at the highest frequency. Choosing an adequate integration time step is very important for the accuracy of the solution. The mentioned criterion can be referred to as the condition (4) and (5) [15]

$$L_e \langle \frac{\lambda_{\text{min}}}{10} \tag{4}$$

$$\Delta t \langle \frac{1}{20\, f\mathrm{max}} \tag{5}$$

3.1 Healthy Plate Simulation

The signals generated from the healthy plate simulation was utilised as the baseline signals. These baseline signals were used as the reference signal for the comparative analysis for damaged plate signals. The sensor was placed at a distance of 10 cm from the actuation point to avoid the superimposed between S_0 and A_0 mode as discussed in Sect. 2. It was observed that there is sufficient time for mode separation at this distance. The sensor was placed at ten different distances for the wave velocity calculation, which is varied, from 10 cm to 28 cm with a graduation of 2 cm, as shown in Fig. 4(a). Figure 5 shows the recorded signals at ten different distances. Two main wave packets were observed in each of the displacement time domain. The velocity of each wave packet was used as the indicator for the type of mode identification and it is depicted in Fig. 6. It was calculated that the velocity of the first wave packet is 5573 m/s and for the second wave packet is 3005 m/s. Referring to the theoretical dispersion curve (Fig. 2), the first wave packet is the S_0 mode and the second wave packet is the A_0 mode. Velocity values for both modes were agreed well with the wave visualisation simulated by the authors in the previous report [16]. The visualisation image was presented in Fig. 7(a). The finding for intact plate visualised that other than the smooth contour of wave propagation, the S_0 mode was moving faster than the A_0 mode.

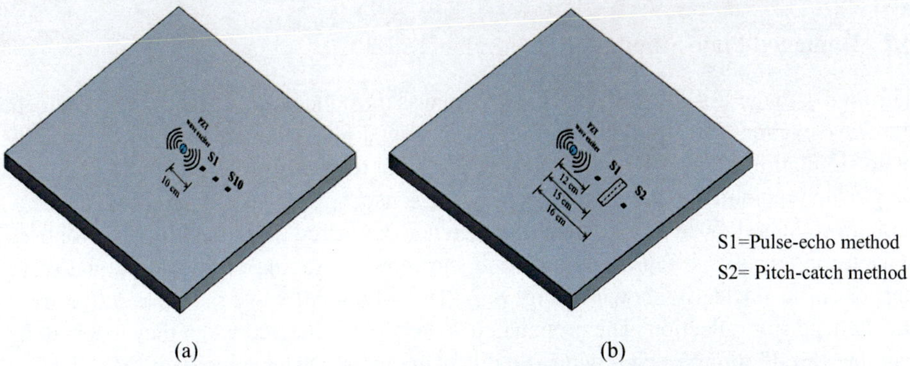

Fig. 4. Simulation configuration for (a) healthy and (b) damaged plate.

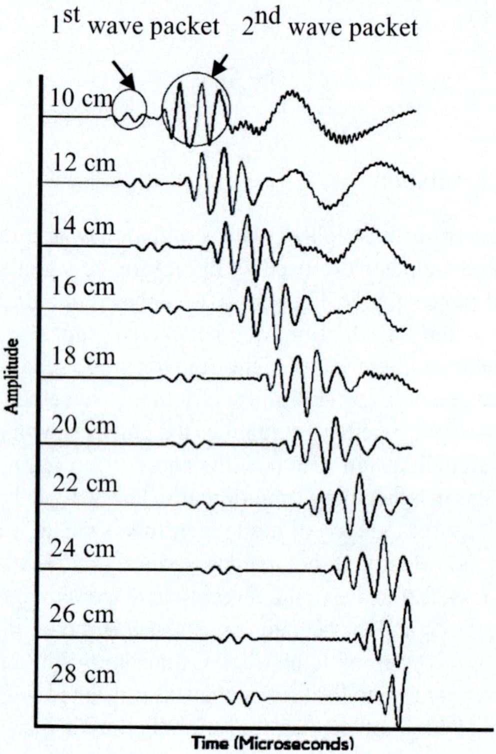

Fig. 5. Displacement signals at ten different positions.

3.2 Damaged Plate Simulation

Through damage slot with a size of 2 × 0.2 cm was introduced at a distance of 15 cm from the wave exciter. The simulation set up for damaged plate was the same as the healthy plate simulation set up. Two sensors were placed at the position for the pulse-echo and pitch-catch method and the arrangement of the sensors is shown in Fig. 4(b). The time-domain signals showed a significant of difference compared to the healthy plate for both sensor configurations. Pulse-echo method showed the existing of one additional wave packet and it is clearly shown in Fig. 8(a). The additional wave packet resulted from the damage slot reflection. The propagated waves were reflected when they reached the damage slot. Figure 7(c) clearly proved this phenomenon. The velocity of the reflection waves was calculated and given as 2703 m/s, which is slower than the waves before reaching the defect due to the energy loss. Short-time Fourier Transform (STFT) was implemented to study in detail the wave interaction with defects. One additional contour has appeared right after the A_0 mode when comparing with a spectrogram of the healthy plate, which is between Fig. 9(a) and (b). This additional contour was observed to have the same frequency spectrum (200 kHz) with the A_0 mode. Thus, the reflected wave from the damage slot is also from *the* A_0 mode family. S_o mode energy is fully attenuating when passing by the damage slot due to its nature which is mainly in-plane motion. It also affected by the large differences in mechanical impedances between the air and the

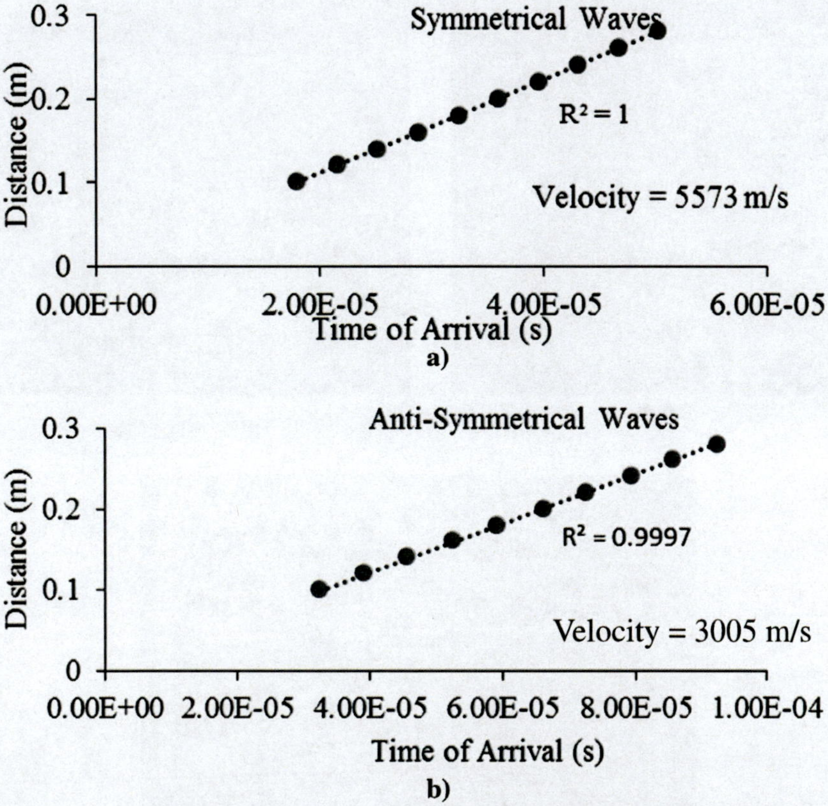

Fig. 6. The group velocity of (a) symmetrical mode and (b) anti-symmetrical mode.

solid-state due to through-hole defects. These factors resulted in only A_0 mode appeared in time-domain signals.

Pitch-catch method showed a different phenomenon. The transmitted waves that are recorded by the sensor placed after the defects show an identical wave pattern with the healthy plate, but with a significant decrement of amplitude range at about 2.8×10^{-10} m which is equal to 48%. This value was obtained from the amplitude difference between healthy and damaged plate signals at the position of 16 cm from the actuator (S2). Besides, the delay for the time of arrival (ToA) of the signals also affected by the introduction of defects for about 1.1 μs as presented in Fig. 8(b). This is due to the scattering of the waves around the defect. The associated spectrogram results in Fig. 9(c) also presents the delay of TOA for both modes which is well agreed with the respective time-domain signal.

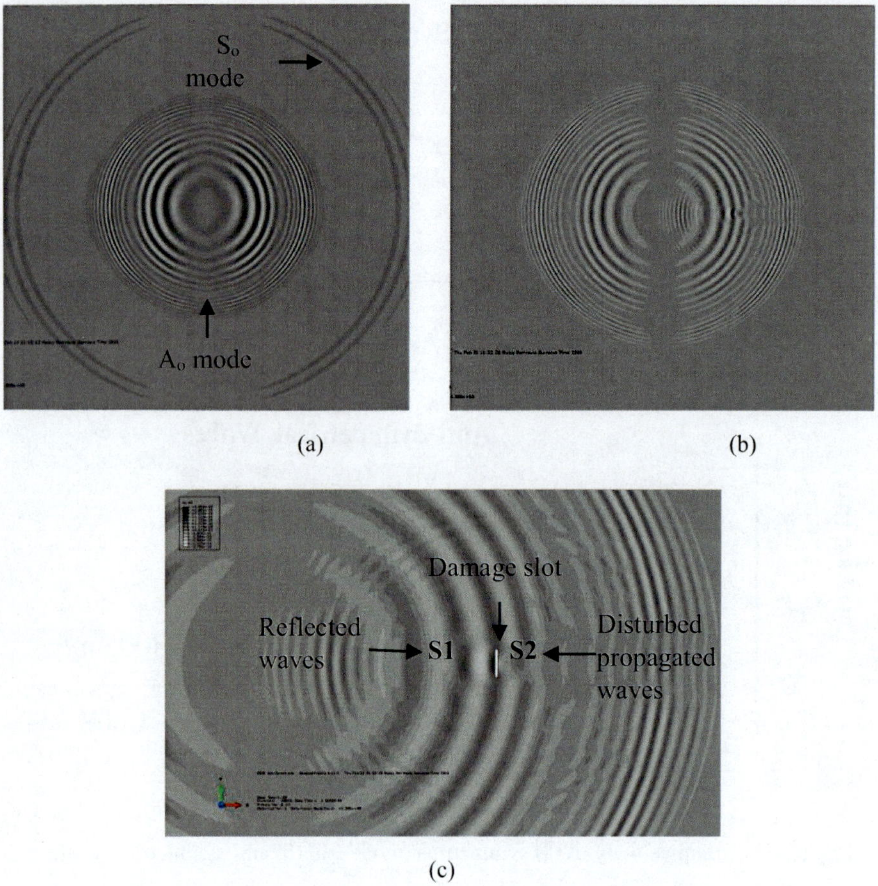

Fig. 7. Visualisation of the wave propagation by N Ismail et al. [16] for (a) intact plate, (b) damaged plate and (c) close-up of the disturbed propagated wave for the damaged plate.

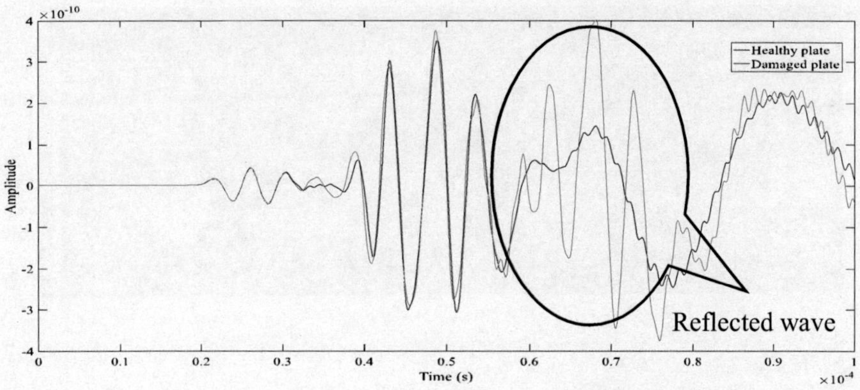

(a)

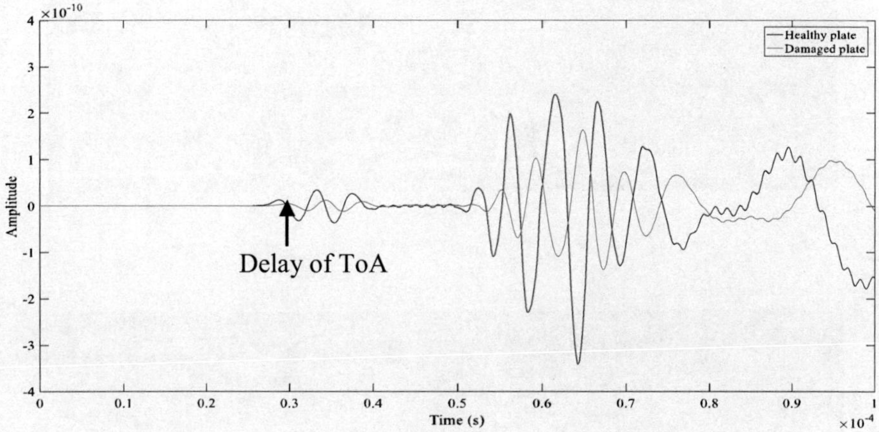

(b)

Fig. 8. Direct comparison between healthy plate and damaged plate for (a) pulse-echo and (b) pitch-catch method.

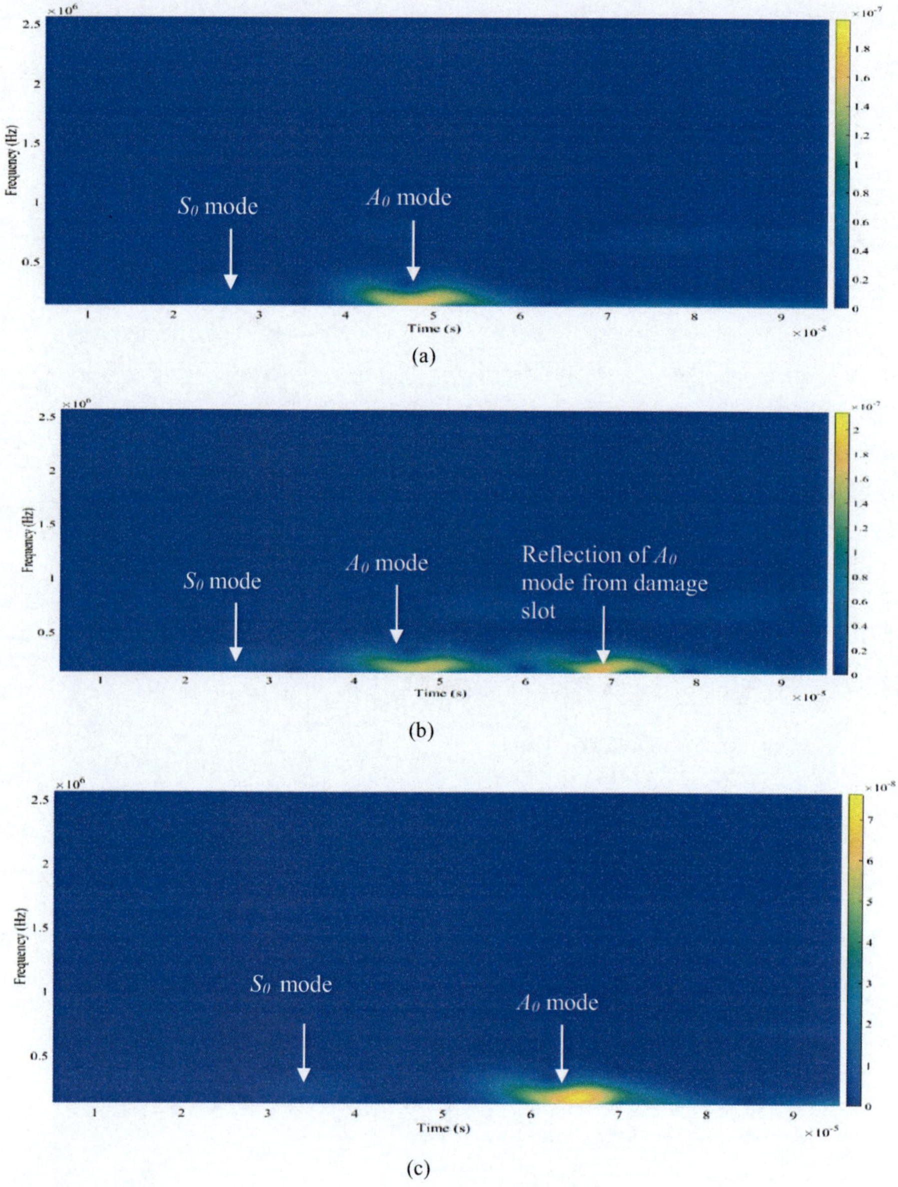

Fig. 9. Spectrogram result for (a) healthy plate, (b) pulse-echo and (c) pitch-catch method

4 Conclusions

Lamb waves are widely used to inspect and monitor the structural integrity of the plates or plate-like structures due to their advantages. However, due to multimode and dispersive nature, a good understanding of their behavior is important for result accuracy in SHM implementation. This study was performed a simulation work to achieve this objective. The additional of the wave packet is the extracted damage features when comparing with the intact plate using the pulse-echo method. The reduction in range amplitude and time delay in the time of arrival is the damage indicators using the pulse-echo method. These parameters were clearly observed in this present study and it proved the applicability of Lamb wave for the diagnostic analysis. Their sensitivity towards any structural changes can be used as an indicator for the condition monitoring of the structure. These defect signatures can be used as the diagnostic parameter for the metallic SHM system.

Acknowledgement. The authors would like to express their gratitude to Universiti Malaysia Pahang through the fund of PGRS180376 for supporting these research activities.

References

1. Mareeswaran, S., Sasikumar, D.T.: The acousto - ultrasonic technique: a review. Int. J. Mech. Eng. Technol. **8**(6), 418–434 (2017)
2. Abbas, M., Shafiee, M.: Structural health monitoring (SHM) and determination of surface defects in large metallic structures using ultrasonic guided waves. Sensors (Basel, Switzerland) **18**(11), 3958 (2018)
3. Giurgiutiu, V.: Structural Health Monitoring with Piezoelectric Wafer Active Sensors, 2nd edn. Academic Press, Oxford (2014)
4. Tian, Z., Yu, L.: Lamb wave frequency–wavenumber analysis and decomposition. J. Intell. Mater. Syst. Struct. **25**(9), 1107–1123 (2014)
5. Staszewski, W.J.: Intelligent signal processing for damage detection in composite materials. Compos. Sci. Technol. **62**(7), 941–950 (2002)
6. Manson, G., Pierce, S.G., Worden, K., Monnier, T., Guy, P., Atherton, K.: Long-term stability of normal condition data for novelty detection. In: SPIE's 7th Annual International Symposium on Smart Structures and Materials, SPIE, vol. 3985 (2000)
7. Liu, Y., Fard, M.Y., Kim, S.B., Chattopadhyay, A., Doyle, D.: Damage detection in composite structures using Lamb wave analysis and time-frequency approach. In: SPIE Smart Structures and Materials + Nondestructive Evaluation and Health Monitoring, SPIE (2011)
8. Ming, Z., Liang, Z., Jing, L., Wentao, W.: Mode identification and extraction of broadband ultrasonic guided waves. Measur. Sci. Technol. **25**(11), 115005 (2014)
9. Nienwenhui, J.H., Neumann, J.J., Greve, D.W., Oppenheim, I.J.: Generation and detection of guided waves using PZT wafer transducers. IEEE Trans. Ultrason. Ferroelectr. Freq. Control **52**(11), 2103–2111 (2005)
10. Liu, W., Giurgiutiu, V.: Finite element simulation of piezoelectric wafer active sensors for structural health monitoring with coupled-filed elements. In: SPIE Smart Structures and Materials + Nondestructive Evaluation and Health Monitoring, SPIE, vol. 6529 (2007)
11. Silva, C., Rocha, B., Suleman, A.: A structural health monitoring approach based on a PZT network using a tuned wave propagation method (2009)
12. Rose, J.L.: Ultrasonic Guided Waves in Solid Media. Cambridge University Press, Cambridge (2014)
13. Rosenthal, A., Razansky, D., Ntziachristos, V.: High-sensitivity compact ultrasonic detector based on a pi-phase-shifted fiber Bragg grating. Opt. Lett. **36**(10), 1833–1835 (2011)
14. Leng, J., Asundi, A.: Structural health monitoring of smart composite materials by using EFPI and FBG sensors. Sens. Actuators, A **103**, 330–340 (2003)
15. Wada, A., Tanaka, S., Takahashi, N.: Optical fiber vibration sensor using FBG fabry–perot interferometer with wavelength scanning and fourier analysis. IEEE Sens. J. **12**(1), 225–229 (2012)
16. Ismail, N., Hafizi, Z.M., Nizwan, C.K.E., Ali, S.: Simulation of Lamb wave interactions with defects in a thin plate. J. Phys. Conf. Ser. **1262**, 012030 (2019)

Automatic Identification and Categorize Zone of RFID Reading in Warehouse Management System

Chun Sern Choong(✉), Ahmad Fakhri Ab. Nasir, Anwar P. P. Abdul Majeed, Muhammad Aizzat Zakaria, and Mohd Azraai Mohd Razman

Innovative Manufacturing, Mechatronics and Sports Lab (iMAMS), Faculty of Manufacturing, University Malaysia Pahang, Pahang, Malaysia
ash6697@gmail.com
http://www.imamslab.ump.edu.my/

Abstract. Radio Frequency Identification (RFID) technology has improved the operational efficiency and process flow in the distribution of warehouse management system (WMS) around the globe. Nonetheless, a moving or missing tag as well as known and unknown tag's location that may occur in the detection could reduce the efficiency of process flow. This study aims at identifying the location of goods in between two RFID reading zones by means of machine learning, particularly Support Vector Machine (SVM). A total of seven statistical features are extracted from the received signal strength (RSS) value from the raw RFID readings. SVM classifier are evaluated by considering the combination of different statistical features namely COMBINE to produce a more effective classification in comparison to individual statistical feature. The performance of the classifier demonstrated a classification accuracy of approximately 94% by considering all features whereas the performance of the classifier by considering individual features alone is below than 90%. This preliminary study establishes the applicability of the proposed automatic identification is able to provide the management of goods as well as supply chain reasonably well without human intervention.

1 Introduction

Warehouse management system (WMS) based on Radio Frequency Identification (RFID) has been widely used in the industry to probe into the logistics and process flow of a given system [1]. WMS for instance, planning, organizing, staffing, directing, and controlling the utilization of warehouse facilities has become more complex in supply chain. Conversely, RFID based WMS system are employed for stock storage, goods locating, order status, cross docking shipment, sorting and cycle counts in order to minimize poor utilization of space in warehouse, miss locating goods, miss identification, fault identification or reidentification of the product [2]. RFID is an automatic identification and data capture (AIDC) system that transmits the identity (in the form of a unique serial number) of an object using electromagnetic fields or electrostatic coupling in wireless communication [3]. The fundamentals of an RFID system consists

M. A. Zakaria et al. (Eds.): *Advances in Mechatronics, Manufacturing, and Mechanical Engineering*, LNME, pp. 194–206, 2021.
https://doi.org/10.1007/978-981-15-7309-5_20

of two major components: optical character readers (also known as transceiver) which transmits the instruction through an antenna using Radio Frequency (RF) waves to and fro the transceiver [4], and tags (or often referred to as transponders) which is located at the goods that is required to be identified [5].

RFID is one of the simple techniques for indoor localization where GPS and other satellite technologies lack in precision or fail entirely [6]. A considerable amount of literature has been published on indoor localization to mitigate the drawback of GPS and other satellite technologies in locating the object in an indoor environment [7, 8]. These studies applied different approaches to improve the accuracy, processing time, hardware architecture and cost development. However, although different methods have been employed in improving the aforesaid technologies, RFID technology have been demonstrated to outperform the technologies particularly with regards to indoor localization. Want et al. [9] has been reported to be one of the earliest that applied infra-red technology to study the indoor localization. The Active Badge system is implemented as a centralized location service for locating staff in the office through a network of sensors through badges. Bahl et al. [10] further improved the idea by exploiting RF map to localize and track workers inside buildings. RADAR based WLAN system is recorded and the Received Signal Strength (RSS) is processed at multiple base stations to cover the full floor to locate the workers from the mobile's position which is registered in the database.

LANDMARC [11] employs the concept of reference tags to reduce number of reader, yet it is still able access every corner in the floor. These existing solutions can be compared with RADAR based WLAN system which requires at least 3 readers to deliver a good geometry for reducing the complexity of system apart from providing a cost-effective solution. Despite such improvements, the drawbacks of RFID still exist, namely the processing time to estimate the object from 3 readers [12] and the effect of other environmental frequencies at the floor [13] amongst others. To date, there are many investigations that have been carried out utilizing SVM for classification that SVM have the highest classification accuracy among the others [14–16].

In this paper, we investigate an indoor positioning system for RFID tagged goods by overlapping the RFID reader antenna. The challenge of this research is to provide a reasonably accurate decision support to the system regarding the correct location of the tagged goods within the RFID reading zone readers. The detected tagged goods are measured by RSS measurement to estimate the reader-tags distance from the intersection reader. Seven statistical features are extracted from received signal strength (RSS) in the RFID readings and the influence of the features combined and sole (individual) towards the performance of a Support Vector Machine (SVM) classifier is investigated. The rest of the paper is organized as follows. Section 2 details on the system architecture of the developed ultra-high frequency (UHF) RFID system in WMS which consists of the front end and back end indoor localization algorithm. Section 3 evaluates the details of classification performance. The reliability of the system has been tested and future work is discussed in Sect. 4.

2 System Structure of RFID in WMS

The physical hardware structure of the UHF RFID system designed for this experimental investigation is illustrated in Fig. 1. The hardware includes the RFID reader antenna, Power over Ethernet (PoE) switch, PoE splitter, RS232 RS485 Ethernet Converter, personal computer (PC) and other peripheral devices. The rationale of utilizing PoE instead Cat5 or Cat6 ethernet cable, primarily on its ability to deliver data and power. Moreover, PoE has been demonstrated to reduce network congestion for multiple reader antenna. A custom software was developed to configure the multiple reader antenna. The UHF RFID system consists of power management, RFID data management and network configuration management that is running on Windows 10 as shown in Fig. 2.

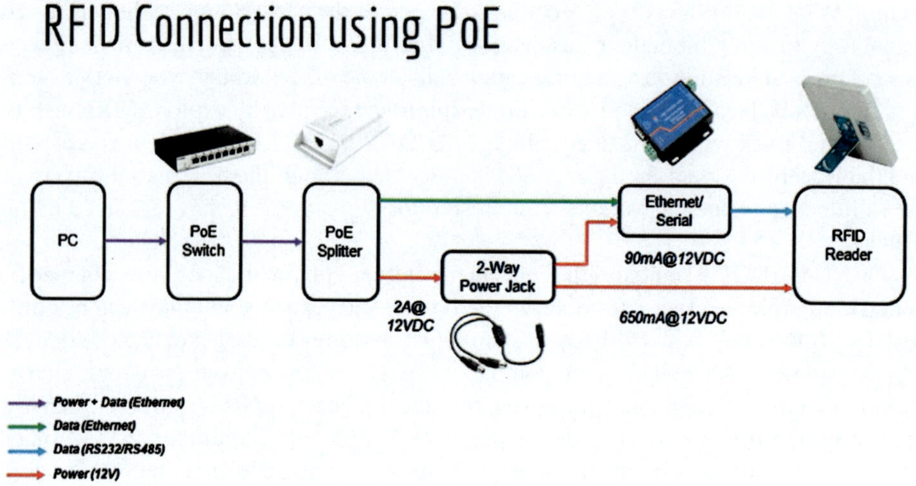

Fig. 1. A connection between the laptop and multiple RFID readers via PoE

The power management is initialized automatically when the PC is powered on while the RFID data management is a lambda architecture of data stream processing methods for handling massive quantities of data. Furthermore, network configuration management supports manually addition of the readers by designating the reader's IP address. After the reader is detected by means of Transmission Control Protocol/Internet Protocol (TCPIP) configuration, it is able to set up the parameter of a new RFID reader that also consists of network properties, frequency of the antenna, antenna's power gain and other different application as depicted in Fig. 3.

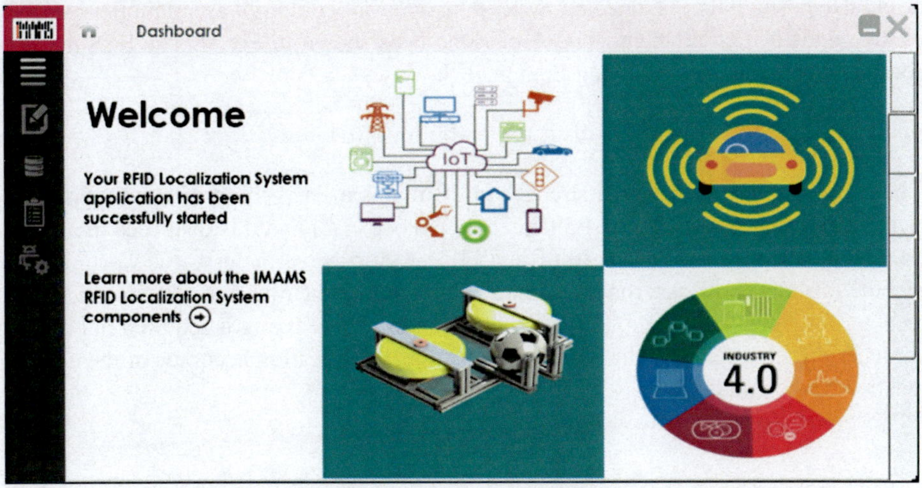

Fig. 2. UHF RFID system application

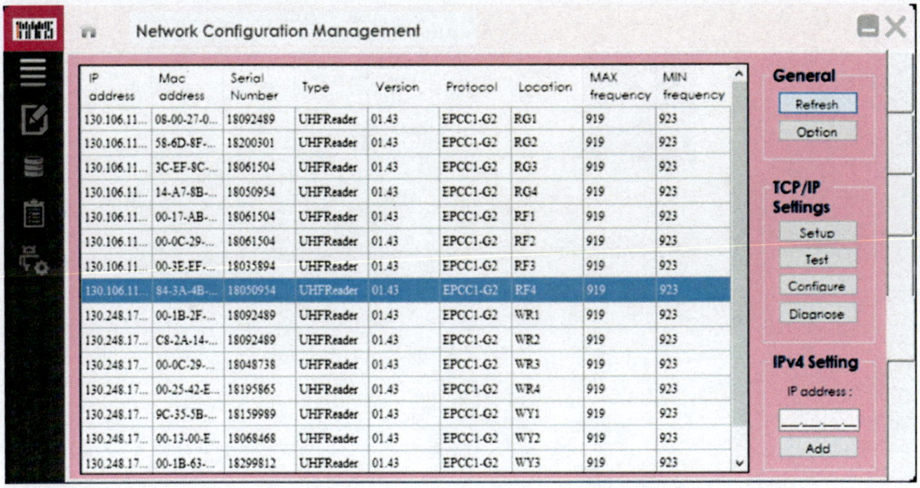

Fig. 3. Network configuration management of UHF RFID system

2.1 UHF RFID Front-End System

RFID tagging only consists of a tag antenna where the foil on the surrounding edges for receive the incoming radio frequency and reflect back to the reader; a special tag chip to generate a unique identifier for each individual tag known as the electronic product code (EPC); and the material of the form factor [17]. The EPC is akin to a bar code technology and it is considered as one of the most common methods used today for storing data on the tags that can be retrieved via either fixed or handheld scanning devices. The front-end of this system provides the decision support for the current location of goods where the goods in RFID reader antenna reading zone (Zone A) is moved, either is false negative that the goods are not detected by the RFID reader as well as if the goods are missing or is false positive that is detected by RFID reader but could not ascertain the location of the goods if it is either in Zone A or Zone B as shown in Fig. 4. The RSS may be computed by using the following Eq. (1) [18].

$$RSS(\text{dBm}) = -10n\ \log_{10}(d) + A \tag{1}$$

where A is the received signal strength in dBm at 1 m, n is the propagation constant or path-loss exponent is 2.7 to 4.3 (Free space has n = 2) and d is distances in meters. Receive Sensitivity: 802.11ac: 30 dBm@11.52 Mbps. Nevertheless, it is worth noting that the formula above is a rough estimate of computing the minimum RSS needed by a receiver for a given data rate. However, further signal processing is required in order to engineer different features that could improve the classification accuracy of the desired location of the goods.

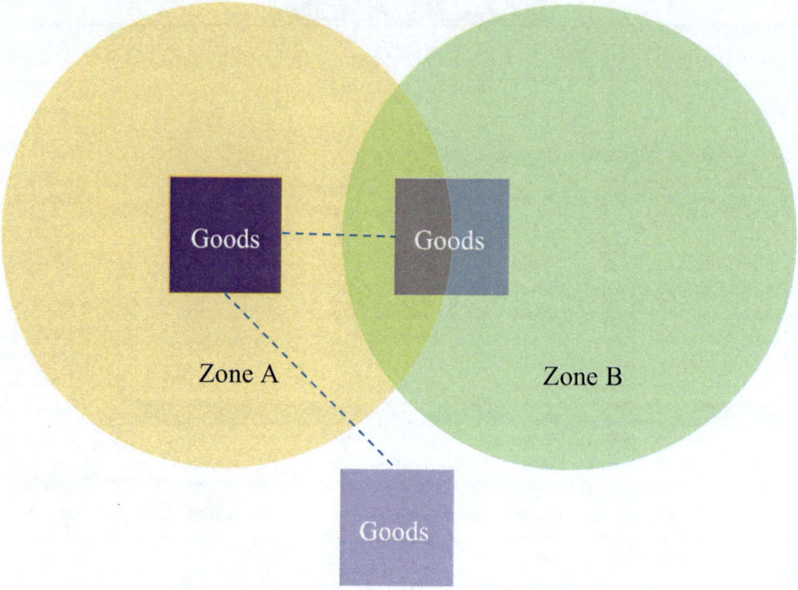

Fig. 4. Reading zone of multiple of the reader antenna

2.2 UHF RFID Back-End System

Back-End server and connectivity are the multiple RFID readers constantly worked on reading events connected with the central server to realize a secure system [19]. The central server provides the functionality for handling database information, for instance, define, manipulate, retrieve and manage data that maintains and updates the tag database, generate alerts when specific performance issues arise. The use of speed layer processes data streams in SQL database engines and a robust PoE TCP/IP stack is a well-versed approach to integrate between the back-end database and the readers (front-end).

As previously mentioned, the RSS computed by aforesaid formula does not provide accurate location of the tagged goods either in Zone A or Zone B due to the non-static nature of it. Therefore, feature engineering or extraction is carried out on the estimated RSS value. Seven statistical features namely Maximum (MAX), Minimum (MIN), Mean, Mode, Median, Variance as well as Range through Received Signal Strength with a sampling time of 10 s obtained from different RFID readings [20]. Table 1 provides an overview of the selected features that will be used.

Table 1. Features extracted from raw RSS value

Features	Description
MAX	The maximum of the RSS value during the data collection. The reading of RSS will increase if the static tag is near to the RFID antenna
MIN	The minimum of the RSS value during the data collection. The reading of RSS selects from the data collection within 1 s
Mean	The average of the RSS value during the data collection. The reading of mean RSS will increase if the static tag is near to the RFID antenna
Mode	The mode of the RSS value during the data collection. The most constant reading of RSS from the data collection within 1 s
Median	The median of the RSS value during the data collection. The midpoint reading of RSS calculates from the data collection within 1 s
Variance	The reading of variance based on the distance between the tag and the antenna if the change of tag's orientation so the larger variance
Range	The difference between the largest RSS and smallest RSS value in the data collection

On the other hand, statistical features extracted from the RSS are also exploited to characterize the RFID readings. These parameters can be calculated at the following.

$$Mean = \frac{\sum_{i=1}^{n} \chi_i}{N} \tag{2}$$

$$Median = L + \left(\frac{\frac{n}{2} - F}{f_m} \right) C \tag{3}$$

$$Variance = \frac{1}{n-1} \sum\nolimits_{i=1}^{n} (\chi_i - \overline{\chi})^2 \tag{4}$$

$$range = MAX - MIN \tag{5}$$

where $\overline{\chi}$ is the average value of the discrete set of numbers, n is the number of a discrete set of χ_i, L is lower boundary of median class, F is size of the median class interval, f_m is frequency corresponding to the median class and C is cumulative frequency preceding median class.

Together these features provide important insights into localizing goods during the system is running. As can see from Fig. 5(a) and Fig. 5(b), the RSS value for the moving tag decreases in the Zone A but appears and increases in the Zone B. In the event that the tag is moved away from Zone A, it is expected that the tag is moved away from the reader antenna of Zone A vice-versa. Moreover, as shown in both figures, the RSS signal shifts to a higher negative value if the tag is moved. Moreover, the distance between the goods and reader antenna changes continuously when the goods transfer to another area, the RSS value of the corresponding goods will change respectively.

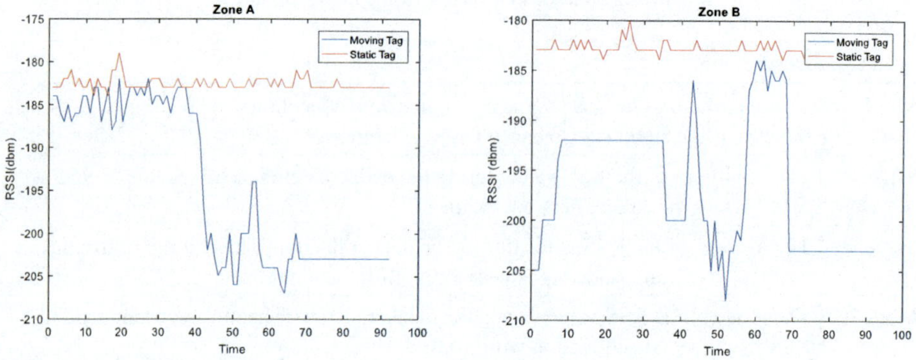

(a) (b)

Fig. 5. RFID RSS distribution for static tag and moving tag (a) Zone A (b) Zone B

2.3 Evaluation Metrics

The performance of the classifier may be described by the confusion matrix tabulated in Table 2. It is apparent that the present investigation is akin to a simple binary classification [21]. The false positive described in this study is essentially an error in the classification in which a test result improperly indicates the location of the presence of a tag, for example, a tag is in between two-zone and the system cannot differentiate which is the correct one. Similarly, a true positive is an outcome in which a test result properly indicates the actual location presence of a tag. While a false negative is an error in which a test result improperly indicates no presence of a tag, such as a tag is gone from the zone but the track recorded mention it is used to be there. And a true negative is an outcome in which a test result correctly predicts none of a tag is shown before in Zone.

The Area under the curve (AUC) is a performance metrics for a binary classifier came from receiver operating characteristic curve (ROC) which is a technique to display the performance of a classification model and compare the performance of different classifiers. This curve is plotting by the true positive rate (TPR) against the false positive rate (FPR) at various threshold settings defined as follows [22]:

$$TPR = \frac{Moving\ tag\ classified\ located\ at\ Designated\ Zone}{Total\ number\ of\ Moving\ tag} \tag{6}$$

Table 2. 2 × 2 Confusion Matrix that depicts all four possible outcomes

	Positive (actual)	Negative (actual)
Positive (predicted)	True Positive (**TP**): **Moving tag** classified located at **Designated zone**	False Positive (**FP**): **Moving tag** classified located at **Unknown zone**
Negative (predicted)	False Negative (**FN**): **Moving tag missing** Classified located at **Out of the zone**	True Negative (**TN**): **Moving tag missing** classified located at **Designated zone**

$$FPR = \frac{Moving\ tag\ classified\ located\ at\ Unknown\ Zone}{Total\ number\ of\ Moving\ tag\ missing} \tag{7}$$

The experiments were carried out with tags moving away from the zone at a relative distance of 4 meters and the rest of the tags at 1 m away. The classification performances have been repeated tests to perform the fraction of predictions model got right. The proportion of correctly classified tags (TP + TN) and the total number of classified tags (TP + TN + FP +FN) represents the Accuracy Index (ACC) [23]:

$$ACC = \frac{Number\ of\ correctly\ classified\ tags}{Total\ number\ of\ classified\ tags} \tag{8}$$

Precision is a measure of statistical variability to determine random errors (False Positive). Talking about this zone detection, FP means that a location that is non-error has been identified as an error. The model might lose decision support to identify the tag's location if the precision is not high for the zone detection model. Precision address in this model are [24]:

$$Precision = \frac{TP}{TP + FP} \tag{9}$$

Recall or sensitivity measures the percentage of actual positives in the model that correctly classified (TP). By understanding Recall in tag detection, the lowest value of FN will be select as the best model from the model metric. The detected tag (designated zone) goes through the test and predicted as a missed tag (out of the zone), the consequence can be inventory loss. It is insignificant to achieve a recall of 100% by detecting all tags in response to the model. As a result of this, the number of missed tags also important to perfect the model and the recall is calculated below [14]:

$$Recall = \frac{TP}{TP + FN} \tag{10}$$

F1 score is the harmonic mean between precision and recall of the test to compute the score. It supports to classify an uneven class distribution which is a large number of the actual negative so F1 score achieves its best value at 100% (ideal precision and recall) will have better performance of the model and the worst will be 0%. Therefore, the high precision but lower recall is giving an extremely accurate, but it misses a large number of instances that are difficult to classify. Mathematically, it can be expressed as [16]:

$$F1 = 2 \times \frac{Recall \times Precision}{Recall + Precision} \tag{11}$$

2.4 Classifier

In this study, the efficacy of a class of supervised classification model of the location of the tag namely Support Vector Machine (SVM) is investigated. The misclassification of the RSS may transpire owing to the influence of physical objects, RF interface, electrical interface and environmental factor amongst others [25, 26]. Therefore, there is a stark difference of the RSS reading between the theoretical calculating methods and the distance from the reader antenna to the tag. The ability of the features extracted either individual or combined effect towards the classification accuracy of the SVM model. SVM discriminates between classes by identifying the optimal hyperplane that is able to distinguish the dataset [27]. It is worth noting that SVM has been demonstrated to be able to classify well on different applications [28]. In this study, the linear kernel based SVM model is evaluated using Eq. 12 with hard margin [30]. The hyperplane pass through the support vector is based on the inner product <x, y> plus an optional constant c. Based on the study in [31, 32], leave-one-out cross-validation (LOOCV) is often better for the smaller train data size but due to the different pattern in the train data, the error of bias and variance will be lower which means the poor ability to differentiate noise and data sampling used in training. Therefore, five-fold cross validation technique is employed in this investigation as it has been established to mitigate the idea of overfitting [33]. The performance of the classifier in classifying the location of the tags is evaluated by means of the evaluation metrics described in the previous subsection.

$$k(x, y) = x^T y + c \tag{12}$$

3 Results and Discussion

The ability of the developed SVM model in classifying the location of the tag is evaluated with regards to the features extracted both individual features as well as the combination of all statistical features. Figure 6 below shows the performance of the SVM model by taking into consideration the individual features. It could be seen that the highest ACC attainable by employing the MODE feature with a classification accuracy of 87.5%. The feature which could not describe the location of the tags is the Variance feature, as it could only provide an ACC of 62.5%. In contrast, it could be observed that the combination of all features yields a classification accuracy of 93.8%, suggesting that it is non-trivial to include all the engineered or extracted features.

It could be observed from the tabulated results that the linear kernel function based SVM model for combination of all features yields classification accuracy in train and

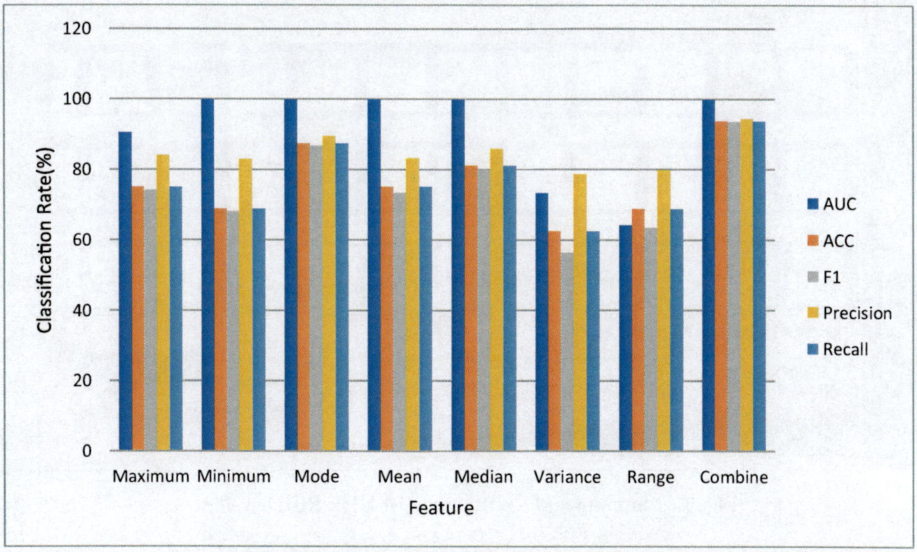

Fig. 6. Classification Rate for training data

test data is shown in Table 3. The comparison in train set demonstrate that the variations provide high accuracy of classification with an error rate of 6.2% that subtracted from the classification accuracy. In additional, the test set in linear-SVM model is perfectly predicting the correct classification of known zone and unknown zone, as it yields the error rate is 0.0%. Based on the received signal strength value, it is obvious through the current research that the linear-based kernel functions SVM model can be used for identification purposes for locate the zone.

Table 3. Confusion Matrix of the known zone and unknown zone in linear-SVM for test and train

Error		Train set		Test set	
		6.2%		0.0%	
Confusion matrix		Known zone	Unknown zone	Known zone	Unknown zone
	Known zone	8	0	3	0
	Unknown zone	1	7	0	3

As pointed out in the confusion matrix of the evaluated combination of all features that the high classification accuracy with the total 16 data sample in training and 6 data sample in testing for classify the zone area is apply into UHF RFID system to locate the tags in real-time. By applying linear-SVM model, the system could classify correct the location of the goods as shown in Fig. 7.

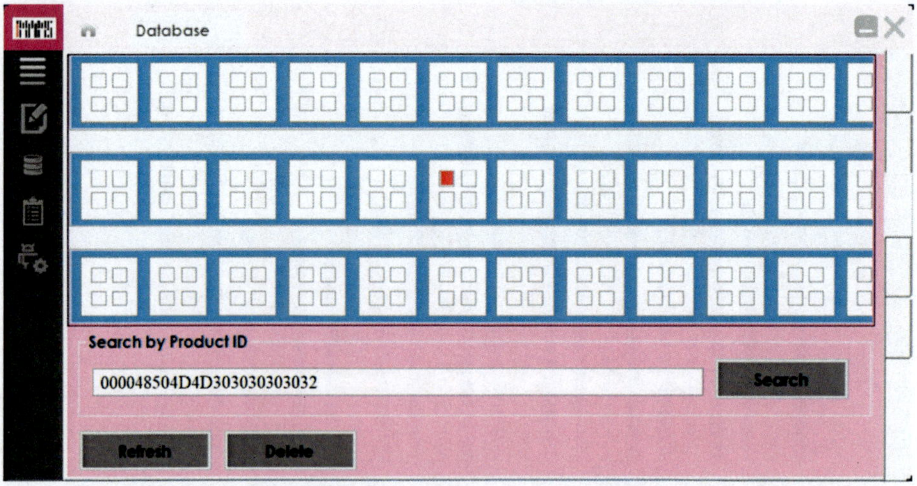

Fig. 7. Mapping and Localization of UHF RFID system

4 Conclusion

In this overview, a new classification method by extracting features from RSS value to discriminate tags for automatic detection of false positive and false negative RFID readings has been presented. The false positive and false negative RFID readings provide the classification accuracy for the decision support to categories the correct location of the tagged goods within the RFID reading zone readers. The classification procedure here proposed exploits combination of seven selected extracted features. It has been demonstrated that the Linear-SVM model by using all the features extracted is able to yield a classification accuracy of 87.5%. Nonetheless, it is worth noting that this study is rather in its preliminary state, further improvement could be carried out by employing a more appropriate feature selection method, for instance Principal Component Analysis, Box Plot Analysis or Singular Value Decomposition techniques amongst others. Moreover, in this study, only the linear kernel is investigated, other kernels such as rbf and polynomial will be further evaluated in order to evaluate its efficacy in classifying the location of the tags.

References

1. Adiono, T., Ega, H., Kasan, H., Harimurti, C.S.: Fast Warehouse Management System (WMS) using RFID based goods locator system. In: 2017 IEEE 6th Global Conference on Consumer Electronics (GCCE), pp. 1–2 (2017)
2. Yan, B., Chen, Y., Meng, X.: RFID technology applied in warehouse management system. In: 2008 ISECS International Colloquium on Computing, Communication, Control, and Management, pp. 363–367 (2008)
3. Karygiannis, T., Eydt, B., Barber, G., Bunn, L., Phillips, T.: Special Publication 800-98 Guidelines for Securing Radio Frequency Identification (RFID) Systems Recommendations of the National Institute of Standards and Technology

4. Kaur, M., Sandhu, M., Mohan, N., Sandhu, P.S.: RFID technology principles, advantages, limitations and its applications. Int. J. Comput. Electr. Eng. **3**(1), 1793–8163 (2011)
5. Ahmad, T.: An improved accelerated frame slotted ALOHA (AFSA) algorithm for tag collision in RFID. Int. J. Mob. Netw. Commun. Telemat. **2**(4), 1–8 (2012)
6. Jeffrey Hightower, G.B.: A survey and taxonomy of location systems for ubiquitous computing. IEEE Comput. **34**, 57–66 (2001)
7. Bekkali, A., Sanson, H., Matsumoto, M.: RFID indoor positioning based on probabilistic RFID map and Kalman filtering. In: Third IEEE International Conference on Wireless and Mobile Computing, Networking and Communications (WiMob 2007), pp. 21–21 (2007)
8. Zafari, F., Gkelias, A., Leung, K.K.: A survey of indoor localization systems and technologies. IEEE Commun. Surv. Tutorials **23**, 2568–2599 (2019)
9. Want, R., Hopper, A., Falcão, V., Gibbons, J.: The active badge location system. ACM Trans. Inf. Syst. **10**(1), 91–102 (1992)
10. Bahl, P., Padmanabhan, V.N.: RADAR: an in-building RF-based user location and tracking system. In: Proceedings IEEE INFOCOM 2000. Conference on Computer Communications. Nineteenth Annual Joint Conference of the IEEE Computer and Communications Societies (Cat. No.00CH37064), vol. 2, pp. 775–784 (2000)
11. Ni, L.M., Liu, Y., Lau, Y.C., Patil, A.P.: LANDMARC: indoor location sensing using active RFID. In: Proceedings of the 1st IEEE International Conference on Pervasive Computing and Communications, PerCom 2003, pp. 407–415 (2003)
12. Griffin, J., Durgin, G.: Radio link budgets for 915 MHz RFID antennas placed on various objects. In: Proceedings of 2005 Texas Wireless Symposium, no. 1, pp. 22–26 (2005)
13. Hassan, M., Ali, M., Aktas, E.: Radio frequency identification (RFID) technologies for locating warehouse resources: a conceptual framework. In: Smart SysTech 2012 - European Conference on Smart Objects, Systems and Technologies (2012)
14. Alfian, G., Syafrudin, M., Yoon, B., Rhee, J.: False positive RFID detection using classification models. Appl. Sci. **9**(6), 1154 (2019)
15. Fan, X., Wang, F., Wang, F., Gong, W., Liu, J.: When RFID meets deep learning: exploring cognitive intelligence for activity identification. IEEE Wirel. Commun. **26**(3), 19–25 (2019)
16. Geigl, F., Moik, C., Hinteregger, S., Goller, M.: Using machine learning and RFID localization for advanced logistic applications, pp. 3–4 (2017)
17. Schwartz, C.W., Khan, J.S., Pfeiffer, G.H., Mustafa, E.: Use of Radio Frequency Identification Tags in Pavements, April 2015
18. Li, G., Geng, E., Ye, Z., Xu, Y., Lin, J., Pang, Y.: Indoor positioning algorithm based on the improved RSSI distance model. Sensors (Switzerland) **18**(9), 1–15 (2018)
19. Mitrokotsa, A., Rieback, M.R., Tanenbaum, A.S.: Classifying RFID attacks and defenses. Inf. Syst. Front. **12**(5), 491–505 (2010)
20. Keller, T.: 25 classification models for RFID-based real-time detection of process events in the supply chain: an empirical study. ACM Trans. Manage. Inf. Syst. **5**, 1–30 (2014)
21. Li, X., Zhang, Y., Marsic, I., Sarcevic, A., Burd, R.S.: Deep learning for RFID-based activity recognition. In: Proceedings of International Conference on Embedded Networked Sensor Systems, vol. 2016, pp. 164–175, November 2016
22. Kamaludin, H., Mahdin, H., Abawajy, J.H.: Filtering redundant data from RFID data streams. J. Sensors **2016**, 1–7 (2016)
23. Brandl, M.: Accuracy of RFID position estimation using SAW compressive receivers. Procedia Eng. **120**, 320–323 (2015)
24. Chang, L., et al.: Towards large-scale RFID positioning: a low-cost, high-precision solution based on compressive sensing. In: 2018 IEEE International Conference on Pervasive Computing and Communications, PerCom 2018 (2018)
25. Meyers, M.: CompTIA Network + N10-007 (2018)

26. Li, Y.Y., Parker, L.E.: Classification with missing data in a wireless sensor network. In: IEEE SoutheastCon 2008, pp. 533–538 (2008)
27. Cortes, C., Vapnik, V.: Support-vector networks. Mach. Learn. **20**(3), 273–297 (1995)
28. Rifkin, R.M.: Everything old is new again: a fresh look at historical approaches in machine learning. Massachusetts Institute of Technology (2002)
29. Taha, Z., Musa, R.M., Abdul Majeed, A.P.P., Alim, M.M., Abdullah, M.R.: The identification of high potential archers based on fitness and motor ability variables: a Support Vector Machine approach. Hum. Mov. Sci. **57**, 184–193 (2018)
30. Kothari, S.C., Oh, H.: Neural networks for pattern recognition. Adv. Comput. **37**(C), 119–166 (1993)
31. Lachenbruch, P.A., Mickey, M.R.: Estimation of error rates in discriminant analysis. Technometrics **10**(1), 1–11 (1968)
32. Luntz, A., Brailovsky, V.: On estimation of characters obtained in statistical procedure of recognition. Tech. Kibern. **3**(6), 6–12 (1969)
33. Taha, Z., et al.: The identification of hunger behaviour of lates calcarifer through the integration of image processing technique and support vector machine. In: IOP Conference Series: Materials Science and Engineering, vol. 319, no. 1, p. 012028, March 2018

Utilization of Ikaz and Direct Quadrature for Transient Test-Based Technique for Leakage Detection Purpose in Pipeline System

Hanafi M. Yusop[1,2], M. F. Ghazali[1,2,3(✉)], W. H. Azmi[1,4], M. F. M. Yusof[1,2], M. A. PiRemli[1,2], and M. Z. Noordin[1,2]

[1] Faculty of Mechanical Engineering, Universiti Malaysia Pahang, 26600 Pekan, Pahang, Malaysia
fairusham@ump.edu.my

[2] Advanced Structural Integrity and Vibration Research (ASIVR), Faculty of Mechanical Engineering, University Malaysia Pahang, Pekan, Pahang, Malaysia

[3] Centre of Excellence for Advanced Research in Fluid Flow (CARIFF), Universiti Malaysia Pahang Lebuhraya Tun Razak, 26300 Gambang, Kuantan, Pahang, Malaysia

[4] Automotive Engineering Centre, Universiti Malaysia Pahang, 26600 Pekan, Pahang, Malaysia

Abstract. There is a plan for a further change to one-point identification, which includes a pipeline-ended measuring platform. The approach makes use of the frequency response and tests using low frequency bandwidth are satisfactory. This article introduces Empirical Mode Decomposition (EMD) as the technique that denotes a noisy pressure transient signal by instantaneous frequency analysis before further analysis by the signal. Thus EMD is the best way to decay from the signal into the IMF (Intrinsic Mode Function). It is nevertheless challenging to choose a fitting IMF. This is why the paper suggested the use of the Integrate Kurtosis-based Algorithm for Z-filter techniques to enable the autonomous selection of suitable IMF. Such research illustrates the mounted leakage simultaneous tubing on a 67.9-m Medium-High-Density Poly-Ethylene (MDPE) sheet. The water pressure of the experiment is about 1–4 in. The findings of the study using Normalized Ikaz have shown that the system can be used in automated IMF selection although the signal noise level ratio is lower. Standardized Ikaz is suggested to be applied as a Direct Square (DQ) study for automated detection of the intrinsic mode feature (IMF).

Keywords: Ikaz · Leak · Pressure transient · EMD · DQ

1 Introduction

The discrepancy between the volume of water generated before and after exposure known as non-revenue waters. This is because of many causes, such as the leakage within the system and the piping. Water loss is a global issue and critical water management concerns [1]. Statistical analysis reveals that Malaysia's normal level of non-revenue water was 29.4% in 2010, resulting in big currency, supply and weight misfortunes and

M. A. Zakaria et al. (Eds.): *Advances in Mechatronics, Manufacturing, and Mechanical Engineering*, LNME, pp. 207–218, 2021.
https://doi.org/10.1007/978-981-15-7309-5_21

excessive use of vitality [2]. In 2017, the rate increased to 35.3% [3]. The consequence can, therefore, take us to the negative effects of governments and the position. High-income water gives the water supply officials a bad impression [4]. It represents the problem of the water supply network's delicacy [5]. The outcome can, therefore, lead to negative impacts on infrastructure and the environment [6]. The water supply providers have a bad feeling about heavy non-revenue sewage. This represents the problem of water supply system instability [7, 8].

So there are various ways to decrease and control the non-revenue water [9]. The most commercial technique is passive control where the issue is typically confirmed by the user or noted by the company's own management. The technique may be tolerated in areas where the supplies are plentiful or good value for money. Furthermore, successful leakage management consists of two approaches which are a routine survey and monitoring of the leakage. Daily control is the method of listening to piping leakage and adjusting, reading the metered flow into the temporary zoned region to track high volume night traffic. Flow management is the monitoring of flow into areas or regions to assess leak and optimize leak detection tasks [10]. However, standard leak detection facilities are sometimes pricey and need to be selected in a developing nation with more developed technique. Therefore the selling strategy has to be something downside.

Analysis of leak detection deserves to be pioneered to identify the leak and to take steps for the industries. However, the issue we have to tackle is the profit incentive from the industry. The running costs have to be lower so they can save money. Further efficiency and precision must be awareness. Repeatability of the process analyzes must be taken to monitor the pipe network system. The transient pressure technique could also analyze the identification of leakage [2]. Transient analysis, for example, may provide essential leak detection, which includes an exact transient system model [11]. Additional specifics are the benefits of transient analysis, and the number of complexities can be catastrophic in one system [12].

The approach for calculating both non-stationary-linear data is empiric mode decomposition (EMD) [13]. This approach is descriptive, simple, retrospective and adaptive, based on and extracted from data decomposition. Empirical mode decomposition, which can decompose any complicated data set into a small and limited amount of intrinsic mode function (IMFs). In the past, it is difficult to pick IMFs because only the experts can evaluate the signals. However, the graduation of data dispersion concerning the data centroid is today easily established with the invention of an Integrated Kurtosis Algorithm for Z-filter (IKAZ) [14]. A direct quadrating (DQ) method is proposed with the periodic solution to resolve Volterra integral equations. The approach is focused on an exponentially modified Gaussian style rule whose parameters depend on the question to replicate the output of the analytical solution. The root means square (RMS) has been studied and direct convergence analysis has been conducted. Several computational tests are applied with other existing methods for contrast [15].

The present paper suggests an integral kurtosis algorithm for the selection of Z-notch filter hybrid for an intrinsic mode function (IMF) by a higher-order statistical method (kurtosis). Since this system is inherently flexible and identifies some shifts in the signals, the Ikaz has been selected [14]. In comparison to most modern studies such as variance, standard deviation and kurtosis, the Ikaz technique was willing, by collecting both the

Ikaz representation and the Ikaz coefficient, to show amplitude and frequency variations at the same time, $Z\sigma$. Description of the test results was addressed with [14] and [16] using the collection of the vibrating signal from spinning part of the machinery (Bearing) and tool wear while turning. The study carried out by Nuawi 2014 et al., The Ikaz was related to a variable, the findings were that both amplitude and frequency variations were not found in the non-stationary signal parameter. Therefore, the Ikaz approach was reliable particularly to the surveillance of the frequencies and amplitude changes of the signal [14].

This paper concentrated on the application of the uniform Ikaz coefficient in automatic mode selection. This approach was tested using transmission line modelling (TLM) through synthetic pressure transient signal simulations. The findings showed a good impact [17]. The results of an experimental analysis using laboratory scaled test rig are as follows in this article.

1.1 Integrated Based Kurtosis Algorithm for Z-Filter Technique to Kurtosis Ratio

Integrated Based Kurtosis Algorithm for Z-Filter Technique

Integrated Kurtosis-based Algorithm for z-filter technique (Ikaz) has been formed based on the statistics dispersing idea throughout the centroid. The raw signal sample frequency was selected for 2.56, regarding the number Nyquist [14]. Some researchers in signal analysis and processing are comfy with the proportion. The maximum frequency range is to avoid the aliasing effect, the maximum frequency span will be of Eq. 1.

$$F\text{max} = \frac{fs}{2.56} \tag{1}$$

Ikaz splits the signal for time field into three frequency levels, describing the x-axis as the medium frequency with a region of higher frequency (LF) 0–0.25 of f_{max}, Y-axis is accompanied by high frequency (HF) with a spectrum of 0.25–0.5 of f_{max}. Finally, the z-axis is a high frequency (VF) of size 0.5 of f_{max}. The 0.25 f_{max} and the 0.5 f_{max} was chosen with respect to the 2nd order of the Daubechies principle as low and high-frequency spectrum limit in signal decomposing [18]. The Ikaz approach is used to provide 3D schematic representations of the calculated signal frequency spectrum with respect to kurtosis. The variance, σ^2 of each frequency band which is σL^2 represent as a low-frequency band, σH^2 represent as a high-frequency band and σV^2 represent as a very-high-frequency band which calculated as in Eq. 2, 3 and 4 to measure the scattering of data distribution.

$$\sigma L^2 = \frac{\sum_{i=1}^{N} (x_i - \mu_L)^2}{n} \tag{2}$$

$$\sigma H^2 = \frac{\sum_{i=1}^{N} (x_i - \mu_H)^2}{n} \tag{3}$$

$$\sigma V^2 = \frac{\sum_{i=1}^{N} (x_i - \mu_V)^2}{n} \tag{4}$$

The I-Kaz coefficient, $Z\sigma$ can be simplified in term of variance, σ as in Eq. 5 [16].

$$Z\sigma = \sqrt{(\sigma L^2)^2 + (\sigma H^2)^2 + (\sigma V^2)^2} \tag{5}$$

Kurtosis

Kurtosis is an indication of the peak analyzes for identification of spikes in a non-stationary signal component, such as a pressure transient, and is, therefore, a good indicator. Kurtosis is said to be;

$$\text{Kurtosis (x)} = \frac{\mathrm{E}\{(\mathrm{x} - \mu)^4\}}{\sigma^4} \tag{6}$$

where μ and σ represent the mean and standard deviation of time series signal respectively. The E demonstrates the process of perceptions. The kurtosis reveals the spiciness and peak distribution of likelihood related to the instantaneous amplitudes of the time series analysis [19]. Therefore, the Ikaz-kurtosis ratio as Eq. 7 was presented.

$$ZK\sigma = \frac{(\sqrt{(\sigma L^2)^2 + (\sigma H^2)^2 + (\sigma V^2)^2})(\sigma^4)}{(\mathrm{E}\{(\mathrm{x} - \mu)^4\})} \tag{7}$$

Algorithm

Automatic Selection of Intrinsic Mode Function (IMF) Based on Pressure Transient Signal

1. Do EMD on a transient signal of pressure.
2. find the local limit of the absolute value for IMF and set the end values as optimum to increase the End Effect (for the gain of both the upper and lower envelopes).
3. Build a maximum Spline Envelope (SE).
4. For this SE section, the straight-line envelope is used when the envelope goes under the data.
5. The signal was decomposed by EMD to the so-called intrinsic mode function (IMF).
6. For all the Z-filter method, estimation of the combined kurtosis algorithm with the coefficient Eq (Ikaz-Kurtosis) 7.
7. The IMF has been identified to reflect the largest Ikaz kurtosis ratio.
8. At the IMF point, Hilbert Transform (HT) and Hilbert Spectrum (HS)include coefficients of Ikaz and Kurtosis.
9. Using SE data: N-data = Data/SE to normalize the data. You can repeat these steps.
10. The Hilberts Transform of N-data determines IF (FM) and Absolute Value (AV).
11. Definition: Index Error = $(AV-1)^2$.
12. Fast Frequency for Calculation for SE (AM).

2 Methodology

In this research, a pressure sensor is used to obtain the reaction of the signal produced from the solenoid valve. The solenoid valve is usually closed down during the procedure and water hammer anomalies are triggered for three consecutive times during the opening and closing of the valve. The test data is collected with the software MatLab.

Experimental Set Up

The test was done in a 67.90-m long ground loop pipeline test rig. The system was designed by implies of Medium-High-Density Polyethylene (MDPE) waterways with an external cross-section of 60 mm, a medium height inside a 55 mm and 2.6 mm cross-section. A false hole check system was put in place at a distance of 19.7 m from the intent of the pipeline study. A surface tank where water was released from the tube locks is connected to the outlet of the tube. This protects rapid expansion phenomena by the pressure waves and reduces the adverse pressure wave because the data in the transducer are affected. The driven sound speed of the test device is 524.3 ms^{-1}.

The pressure sensor and the solenoid valve that was fitted 10 m away from the electric motor to prevent the concussion during information accumulation were presented in Fig. 1. This is due to the characteristic of the wave as a turbulent flow when the water leaves the pump, due to the distance it travels, it is almost laminar or steady. The result of friction caused by water movement and moisture from the wall tubing. This method is used in reservoir exploration techniques regularly. The thought was received from the use of pressure transient progression of water in the pipeline system. Hypothetically, the pressure transient reaction happens when an abrupt change in the progression of water by shutting or opening the valve in the pipeline system. The wave propagation of the water will be generated in combination with the pipeline network, as well as the wave propagation characteristics used as a leak detection tool, together with the pipeline network. Depending on the size of the leak, leak width, the tube style function and pipe's diameter, the wave within the pipeline network travels with different signals.

Figure 1 shows the experimental test platform at the University of Malaysia Pahang in Malaysia's Fluid Mechanic Laboratory. To order to fix the valve and the pressure sensor for the solenoid field test, the fire hydrate cap was designed and made. The solenoid valve is used to generate water hammer phenomena in both parts of the system and a pressure sensor is used for the acquisition of the signal. With the sudden opening or closure of the valve and the failure of the pump, the system will in a few instances be shut down. This creates a "water hammer" on the pipeline system. Both pressurized tube structures also suffered from this phenomena and induced heavy vibrations and failure of the tube system [20]. This effect is created through the solenoid valve (Fig. 1) used to build a hammer and pipeline device. The same condition is generated in real life when the underground pipe has a pressure disorder [21].

3 Results and Discussion

The process starts with the pressure transient signal obtained by using Matlab from the test rig. The obtained signal restores the properties of the whole network of the pipeline.

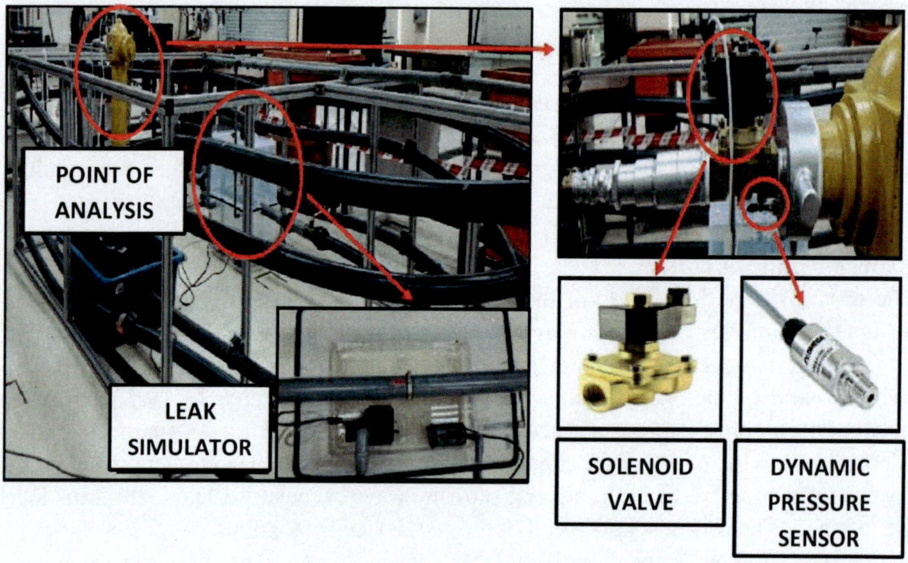

Fig. 1. Experimental test rig in laboratory scale

The transient pressure waves spread in both directions alongside the pipeline system and from the explosive source through the sound speed of a water distribution system. DQ and Empirical Mode Distribution (EMD) are used as methods in this research to analyze the pressure transient signal as the pre-processing technique and the DQ function post-processing technique.

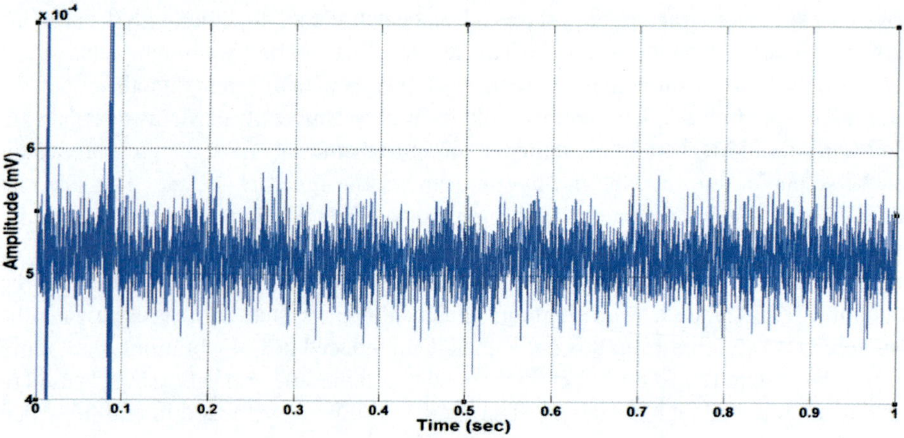

Fig. 2. Raw pressure transient signal acquire from test rig

Figure 2 shows the raw pressure transient signal response acquire from the test rig. The first two most intense amplitudes in the figure indicate the begin and shutting down of the solenoid valve that delivers the pressure transient signals, transmitted together with the pipeline network and reflected back due to the troubling effect. There have been two separate pressure levels of the first reaction: 1 bar and 2 bar. There were different results with the signal gain at an alternative level of pressure. Higher pressure created a lower signal frequency, which describes the higher signal amplitude. To get the details in the signal, the original data must be analysed. DQ has been used as a signal analysis in this research and Normalized Ikaz is being used for the automatic selection of the appropriate IMF. The study phase beginning with empirical mode decomposition (EMD) decomposes the signal to inherent mode feature level 13 (Fig. 3). A category of the higher frequency signal, the noise signal, is the first stage of IMF. This last level was a lower frequency signal reserve. In addition to the analysis of frequency noise at these levels, the first and second level of IMF are generally not considered. In the meantime, IMF level 7 and the remains contain the network's basic response. Consequently, all these IMFs have been discarded. The rest of the IMF 3–6 were recombined to produce a noise-free signal [22]. The structured Ikaz acts as an advanced statistical tool to autonomously select the right IMF to be used to correctly identify the right IMF without the professional user's visual intervention. The value of the standardized Ikaz coefficient calculates the value for each IMF level and the following.

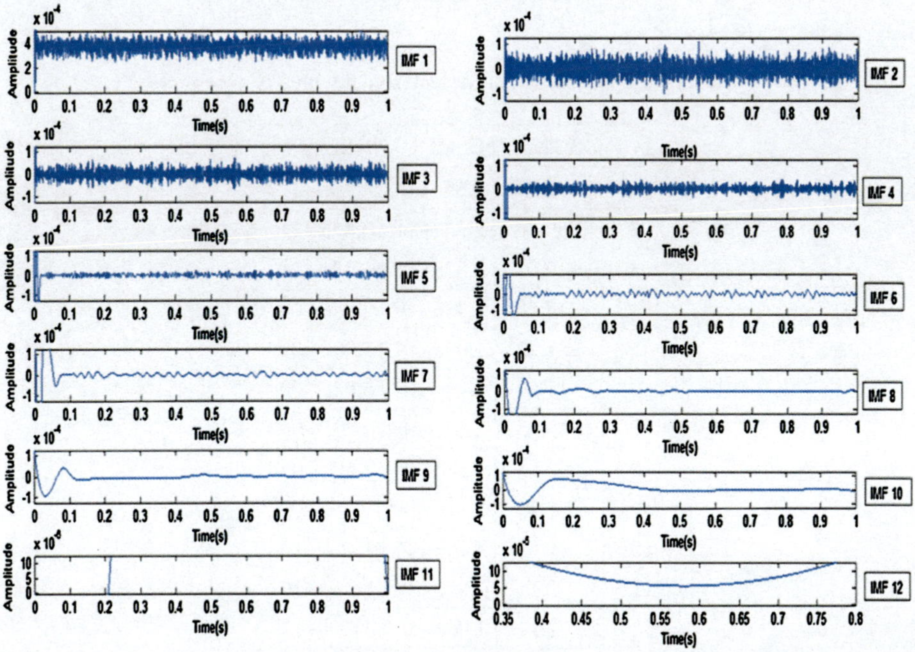

Fig. 3. Intrinsic Mode Function (IMF) from level 1-1

Table 1 and Table 2 showed the correlation of the kurtosis, Ikaz and Normalized Ikaz coefficient for every degree of IMF. From the perception, IMF level 1 has the most elevated kurtosis and Ikaz coefficient. In relation to [20], the first and second degree of IMF contains the signal with noise. In this manner, in this examination, Ikaz and kurtosis

Table 1. Normalized Ikaz coefficient no leak signal with 2 bar pressurised pipe

IMF	Kurtosis	IKAZ	NIKAZ
1	**2803.53**	**2.69e−09**	9.59e−13
2	888.82	1.64e−09	1.85e−12
3	653.63	3.51e−10	5.37e-13
4	118.31	9.06e−10	7.65e−12
5	35.74	1.59e−09	**4.46e−11**
6	25.34	4.80e−10	1.89e−11
7	13.14	1.12e−10	8.52e−12
8	13.61	6.13e−11	4.50e−12
9	9.44	2.22e−12	2.35e−13
10	6.59	4.48e−14	6.80e−15
11	3.51	1.00e−14	2.86e−15
12	1.53	2.78e−14	1.80e−14
13	3.00	1.66e−12	5.54e−13

Table 2. Normalized Ikaz coefficient for leak signal with 2 bar pressurised pipe

IMF	Kurtosis	IKAZ	NIKAZ
1	**3077.94**	**2.59e−07**	8.43e−12
2	669.46	2.32e−08	3.47e−11
3	112.02	3.83e−08	3.42e−10
4	48.59	1.45e−07	2.99e−09
5	21.22	1.05e−07	4.98e−09
6	11.34	5.75e−08	**5.06e−09**
7	5.052	2.17e−08	4.29e−09
8	2.85	7.41e−09	2.59e−09
9	2.49	3.91e−09	1.56e−09
10	1.55	3.78e−10	2.42e−10
11	2.14	4.01e−16	1.87e−16
12	2.14	1.04e−10	4.88e−11
13	2.39	9.66e−10	4.02e−10

coefficient is not appropriate to be utilized for programmed determination of IMF. Other than that, IMF 5 (Table 1) and IMF 6 (Table 2) contains the most outstanding Normalized Ikaz coefficient and by alluding to [20], this level is in the scope of IMF 3–6 that ought to contain the signal without noise. To guarantee IMF 5 and 6 have the most noteworthy Normalized Ikaz coefficient for thought as the correct IMF for further examination, IMF 5 and 6 were exposed to facilitate investigation utilizing DQ. The aftereffect of DQ is introduced in Fig. 4 and 5.

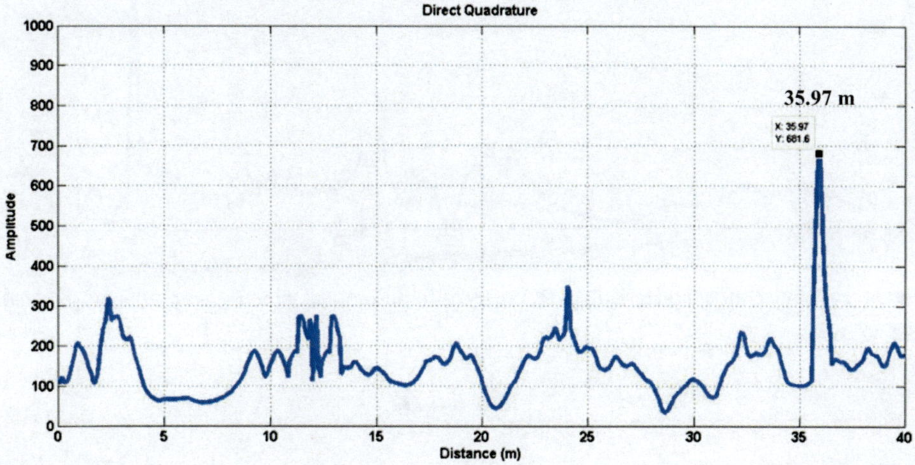

Fig. 4. Instantaneous characteristic of DQ analysis for the no-leak pipeline with 2 bar pressurized pipe

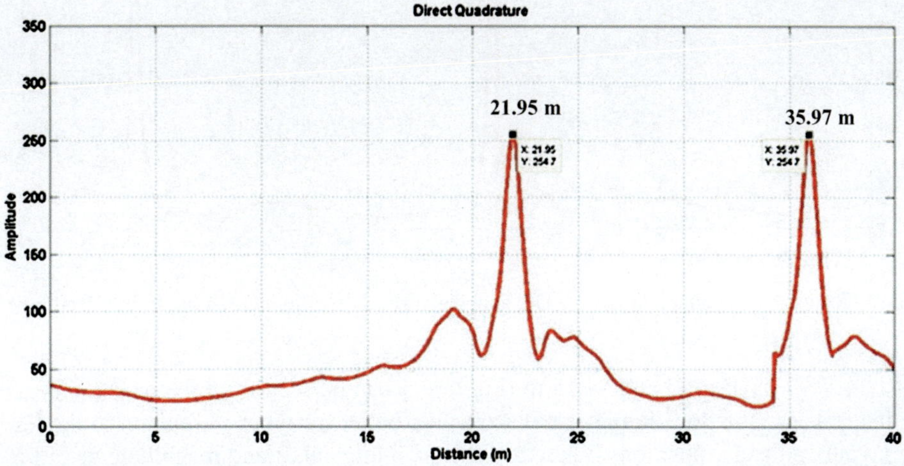

Fig. 5. Instantaneous characteristic of DQ analysis for leak pipeline with 2 bar pressurized pipe

Figure 4 and 6 show the result of DQ for no-leak pipeline system while Fig. 5 and 7 for leak pipeline system with 2 and 4 bar pressurized pipe respectively. In this context, it is more useful to describe a signal as to its instantaneous frequency by using the non-steady signal, which shifts frequency value at any moment. Instant frequency defines the frequency which matches the signal locally [20]. The maximum spikes and magnitude

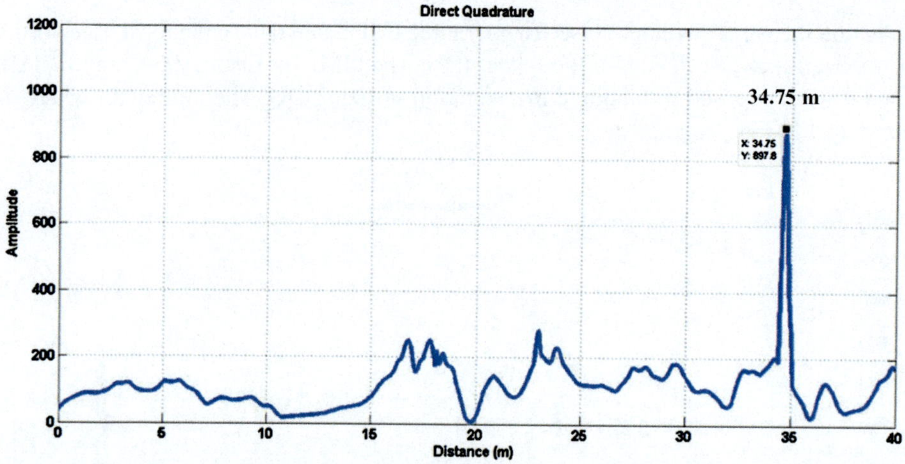

Fig. 6. Instantaneous characteristic of DQ analysis for the no-leak pipeline with 4 bar pressurized pipe

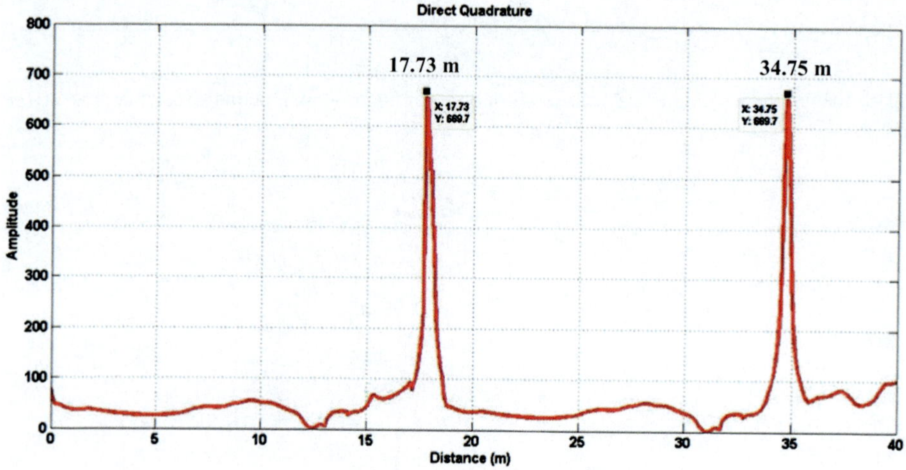

Fig. 7. Instantaneous characteristic of DQ analysis for leak pipeline with 4 bar pressurized pipe

exist at 35.97 m (Fig. 4) and 34.75 m (Fig. 6), as can be seen from the measurements in this respect. This location is calculated as the outlet for pipes in relation to the test rig layout and exact position. Whereas, the maximum spikes and magnitude at 21.95, 35.97, 17.73 and 36.88 m respectively, are shown by Figs. 5 and 7. In reference to the research plant blueprint, the leak was detected at a distance of 19.7 metres, while the exit pipe was found at a distance of 36.6 metres. The longitudinal work and study have been replicated under different pressure and the outcomes are tabled in Table 3.

The contrasts between statistical estimation and the experimental location shown in Table 3. The location of the leak indicates a defect of 9.75% in leak details and 1.72% in the direction of the source. The final part of the DQ review clearly demonstrates

Table 3. Comparison between simulated and experimental

Water pressure	Signal response	Pipe feature	Measured position (m)	IMF contains outstanding normalized Ikaz coefficient	Analysis position (m)	Error %
2 bar	Leak data	Leak	19.7	5	21.95	9.75
		Outlet	36.6		35.97	1.72
	No leak data	Outlet	36.6	4	35.97	1.72
4 bar	Leak data	Leak	19.7	5	17.73	11.35
		Outlet	36.6		34.75	5.05
	No leak data	Outlet	36.6	4	34.75	5.05

the location of the leaked pipe function and the pipe outlet is in the same place as for each numerical measurement and the original locations. It also indicates that the DQ analytics were able in the non-stationary pressure transient signal to identify and position the transient results.

4 Conclusion

This article worked with the automatic form of selection for IMF selection by means of the empirical mode decomposition (EMD) and DQ analysis. The result shows that Ikaz-kurtosis ratios are the correct and suggested automatic filtering tool for selecting a large IMF. The automatic self-decision process of IMF by DQ was developed and the question of the selection of IMF was statistically analyzed by the suing of Ikaz-kurtosis ratio. Therefore the degree of automation for EMD + DQ was improvised to identify leakage in the live water distribution system by using pressure transient signal.

Acknowledgement. University Malaysia Pahang fully supports the facilities and resources for this research. The authors would like to acknowledge the support from the University Malaysia Pahang internal grant RDU170386, RDU190386 and Ministry of Higher Education (MOHE), Malaysia grant FRGS/1/2017/TK03/UMP/02/1(RDU170121).

References

1. Yusop, H.M., et al.: Improvement of cepstrum analysis for the purpose to detect leak, feature and its location in water distribution system based on pressure transient analysis. J. Mech. Eng. **4**(4), 103–122 (2017)

2. Yusop, H.M., et al.: Pipe leak diagnostic using high frequency piezoelectric pressure sensor and automatic selection of intrinsic mode function. In: IOP Conference Series: Materials Science and Engineering. IOP Publishing (2017)
3. Ministry of Water, L.a.N.R.K., Non-Revenue Water (NRW) Statistics - Datasets - MAMPU (2018). http://www.data.gov.my/data/en_US/dataset/non-revenue-water-nrw-statistics
4. See, K.F., Ma, Z.: Does non-revenue water affect Malaysia's water services industry productivity? Util. Pol. **54**, 125–131 (2018)
5. Alkasseh, J.M., et al.: Applying minimum night flow to estimate water loss using statistical modeling: a case study in Kinta Valley, Malaysia. Water Resour. Manage **27**(5), 1439–1455 (2013)
6. Liemberger, R., Wyatt, A.: Quantifying the global non-revenue water problem. Water Supply **19**(3), 831–837 (2019)
7. Jang, D., Choi, G.: Estimation of non-revenue water ratio using MRA and ANN in water distribution networks. Water **10**(1), 2 (2018)
8. Pickard, B.D., et al.: Reducing non-revenue water: a myriad of challenges. Florida Water Resour. J. 26–32 (2008)
9. Farley, M.: Non-revenue water–international best practice for assessment, monitoring and control. In: 12th Annual CWWA Water, Wastewater & Solid Waste Conference (2003)
10. Taghvaei, M., Beck, S:, Staszewski, W.: Leak detection in pipelines using cepstrum analysis. Meas. Sci. Technol. **17**(2), 367 (2006)
11. Colombo, A.F., Lee, P., Karney, B.W.: A selective literature review of transient-based leak detection methods. J. Hydro-Environ. Res. **2**(4), 212–227 (2009)
12. Yusop, H.M., et al.: fault identification in pipeline system using normalized hilbert huang transform and automatic selection of intrinsic mode function. In: MATEC Web of Conferences. EDP Sciences (2018)
13. Huang, N.E.: Review of empirical mode decomposition. In: Wavelet Applications VIII. International Society for Optics and Photonics (2001)
14. Nuawi, M.Z., et al.: Development of integrated kurtosis-based algorithm for z-filter technique. J. Appl. Sci. **8**(8), 1541–1547 (2008)
15. Cardone, A., et al.: Ef-Gaussian direct quadrature methods for Volterra integral equations with periodic solution. Math. Comput. Simul. **110**(Suppl. C), 125–143 (2015)
16. Rizal, M., et al.: A comparative study of I-kaz based signal analysis techniques: application to detect tool wear during turning process. Jurnal Teknologi **66**(3), 99–105 (2013)
17. Yusop, H.M., et al.: Application of Ikaz and direct quadrature for solving leakage in pipeline distribution by using transmission line modelling. In: IOP Conference Series: Materials Science and Engineering. IOP Publishing (2019)
18. Daubechies, I.: Ten lectures on wavelets, vol. 61. SIAM (1992)
19. Wang, Y., et al.: Spectral kurtosis for fault detection, diagnosis and prognostics of rotating machines: a review with applications. Mech. Syst. Signal Process. **66**, 679–698 (2016)
20. Amin, M., et al.: Leak detection in medium density polyethylene (MDPE) pipe using pressure transient method. In: IOP Conference Series: Materials Science and Engineering. IOP Publishing (2015)
21. Yusop, H.M., et al.: A Study of Ikaz and Normalized Hilbert transform for solving faulty in pipeline distribution system using transmission line modelling. In: IOP Conference Series: Materials Science and Engineering. IOP Publishing (2019)
22. Ghazali, M., et al.: Comparative study of instantaneous frequency based methods for leak detection in pipeline networks. Mech. Syst. Signal Process. **29**, 187–200 (2012)

Analysis on Dimensional Accuracy of 3D Printed Parts by Taguchi Approach

Mohd Nazri Ahmad(✉) and Abdul Rashid Mohamad

Faculty of Mechanical and Manufacturing Engineering Technology, Universiti Teknikal Malaysia Melaka, Hang Tuah Jaya, Durian Tunggal, 76100 Melaka, Malaysia
mohdnazri.ahmad@utem.edu.my

Abstract. The part produced by 3D printing has its accuracy varies with the changes in the process parameters of the machine such as quality, temperature, speed, and some others. Taguchi method was applied in order to achieve the optimum accuracy of the printed part. It is a simplified yet powerful method for experimental design using the orthogonal array method. In this project, an orthogonal array of L'9 is used with four parameters which are a print pattern, orientation, support angle, and sidewalk offset. The experiment was set up by involving the process of part design by using SolidWorks. Then nine sample has undergone a process of measuring by Coordinate Measuring Machine (CMM). Thus, the result of a measurement of nine samples has to be compared to CAD data to get the percentage of accuracy. Besides that, the measurement result will have analysed using Taguchi method to determine the default and optimum parameter condition. Then we compare the result from both parameter data. The result shows that optimum parameter condition for this cube pro process is honeycomb for the print pattern, 0 y-axes for orientation, 90° for support angle and 0.4 for sidewalk offset. The print pattern has the most significant effect compared to the orientation, support angle, and sidewalk offset on the dimension accuracy of the part. The shrinkage data value between default and optimum parameter shows that the shrinkages reduce from 0.619 mm to 0.429 mm. An improvement of 0.190 mm with 44.3% of percentage has achieved by applying the optimized parameter.

Keywords: 3D printing · Taguchi · Orthogonal array · Dimensional accuracy

1 Introduction

According to the Sood et al. [1] the term speedy prototyping relates to a speedily growing vary of machine-controlled machines or approaches like stereolithography (SL), coalesced Deposition Modelling (FDM), selective optical maser sintering (SLS), laminated object producing (LOM), and so on. The fabrication of three dimensional solid object from CAD info robotically while not victimization tooling and bottom human intervention. Rayegani et al. [2] states that fused deposition modeling (FDM) is a fast-growing rapid prototyping (RP) technology due to its ability to build functional parts having complex geometrical shapes in reasonable build time.

M. A. Zakaria et al. (Eds.): *Advances in Mechatronics, Manufacturing, and Mechanical Engineering*, LNME, pp. 219–231, 2021.
https://doi.org/10.1007/978-981-15-7309-5_22

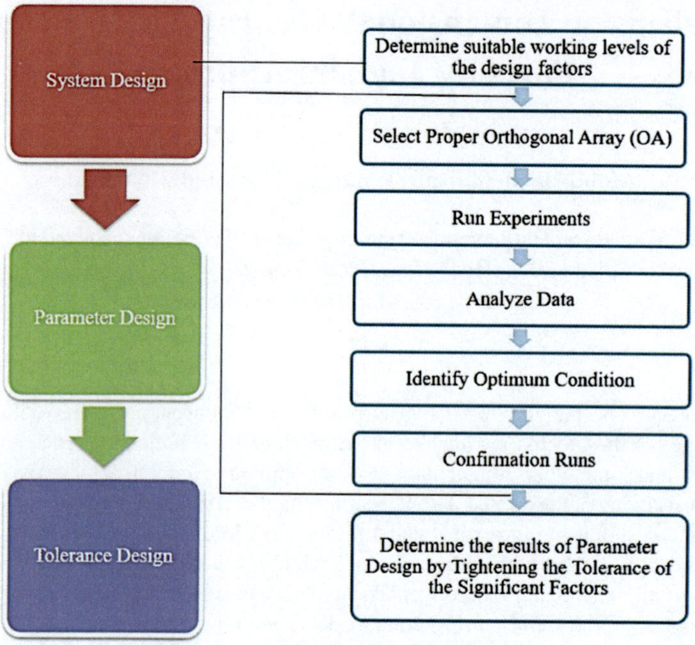

Fig. 1. Taguchi design procedure [4].

Osman et al. [3] has carried out a research to determine the optimization parameter for hole diameter accuracy in dry drilling process of AISI D2 Tool Steel by using Taguchi method. One method presented in this study is an experimental design process called the Taguchi design method. Taguchi design, developed by Dr. Genichi Taguchi, is a set of methodologies by which the inherent variability of materials and manufacturing processes has been taken into account at the design stage [4]. The procedure of the Taguchi design is described in Fig. 1. The complete procedure in Taguchi design method can be divided into three stages: system design, parameter design, and tolerance design. Taguchi has developed a large number of SN ratios for use with differing experimental targets. There has been extensive debate approximately the usefulness of studying SN ratios [5–7].

The goal feature of this work is a reduction, the quantitative relation of S/N quantitative relation delineate in keeping with the Taguchi approach: Smaller-is-good as in Eq. 1:

The goal feature of this work is a reduction, the quantitative relation of S/N quantitative relation delineate in keeping with the Taguchi approach: Smaller-is-good as in Eq. 1:

$$SNs = -10 \log\left(\frac{1}{n}\sum\nolimits_{i-1}^{n} yi^2\right) \tag{1}$$

Nominal-is-better (to reducing variability around a target) as in Eq. 2.

$$\mathrm{SNT} = 10\log\left(\frac{\mu^2}{\sigma^2}\right) \tag{2}$$

Larger-is-better (for making the system response as large as possible) as in Eq. 3

$$\mathrm{SNL} = -10\log\left(\frac{1}{n}\sum\nolimits_{i-1}^{n}\frac{1}{yi^2}\right) \tag{3}$$

Where the $\boldsymbol{\mu}^2$ the average of the observed data is, $\boldsymbol{\sigma}^2$ is the variance of y (the observed data) and $\boldsymbol{n}$ is the number of observation.

Fused Deposition Modeling is an Additive Manufacturing technology able to fabricate prototypes, tooling and functional parts without geometrical complexity limitations. Despite of the potential advantages of this technology, a limiting aspect of its industrial diffusion is the obtainable accuracy [8]. Mahesh et al. [9] proposed a geometry characterized by free form surfaces finding deviations from the nominal dimensions ranging from 5% to 15%. In one case the shape de-formation led to a deviation of 2.5 mm. Sood et al. [1] states the effects of several factors on the dimensional accuracy of FDM parts have been studied. Taguchi Grey method has been used to find significant factors and optimum factors level in order to minimize changes in length, width and thickness. Observed results showed that the FDM accuracy is influenced in a highly nonlinear manner, hence an artificial neural network modeling has been proposed. Tong et al. [10] has developed a compensation method of FDM that based on correcting slice files. This methodology employs the construction of an artifact by the machine considered for the error compensation; then by a CMM a point cloud is acquired to gain coefficients for the model; finally, the STL file is modified according to the formulation. Hafsa et al. [12] has carried out evaluation on Acrylonitrile Butadiene Styrene (ABS) and Polylactic acid (PLA) part produced from Fused Deposition Modeling (FDM) as a master pattern for Investment Casting (IC) process. Results show that model fabricated with hollow internal pattern structure (ABS material) that produced by low layer thickness is better than other models in terms of its dimensional accuracy (−0.19666 mm) and surface roughness (1.41 µm). Even though the ABS built part performed better as the model, the PLA build part produces better overall casting result. The previous study shows that the surface finish of the components is greatly improved by this vapor treatment process with minimal variations in part geometric accuracy after the treatment [13]. Dimensional accuracy refers to fidelity of part geometry to the computer design and is typically quantified through statistical measures in comparison to a design tolerance. This is distinct from print resolution which refers to the minimum obtainable feature size [15, 16].

3D printing is a method similar to "standard" printing with multiple layers of "ink" printed on top of one another to produce a solid object. In practice, conventional inks are too thin, so most printers use plastic material. Regardless of the control geometry used, the method of printing is the same for both types of FDM printers. The print head is a metal tube with a heating element and thermistor to control the temperature. The plastic substrate is melted by the heat of the print head, and pressure is applied by forcing more plastic in, causing some of the liquid plastic to extrude through a small nozzle that ranges from 0.2 mm to 0.5 mm in size [11]. With FDM, 3D plastic parts can be built layer by layer, depositing build materials and support materials together, both of which are heated and extruded separately through two extrusion nozzles on the liquefier head [14]. Thus, it is important to determine the right parameters of the 3D Printing machine in order to produce a part which can fulfill the desired specifications. Thus, the aim of this study are to determine the optimal process parameter of FDM by using Taguchi approach and evaluate the shrinkage value of printed parts on dimensional accuracy for Acrylonitrile Butadiene Styrene (ABS).

2 Research Methodology

The CubePro, 3D printer types FDM was used in this study to print the samples that base on ASTM D635. The technical specifications of this 3D printer can be seen in Table 1. The overall method of study as shown in Fig. 2. In order to run the experiment, factor and level have to be determined in early stage. Next, printed samples will undergo the measurement process by Coordinate Measuring Machine (CMM). Three linear dimensions were measured which are total length, width and gage length. Lastly, prediction of shrinkage has been made by computer simulation which is Minitab.

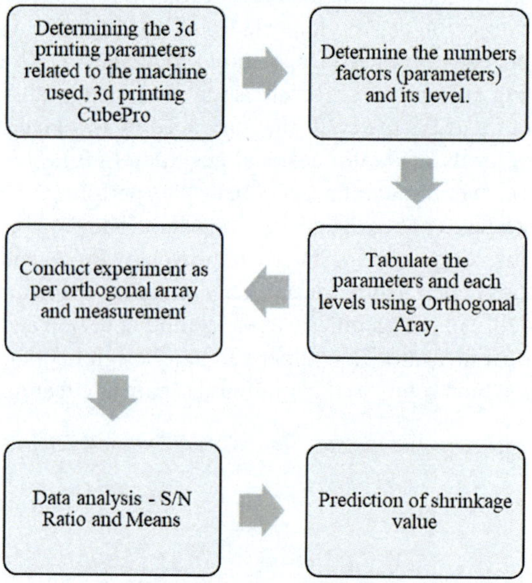

Fig. 2. Process flow of research

Table 1. CubePro 3D printer specification

No.	Items	Specification
1	Technology	Plastic Jet Printing (PJP)
2	Print jets	Single, dual or triple jets
3	Bed size	10.8 × 10.45 × 9.5″, 275 × 265 × 240 mm
4	Raw material	PLA/ABS
5	Layer thickness	75/200/350 μ

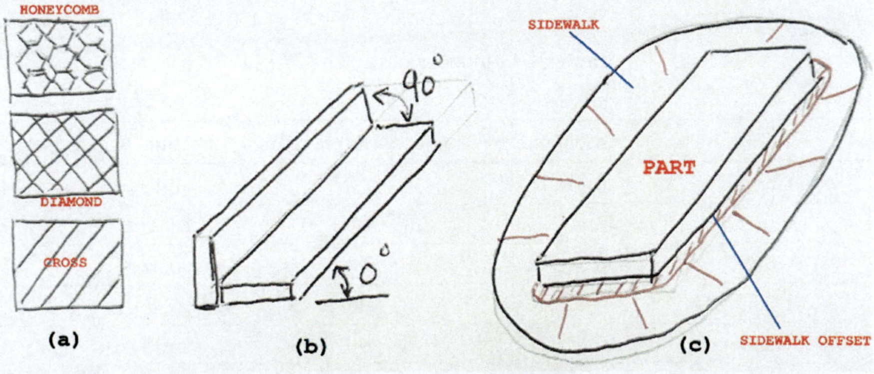

Fig. 3. Printing parameters (a) patterns (b) orientations (c) sidewalk offset

Design of Experiment (DOE) method was used during this stage which is the Taguchi Method. This method will help to design an appropriate experiment with the selected process parameters. Four factors with three levels (4^3) for each were considered, are shown in Table 2. The L-9 orthogonal array is selected by using an orthogonal array. Table of orthogonal array with nine samples as shown in Table 3. Meanwhile, Fig. 3 shows the illustrations of printing parameter for types of patterns, part orientation and sidewalk offset.

Minitab software provides a technique and the same time investigate the consequences of multiple variables on associate in nursing output variable (response). Once conducting the experiments and enter the result, Minitab provides an analytical and graphing tool like main impact graph to assist in understanding the results.

3 Results and Discussion

The first sample was a combination of parameter print pattern (cross), orientation (Y axis) is 0°, support angle (0°) and sidewalk offset is 0.15 mm. Then for second sample was print pattern (cross), sidewalk offset (0.25 mm), orientation (Y axis) and support angle is 45°. Sample number 3 was print pattern (cross) sidewalk offset 0.4 mm and 90° for the orientation (Y axis) and support angle. The combination parameter for the

Table 2. Factors and their levels

Serial	Factor	Levels		
		1	2	3
A	Print pattern	Cross	Diamonds	Honeycomb
B	Orientation	0	45	90
C	Support angle	0	45	90
D	Side walk offset	0.15	0.25	0.4

Table 3. Orthogonal array $L_9\left(3^4\right)$

Trial	Print pattern	Orientation (Y-axis)	Support angle	Side walk offset
1	Cross	0	0	0.15
2	Cross	45	45	0.25
3	Cross	90	90	0.4
4	Diamonds	0	45	0.4
5	Diamonds	45	90	0.15
6	Diamonds	90	0	0.25
7	Honeycomb	0	90	0.25
8	Honeycomb	45	0	0.4
9	Honeycomb	90	45	0.15

sample number 4, 5 and 6 for the print pattern was diamond and other orientations were (Y axis) 0°, 45° and 90°. Meanwhile for the support angle 45°, 90° and 0°. The parameter sidewalk offset is 0.4 mm, 0.15 mm and 0.25 mm for the sample number 4, 5 and 6. Next, print pattern (honeycomb), orientation (Y axis) is 0°, support angle (90°) and sidewalk offset (0.25 mm) for the sample number 7. For the combination parameter of sample number 8 is print pattern (honeycomb), orientation (Y axis) is 45°, support angle (0°) and sidewalk offset (0.4 mm). The last is sample number 9 used the combination parameter is print pattern (honeycomb), orientation (Y axis) is 90°, support angle (45°) and sidewalk offset is 0.15 mm. The result shown in Table 4 shows the result of measurement for dimension of section A, B, and C and also the shrinkage value and percentage of error. The result shows that the section C is the highest percentage of error of the printed part. An average of 3.28%, the percentage of error is obtained for section C. Followed by section B with an average of 1.25% for the percentage of error. Lastly, section A has the lowest percentage of error compare to others section of the parts. An average of 0.36%, the percentage of error is obtained for section A for nine samples. Thus, section A is selected to be optimized because it has the smallest percentage of error.

Table 4. Result of measurement

Run	Measurement	Section		
		A	B	C
		Cad data		
		150	20	10
1	Dimension	149.447	19.7786	9.689
	Shrinkage	0.553	0.2214	0.311
	% Error	0.37	1.12	3.21
2	Dimension	149.377	19.8	9.682
	Shrinkage	0.623	0.2	0.318
	% Error	0.42	1.01	3.28
3	Dimension	149.453	19.859	9.702
	Shrinkage	0.547	0.141	0.291
	% Error	0.37	0.71	3.00
4	Dimension	149.498	19.843	9.709
	Shrinkage	0.502	0.157	0.291
	% Error	0.34	0.79	3.00
5	Dimension	149.456	19.526	9.643
	Shrinkage	0.544	0.474	0.357
	% Error	0.36	2.43	3.70
6	Dimension	149.433	19.635	9.62
	Shrinkage	0.567	0.365	0.46
	% Error	0.38	1.86	4.78
7	Dimension	149.54	19.816	9.748
	Shrinkage	0.46	0.184	0.296
	% Error	0.31	0.93	3.04
8	Dimension	149.492	19.634	9.704
	Shrinkage	0.508	0.366	0.296
	% Error	0.34	1.86	3.05
9	Dimension	149.449	19.891	9.762
	Shrinkage	0.551	0.109	0.238
	% Error	0.37	0.55	2.44

Based on Fig. 4, the deviation of section C shows 6th sample has the highest values among all nine samples. The deviation value of the 6th sample is 0.478. Meanwhile, sample nine in section C has the lowest deviation which the value is 0.24. For section B, the graph shows that the 8th sample also has the highest deviation value of all 9 samples. The deviation value for the 8th sample is 0.186. Meanwhile, sample two has the lowest deviation value for section B with the value of 0.0101. For section A, 2nd sample has the highest value of deviation which is 0.042 and 7th sample has the lowest deviation value of 0.031. Section A is the lowest deviation value among other two sections and it needs to be optimized.

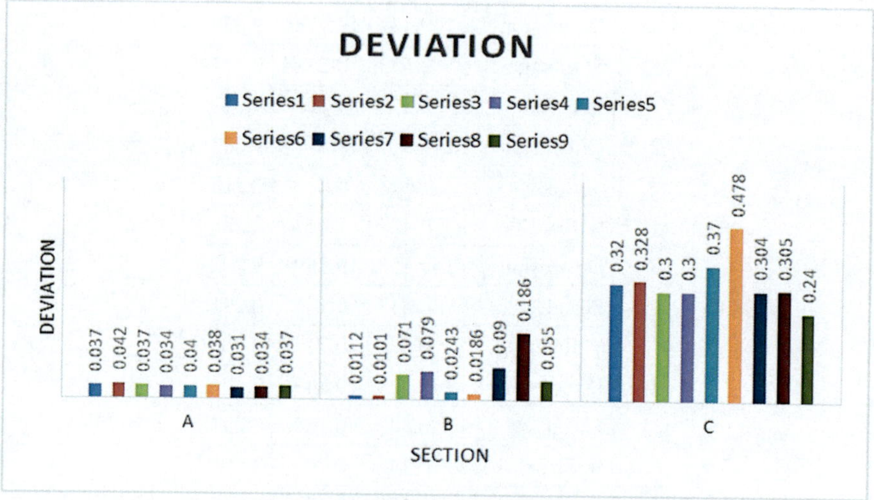

Fig. 4. Result of deviation

In order to optimize the parameter, only the major dimension section being analysed as shown in Fig. 5. For this part, the major section was A. Minitab software has been used to do the analysis of the data. The values of Signal to Noise Ratio (S/N Ratio) as in Eq. 1 and means were obtained throughout the analysis. The analysis uses formula 'smaller is better' for the signal to ratio, signal to noise ratios and means.

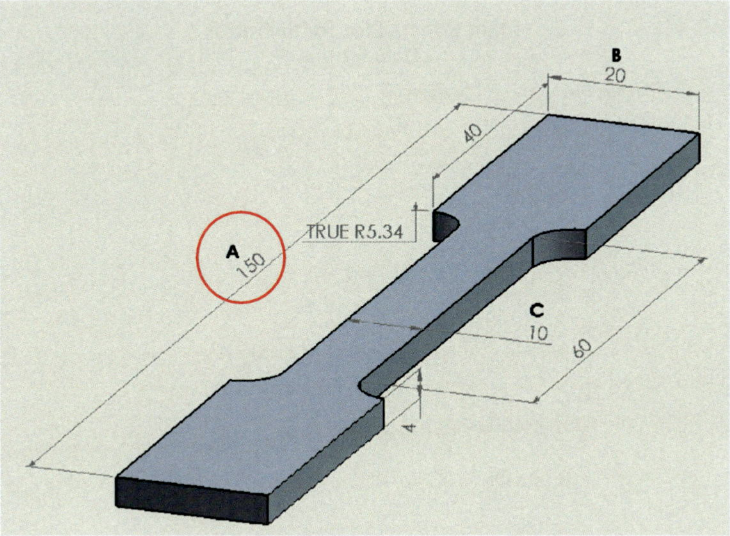

Fig. 5. Section 'A' as the major section

3.1 Main Effects Plots for S/N Ratio and Means

The graph has been plotted for both Signal to Noise Ratio and Mean for shrinkage value. By referring to the graph in Fig. 6 and 7, the analysis is made. In the SN ratio graph, the optimum parameter is observed by the highest level plotted. This is because the highest value of SN ratio means it has lower effect of the noise factor. For the means graph, the analysis on is observed by the lowest level plotted. The graph of means shows the average of variation, meaning that the lowest value of means shows that the variation is the smallest. Main Effect Plot represents the variation in the response variable with the variation in control factors and is used to examine differences between level means for factors. If the line plotted in the Main Effect Plot is horizontal, means that the factor is non-significant since there is no change in responses with the factor. If the line has a higher slope, then it shows that the factor is significantly affecting the response.

Based on Section A main effect plot for S/N ratios, the highest point plotted for the print pattern is at level 3 (Honeycomb). For orientation, the highest point is at level 1 (0). Meanwhile, for support angle, the highest point is at level 3 (90°). Finally, for sidewalk offset, the highest point is at level 3 (0.4). The aim was to minimize the means; it can minimize the variability in the part. Based on Fig. 6, the lowest point plotted for the print pattern is at level 3 (Honeycomb). For orientation, the lowest point is at level 1 (0). Meanwhile, for support angle, the lowest point is at level 3 (90°). Finally, for sidewalk offset, the lowest point is at level 3 (0.4). Hence, the combination for optimum parameter condition is A3B1C3D3 (Honeycomb, 0 y-axis, 90°, 0.4) that will give the optimum shrinkage value for section A dimension.

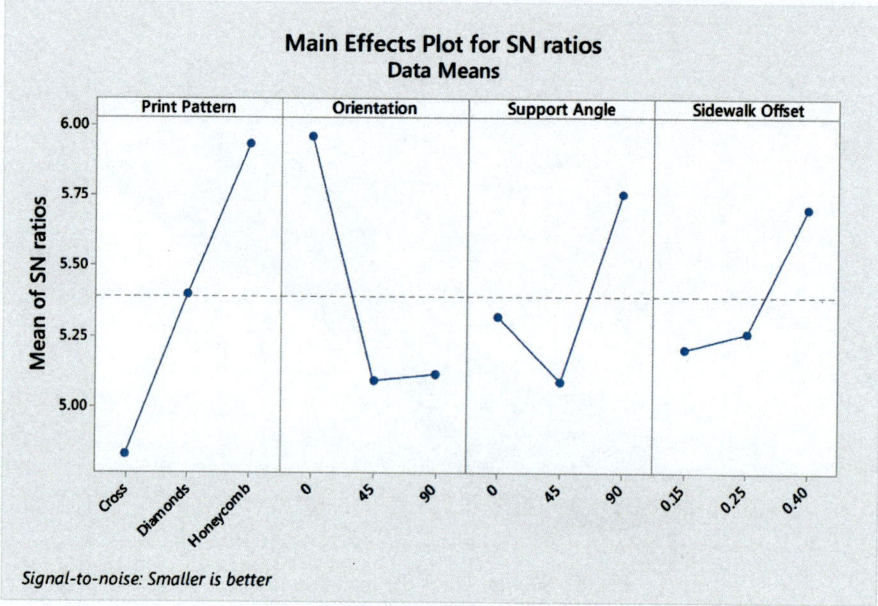

Fig. 6. Main effects plot for SN ratios

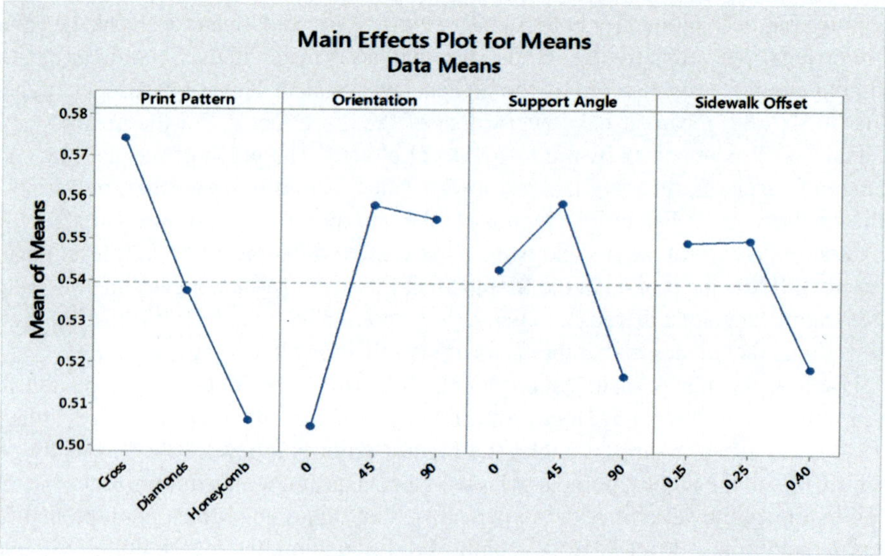

Fig. 7. Main effect plot for means

3.2 Response Table for S/N Ratio and Means

Response table for Signal to Noise Ratio and Means as in Table 5 and 6. The values of Delta are the difference between the maximum and minimum average of SN ratio for the factor. In the response table, it also shows that the highest value of delta is the highest in rank. The highest rank shows it is having the most significant factor effect contributed to this experiment.

Table 5. Section A response table for S/N ratios

Level	Print pattern	Orientation	Support angle	Sidewalk offset
1	4.832	5.959	5.319	5.203
2	5.401	5.094	5.091	5.261
3	5.935	5.115	5.758	5.703
Delta	1.103	0.865	0.667	0.499
Rank	1	2	3	4

Table 6. Section A response table for means

Level	Print pattern	Orientation	Support angle	Sidewalk offset
1	0.5743	0.5050	0.5427	0.5493
2	0.5377	0.5583	0.5587	0.5500
3	0.5063	0.5550	0.5170	0.5190
Delta	0.0680	0.0533	0.0417	0.0310
Rank	1	2	3	4

Based on Table 4, the print pattern is the first rank at (1.103), followed by orientation (0.865) at second. Then support angle (0.667) at the three places and sidewalk offset (0.499) at the last place. By referring to 5, the print pattern is rank at first (0.0680), followed by orientation (0.0533) at second. Then support angle (0.0417) at the three places and sidewalk offset (0.0310) in the last place. It indicates the print pattern has the most significant effect compare to the orientation, support angle and sidewalk offset on the dimension accuracy of the part.

3.3 Predicted Shrinkage Value

The used of the combination of optimum parameter condition, the predicted value for shrinkage was 0.429 mm. The shrinkage value improves by 0.190 mm as in Table 7. Based on the comparison, there is an improvement between the default and optimum parameter. From that, the shrinkage value was improved by 44.3% and this means the percentage is in good range for the improvement.

Table 7. Improvement value for section 'A'

Sections	Default parameter (mm)	Optimum parameter (mm)	Improvement (mm)
A	0.619	0.429	0.190

4 Conclusions

The study shows the determination of each sample dimensional accuracy of a printed part by Cube Pro 3D printer. Taguchi method was used to determine the optimum parameter condition for printing process. Based on the result, some conclusions can be made:

i. The percentage of error for the part was small for each section of the part. This is due to the small shrinkage for each section of the part. Each level of experimental design table L9 did not have an abnormal effect on the dimension of the part. Therefore, the chosen parameter level is acceptable for the experiment.
ii. The optimum parameter condition for this cube pro process was honeycomb for the print pattern, 0 y-axes for orientation, 90° for support angle and 0.4 for sidewalk offset. The print pattern has the most significant effect compared to the orientation, support angle, and sidewalk offset on the dimension accuracy of the part.
iii. The shrinkage data value between default and optimum parameter shows the shrinkages reduce from 0.619 mm to 0.429 mm. An improvement of 0.190 mm was achieved by applying the optimized parameter.

Acknowledgment. The authors wish to thank Universiti Teknikal Malaysia Melaka (UTeM) for supporting this research. Special appreciation and gratitude to Centre of Research and Innovation Management (CRIM) and Faculty of Mechanical and Manufacturing Engineering Technology for giving the full cooperation towards this research.

References

1. Sood, A.K., Ohdar, R.K., Mahapatra, S.S.: Improving dimensional accuracy of fused deposition modelling processed part using grey Taguchi method. Mater. Des. **30**(10), 4243–4252 (2009)
2. Rayegani, F., Onwubolu, G.C.: Fused deposition modelling (FDM) process parameter prediction and optimization using group method for data handling (GMDH) and differential evolution (DE). Int. J. Adv. Manuf. Technol. **73**(1–4), 509–519 (2014)
3. Osman, M.H.B., Bin Ab Rahman, M.H., Bin Ahmad, M.N., Bin Wahid, M.K., Binti Maidin, N.A.: Optimization of drilling parameters on diameter accuracy in dry drilling process of AISI D2 tool steel. Int. J. Appl. Eng. Res. **12**(20), 9644–9652 (2017)
4. Zhang, J.Z., Chen, J.C., Kirby, E.D.: Surface roughness optimization in an end-milling operation using the Taguchi design method. J. Mater. Process. Technol. **184**(1–3), 233–239 (2007)

5. Nair, V.N., Abraham, B., MacKay, J., Box, G., Kacker, R.N., Lorenzen, T.J., Lucas, J.M., Myers, R.H., Vining, G.G., Nelder, J.A., Phadke, M.S.: Taguchi's parameter design: a panel discussion. Technometrics **34**(2), 127–161 (1992)
6. León, R.V., Shoemaker, A.C., Kacker, R.N.: Performance measures independent of adjustment: an explanation and extension of Taguchi's signal-to-noise ratios. Technometrics **29**(3), 253–265 (1987)
7. Steinberg, D.M., Bursztyn, D.: Dispersion effects in robust-design experiments with noise factors. J. Qual. Technol. **26**(1), 12–20 (1994)
8. Boschetto, A., Bottini, L.: Design for manufacturing of surfaces to improve accuracy in Fused Deposition Modeling. Robot. Comput.-Integr. Manuf. **37**, 103–114 (2016)
9. Mahesh, M., Wong, Y.S., Fuh, J.Y.H., Loh, H.T.: Benchmarking for comparative evaluation of RP systems and processes. Rapid Prototyp. J. **10**(2), 123–135 (2004)
10. Tong, K., Joshi, S., Amine Lehtihet, E.: Error compensation for fused deposition modeling (FDM) machine by correcting slice files. Rapid Prototyp. J. **14**(1), 4–14 (2008)
11. Griffey, J.: The types of 3-D printing. Lib. Technol. Rep. **50**(5), 8–12 (2014)
12. Hafsa, M.N., Ibrahim, M., Wahab, M., Zahid, M.S.: Evaluation of FDM pattern with ABS and PLA material. In: Applied Mechanics and Materials, vol. 465, pp. 55–59. Trans Tech Publications (2014)
13. Garg, A., Bhattacharya, A., Batish, A.: On surface finish and dimensional accuracy of FDM parts after cold vapor treatment. Mater. Manuf. Processes **31**(4), 522–529 (2016)
14. Ahmad, M.N., Wahid, M.K., Maidin, N.A., et al.: Mechanical characteristics of oil palm fiber reinforced thermoplastics as filament for fused deposition modeling (FDM). Adv. Manuf. **8**, 72–81 (2020). https://doi.org/10.1007/s40436-019-00287-w
15. Bakar, N.S.A., Alkahari, M.R., Boejang, H.: Analysis on fused deposition modelling performance. J. Zhejiang Univ.-Sci. A **11**(12), 972–977 (2010)
16. Gibson, I., Rosen, D.W., Stucker, B.: Additive manufacturing: rapid prototyping to direct digital manufacturing (2010)

Check for updates

Magnetohydrodynamic Flow of Casson Nanofluid in a Channel Filled with Thermophoretic Diffusion Effect and Multiple Slips

Sidra Aman[1(✉)], Zulkhibri Ismail[1(✉)], Mohd Zuki Salleh[1], and Ilyas Khan[2]

[1] Centre for Mathematical Sciences, Universiti Malaysia Pahang, UMP, 26300 Kuantan, Pahang, Malaysia
sidraaman9@gmail.com, zulkhibri@ump.edu.my

[2] Basic Engineering Sciences Department, College of Engineering, Majmaah University, Majmaah 11952, Saudi Arabia

Abstract. This article studied the impacts of thermophoretic diffusion and chemical reaction on MHD (magnetohydrodynamic) CNTs-Sodium alginate nanofluid in a channel filled with porous medium. Two types of nanotubes i.e. single walls carbon nanotubes (SWCNTs) and multiple walls carbon nanotubes (MWCNTs) are suspended in SA-base fluid. The focus of this work is to examine Multiple slips (hydrodynamic, thermal and mass slip) effects on the flow of a casson nanofluid in a vertical channel. The problem is transformed to nondimensional form and then tackled by perturbation method. Effect of diameter and length of CNTs on nanofluids' thermal conductivity is evaluated numerically using modified Yu-Choi model. The results are plotted via MathCAD software. Fluid velocity decreases by increasing the volume fraction of CNTs. Thermal conductivity of CNTs with diameter 500 pm is greater than those whose diameter is 50 mm. The CNTs with length 50 pm has the highest thermal conductivity while lowest for CNT with 0.1 m length.

Keywords: Casson fluid · Nanofluid · Multiple slips · Perturbation technique · Heat and mass transfer

1 Introduction

Researchers have realized that no-slip condition may no longer exist and needed. Instead, the slip conditions are of great interest for researchers and engineers in experimental and theoretical studies. Motion of particles such as bubbles involves slip condition at the surface, technology (to polish artificial heart valves, micro and Nano channels, molten polymer capillary flows etc.). Theoretically, the fluid's non-adherence to a solid boundary is known as velocity slip. No-slip conditions are not applicable in MEMS and NEMS fluid flows. Therefore, slip condition is a point of interest for such kind of applications. Near the boundary surface, maybe some fluids particles feel bounce, so no-slip condition may

M. A. Zakaria et al. (Eds.): *Advances in Mechatronics, Manufacturing, and Mechanical Engineering*, LNME, pp. 232–246, 2021.
https://doi.org/10.1007/978-981-15-7309-5_23

not be fruitful anymore and allow us to take slip condition into account. Simulations at fluids with low-viscosity such as water in huge vessels where the capillary effects are insignificant, a global slip condition is adequate; no complicated wetting model is required and the free surface to slides freely along the wetted boundary. It is a union of Dirichlet and Neumann conditions. The slip flow model proposes a correlation between the tangential components of velocity at the surface and the velocity gradient normal to the surface [1–3]. Hamad et al. [4] observed that wall heat transfer and concentration increase while slip velocity decreases with hydrodynamic slip parameter. Bhattacharyya et al. [5] observed an elevation in heat transfer rate with slip parameter while studying MHD heat transfer boundary layer flow over a flat plate. Some of the most interesting results have been obtained in [6–11].

Non-Newtonian fluids have unique characteristics; they exhibit both liquid and solid properties, as the correspondence between shear stress and shear rate becomes non-linear. Non-Newtonian fluid is used inside the flexible military suit, when the soldier stands still, moves, or runs; it remains liquid but when the bullet hits it converts to solid state as a bullet proof armor. Their viscosity depends on the rate of applied stress. Some of non-Newtonian fluids are, jelly, honey, soup, human blood, toothpaste, paint, shampoo, ketchup, foods some molecular liquids such as sodium alginate (SA) [11, 12].

Sodium Alginate (Na-Alg) is a natural polysaccharide derived from algae. It has been used as an encapsulating matrix in drug delivery. It has wound healing, antimicrobial properties and the ability to absorb water 200–300 times of its weight. Thus, it is broadly encountered in biological applications such as wound dressing and encapsulation Liakos et al. [13]. Heat transfer and flow analysis for SA-TiO_2 non-Newtonian nanofluid was investigated by Hatami and Ganji [14]. Sheikholeslami and Rashidi [15] investigated heat transfer treatment of ferrofluid in presence of magnetic field. Pawar and Sunnapwar [16] considered Sodium Alginate in their experimental investigation on heat transfer flow in helical coils.

Due to small thermal conductivities of Non-Newtonian fluids in cooling process and other applications, they are regarded as poor heat transfer liquids. To enhance their thermal conductivities different types of nanoparticles are added to these fluids in the practical fields. Yu and Choi [17] investigated the function of interfacial layers to enhance nanofluids' thermal conductivity by their own model. However later, Jiang et al. [18] observed from experimental results that the effect of length and diameter of CNTs must have significant reflection on their thermal conductivities so they modified Yu-Choi model which will be used in this attempt to find thermal conductivity of CNTs. Their thermal properties are directly related to their structure and size which makes them attractive for researchers due to their interesting applications in textile and acoustics. Xiao et al. [19] made practical CNTs thin film loudspeakers as they could emit loud sounds. Asgari et al. [20] surveyed stretching and slip boundary conditions on nonlinear vibrations of nanotube conveying fluid. Kandasamy et al. [21] investigated influence of chemical reaction on Cu, Al_2O_3 and SWCNTs-nanofluid flow with slip conditions. Sheikholeslami and Sadoughi [22] examined simulation of CuO-water nanofluid heat transfer enhancement in presence of melting surface. Masood et al. [23] investigated non-linear radiative flow of three-dimensional Burgers nanofluid with new mass flux effect. Recently, sand particle erosion in long radius elbow for multiphase flow was numerically investigated by Khan et al. [24]. Some appealing related references can be found in [25–29].

Despite of a vast range of their implementations in thermal engineering, mostly experimental or numerical research work is reported in this domain hence a lack of analytical studies in this area has been found particularly in channel flows. The above literature concluded that no study has been reported on analytical work for analysis of a casson nanofluid for multiple slip conditions with radiation effects. However, some of numerical or semi numerical studies have considered multiple slip conditions. Recently Sekhar et al. [30] examined 3D magnetic-Casson nanofluid for multiple slip effects by employing R-K-Feldberg integration method. They found an enhancement due to nanoparticles presence inside the fluid. Some interesting numerical studies on multiple slip for flow of fluids can be found in [31–35] and the references cited in. Yet, the exact analytic studies in this direction are scarce. Hence, the present article will be considered as a pioneering work for exact analytical investigators. Further, the results obtained here can be used as a benchmark by numericalists and experimentalists.

This article aims to study the impacts of multiple slips (velocity, thermal and solutal slip) on MHD flow of SA-CNTs casson nanofluid in a vertical channel. The effect of diameter and length of CNTs on its thermal conductivity is discussed and observed for temperature and velocity of nanofluids. The equations are formulated and transformed to non-dimensional forms together with imposed boundary conditions. Perturbation method has been employed to acquire the solution of velocity, concentration and temperature fields. The results are mapped graphically for important parameters and analysed thoroughly.

2 Formulation and Solution of the Problem

Consider MHD flow of SA-CNTs non-Newtonian nanofluid in a vertical channel with width a, with oscillating and static walls at $y = 0$ and $y = d$ respectively. Three various slip conditions (velocity, thermal and concentration slip) are considered. External pressure gradient together with buoyancy force induces the mixed convection. A uniform magnetic field, B_0 is exerted perpendicular to the axis of flow. In vertical channel, T_0 and T_d show left and right plate temperatures whereas C_0 and C_d show their concentrations. The physical geometry for the problem is given in Fig. 1. The governing equations of momentum, energy and concentration are:

$$\rho_{nf}\frac{\partial u}{\partial t} = -\frac{\partial p}{\partial x}+\mu_{nf}\left(1+\frac{1}{\beta}\right)\frac{\partial^2 u}{\partial y^2} - \left(\sigma_{nf}B_0^2+\frac{\mu_{nf}}{k_2}\right)u + (\rho\beta_T)_{nf}g(T-T_d) + (\rho\beta_T)_{nf}g(C-C_d), \quad (1)$$

$$(\rho c_p)_{nf}\frac{\partial T}{\partial t} = k_{nf}\frac{\partial^2 T}{\partial y^2} + 4\alpha_0^2(T-T_d). \quad (2)$$

$$\frac{\partial C}{\partial t} = D_{nf}\frac{\partial^2 C}{\partial y^2} + \frac{D_{nf}K_T}{T_m}\frac{\partial^2 T}{\partial y^2} - k_r(C-C_d), \quad (3)$$

Where u, T, ρ_{nf}, μ_{nf} , σ_{nf} , β_0, β_T, g, $\left(\rho c_p\right)_{nf}$, k_{nf} , α_0, D_{nf}, K_T, k_r are fluid velocity in the $x-$ direction, temperature, density, the dynamic viscosity, electrical conductivity, thermal expansion coefficient, gravitational acceleration, heat capacitance, thermal conductivity, radiation absorption coefficient, diffusivity, thermophoretic parameter and chemical reaction parameter. The boundary conditions are:

$$u(0,\,t) - \beta_{\text{vel}}\left(1+\frac{1}{\beta}\right)\frac{\partial u(0,t)}{\partial y} = u_0\varepsilon\exp(iwt),\quad u(d,\,t)=0, \tag{4}$$

$$T(\,0\,,t) - \beta_{temp}\,\frac{\partial T(0\,,t)}{\partial y} = T_0,\quad T(d,\,t)=T_d, \tag{5}$$

$$C(0,\,t) - \beta_{\text{conc}}\frac{\partial C(0,t)}{\partial y} = C_0,\quad C(d,\,t)=C_d. \tag{6}$$

where β_{vel}, β_{temp} and β_{conc} are respectively the velocity, temperature and Solutal slips. The external pressure gradient is taken as $-\partial p/\partial x = \lambda_0 + \lambda\varepsilon exp(i\omega t)$ with constant λ and oscillation frequency ω.

Yu and Choi model [17] is used in this problem for thermal conductivity of CNTs nanofluid as under:

$$k_{nf} = \left(1+\frac{n\phi A}{1-\phi A}\right)k_f, \tag{7}$$

where

$$n = 3\psi^{-\alpha} \tag{8}$$

The density ρ_{nf}, thermal expansion $(\rho\beta)_{nf}$, heat capacitance $\left(\rho c_p\right)_{nf}$, thermal conductivity σ_{nf} defined as in Kandasamy et al. [21]:

$$(\rho\beta)_{nf} = (1-\phi)\,(\rho\beta)_f + \phi(\rho\beta)_{CNT},\ \mu_{nf} = \frac{\mu_f}{(1-\phi)^{2.5}},\ \rho_{nf} = (1-\phi)\,\rho_f + \phi\rho_{CNT},$$
$$\left(\rho c_p\right)_{nf} = (1-\phi)\,(\rho c_p)_f + \phi(\rho c_p)_{CNT},\ \ \sigma_{nf} = \sigma_f\left(1+\frac{3\,(\sigma-1)\phi}{(\sigma+2)-(\sigma-1)\phi}\right), \tag{9}$$

where ϕ is the nanoparticles' volume fraction, ρ_f and ρ_{CNT} is the density, β_{CNT} and β_f the thermal expansions coefficient, $\left(c_p\right)_{CNT}$ and $\left(c_p\right)_f$ are specific heat capacities of CNTs and base fluids respectively. Introducing below dimensionless quantities:

$$x^* = \frac{x}{d},\ y^* = \frac{y}{d},\quad p^* = \frac{d}{\mu U_0}p,\ t^* = \frac{tU_0}{d},\ u^* = \frac{u}{U_0},\ T^* = \frac{T-T_d}{T_0-T_d},$$
$$C^* = \frac{C-C_d}{C_0-C_d},\ \omega^* = \frac{d\omega}{U_0},\quad -\frac{\partial p^*}{\partial x^*} = \lambda_0 + \lambda\varepsilon exp\left(i\omega^* t^*\right), \tag{10}$$

From Eqs. (1), (2), (3), (4), (5), and (6), we get

$$a_o\frac{\partial u}{\partial t} = \lambda_0 + \lambda\varepsilon e^{i\omega t} + a_1\frac{\partial^2 u}{\partial y^2} - a_2 u + a_3 T + a_4 C, \tag{11}$$

$$u(0,t) - \beta_1\left(1 + \frac{1}{\beta}\right)\frac{\partial u(0,t)}{\partial y} = \varepsilon\exp(iwt), \quad u(1,t) = 0, \tag{12}$$

$$b_0^2\frac{\partial T}{\partial t} = \frac{\partial^2 T}{\partial y^2} + b_1^2 T, \tag{13}$$

$$T(0,t) - \beta_2\frac{\partial T(0,t)}{\partial y} = 1, \quad T(1,t) = 0, \tag{14}$$

$$b_2^2\frac{\partial C}{\partial t} = \frac{\partial^2 C}{\partial y^2} + b_3^2\frac{\partial^2 T}{\partial y^2} + b_4^2 C, \tag{15}$$

$$C(0,t) - \beta_3\frac{\partial C(0,t)}{\partial y} = 1, \quad C(1,t) = 0, \tag{16}$$

where

$$\mathrm{Re} = \frac{U_0 d\rho_f}{\mu_f}, \; M = \frac{\delta\beta_0^2 d^2}{\mu_f}, \; Ri_T = \frac{gd(\beta_T)_f(T_0 - T_d)}{U_0^2}, \; Ri_c = \frac{gd(\beta_T)_f(C_0 - C_d)}{U_0^2},$$

$$\beta_2 = \frac{\beta_{\text{temp}}}{d}, \; \beta_3 = \frac{\beta_{\text{conc}}}{d}, \; Pe = \frac{U_0 d(\rho c_p)_f}{k_f}, \; N^2 = \frac{4\alpha^2 d^2}{k_f}, \; b_1^2 = \frac{N^2}{\lambda_{nf}}, \; \lambda_{nf} = \frac{k_{nf}}{k_f},$$

$$\phi_1 = (1-\phi) + \phi\left(\frac{\rho_{CNT}}{\rho_f}\right), \; \phi_2 = \frac{1}{(1-\phi)^{2.5}}, \; \phi_3 = \left(1 + \frac{3(\sigma-1)\phi}{(\sigma+2)-(\sigma-1)}\right), \; a_0 = \phi_1\mathrm{Re},$$

$$\phi_4 = (1-\phi) + \frac{\phi(\rho\beta_T)_{CNT}}{(\rho\beta_T)_f}, \; \phi_5 = (1-\phi) + \phi\frac{(\rho c_p)_{CNT}}{(\rho c_p)_f}, \; a_1 = \phi_2\left(1 + \frac{1}{\beta}\right), \; a_2 = \phi_3 M + \frac{\phi_2}{k},$$

$$Sc = \frac{U_0 d}{D_f}, \; a_3 = Ri_T Re\phi_4, \; a_4 = \phi_4 ReRi_c, \; b_0^2 = \frac{\phi_5 Pe}{\lambda_{nf}}, \; Sr = \frac{D_f k_T T_a}{(C_0 - C_d)T_m\mu_f}, \; \beta_1 = \frac{\beta_{\text{vel}}}{d},$$

where Re, M, Ri_T, Ri_c, β_1, β_2, β_3, Pe, N are Reynolds number, magnetic parameter, Richardson number, dimensionless velocity, thermal and solutal slips, radiation parameter and Peclet number respectively. To interpret Eqs. (14, 16 and 18) with boundary conditions (15), (16), (17), (18), and (19), the perturbed solutions of the following forms are taken:

$$u(y,t) = u_0(y) + \varepsilon\exp(i\omega t)u_1(y), \tag{17}$$

$$T(y,t) = T_0(y) + \varepsilon\exp(i\omega t)T_1(y), \tag{18}$$

$$C(y,t) = C_0(y) + \varepsilon\exp(i\omega t)C_1(y). \tag{19}$$

Using Eqs. (17), (18), and (19) into (11), (12), (13), (14), (15), and (16), we acquire the following system of ODE's:

$$\begin{aligned}
&\frac{\partial^2 u_0(y)}{\partial y^2} - a_5^2 u_0(y) = -\frac{a_3}{a_1}T_0 - \frac{a_4}{a_1}C_0 - \frac{\lambda_0}{a_1}, \\
&a_1\frac{\partial^2 u_1(y)}{\partial y^2} - a_2 u_1(y) - a_0 i\omega t = -a_3 T_1 - a_4 C_1 - \lambda_1,
\end{aligned} \tag{20}$$

$$\frac{\partial^2 T_0(y)}{\partial y^2} + b_1^2 T_0(y) = 0\,,$$
$$\frac{\partial^2 T_1(y)}{\partial y^2} + b_2^2 T_1(y) = 0\,, \tag{21}$$

$$\frac{\partial^2 C_0(y)}{\partial y^2} + b_4^2 C_0(y) = -b_3^2 \frac{\partial^2 T_0}{\partial y^2}\,,$$
$$\frac{\partial^2 C_1(y)}{\partial y^2} + b_4^2 C_1(y) - b_2^2 i\omega C_1(y) = -b_3^2 \frac{\partial^2 T_1}{\partial y^2}\,, \tag{22}$$

with the boundary conditions

$$u(0,t) - \xi_1 \frac{\partial u(0,t)}{\partial y} = \varepsilon \exp(iwt),\;\; u_1(0) = 0,\;\; u_1(1) = 0\,,$$
$$T(0,t) - \xi_2 \frac{\partial T(0,t)}{\partial y} = 1,\;\; T_0(1) = 0\,,\;\; T_1(1) = 0,$$
$$C(0\,,t) - \xi_3 \frac{\partial C\,(0,t)}{\partial y} = 1,\;\; C_0(1) = 0\,,\;\; C_1(1) = 0, \tag{23}$$

where

$$a_5^2 = \frac{a_2}{a_1},\; b^2 = (b_1^2 - b_0^2 i\omega),\; b_2^2 = (ReSc),\; b_3^2 = (SrSc),\; b_4^2 = \left(Ri_T^2 \gamma Sc\right),$$
$$b_5^2 = (b_4^2 - b_2^2 i\omega),\; b_6^2 = \left(\frac{a_2 + a_0 i\omega}{a_1}\right),\; \xi_1 = \beta_1\left(1 + \frac{1}{\beta}\right),\;\; \xi_2 = \beta_2,\;\; \xi_3 = \beta_3\,.$$

Solutions of Eq. (20), (21), and (22), under the boundary conditions in (23) are

$$\begin{aligned} u(y,t) = {} & c_1 \sinh a_5 y + c_2 \cosh a_5 y + m_1 m_6 \cos b_1 y - m_2 m_6 \sin b_1 y - m_3 m_7 \sin b_4 y \\ & + m_4 m_7 \cos b_4 y - m_3 m_8 \sin b_4 y - m_3 m_9 \sin b_4 y + m_4 m_8 \cos b_4 y \\ & + m_4 m_9 \cos b_4 y + \frac{m_2 m_9 \sin b_1 y}{m_1} - m_9 \cos b_1 y + \frac{\lambda_0}{a_1 a_5^2} \\ & + \varepsilon e^{i\omega t}\left(\frac{\sinh b_6 \sinh b_6 y}{m_5}\left(\frac{\lambda_1 \cosh b_6}{a_1 b_6^2} - \cosh b_6 - \frac{\lambda_1}{a_1 b_6^2}\right) + \cosh b_6 y\right) \\ & + \varepsilon e^{i\omega t}\left(\left(m_5 + \frac{\lambda_1 m_5}{a_1 b_6^2 \cosh b_6} - \frac{\lambda_1 m_5}{a_1 b_6^2} - \frac{\lambda_1}{a_1 b_6^2 \cosh b_6}\right) + \frac{\lambda_1}{a_1 b_6^2}\right). \end{aligned} \tag{24}$$

$$T(y,t) = \left[-m_2 \sin b_1 y + m_1 \cos b_1 y\right]. \tag{25}$$

$$C(y,t) = (m_7 + m_8 + m_9)\frac{m_3}{m_7}\left[\frac{\sin b_4 \cos b_4 y}{\cos b_4} - \sin b_4 y\right] + b_1^2 b_3^2 (m_2 \sin b_1 y - m_1 \cos b_1 y), \tag{26}$$

where

$$c_1 = \left[\begin{array}{l} (\sin b_4 - 1) m_9 m_{10} + \dfrac{\lambda_0 m_{10}}{a_1 a_5^2} + (m_1 m_6 + m_4 m_7 + m_4 m_8) m_{10} + b_1 \xi_1 m_2 m_6 m_{10} \\ \quad +(m_7 + m_8 + m_9) b_4 \xi_1 m_3 m_{10} - \dfrac{\xi_1 m_8 m_{10}}{\xi_3} + (a_5 \xi_1 m_{10} - 1)\dfrac{\lambda_0}{a_1 a_5^2 \sinh a_5} \end{array}\right]$$

$$c_2 = \left[\begin{array}{c} (1 - \sin b_4)m_9 m_{11} - (m_1 m_6 + m_4 m_7 + m_4 m_8)m_{11} - b_1 \xi_1 m_2 m_6 m_{11} \\ -\left(1 + \frac{a_5 \xi_1}{\sinh a_5}\right)\frac{\lambda_0 m_{11}}{a_1 a_5^2} - (m_3 m_7 + m_8 m_3 + m_3 m_9) b_4 \xi_1 m_{11} + \frac{\xi_1 m_8 m_{11}}{\xi_3} \end{array} \right]$$

and

$$m_1 = \frac{\sin b_1}{(\sin b_1 + \xi_2 b_1 \cos b_1)},\ m_2 = \frac{\cos b_1}{(\sin b_1 + \xi_2 b_1 \cos b_1)},\ m_3 = \frac{\cos b_4}{(\sin b_4 + \xi_3 b_4 \cos b_4)},$$

$$m_4 = \frac{\sin b_4}{(\sin b_4 + \xi_3 b_4 \cos b_4)},\ m_5 = \frac{sinhb_6}{(\sinh b_6 - \xi_1 b_6 \cosh b_6)},\ m_6 = \frac{a_3}{a_1\left(b_1^2 + a_5^2\right)},$$

$$m_7 = \frac{a_4}{a_1\left(b_1^2 + a_5^2\right)},\ m_8 = \frac{\xi_3 b_1^3 b_3^2 m_2 m_7}{\left(b_1^2 - b_4^2\right)},\ m_9 = \frac{b_1^2 b_3^2 m_1 m_7}{\left(b_1^2 - b_4^2\right)},$$

$$m_{10} = \frac{cosha_5}{(\sinh a_5 + \xi_1 a_5 \cosh a_5)},\ m_{11} = \frac{\sinh a_5}{(\sinh a_5 + \xi_1 a_5 \cosh a_5)}$$

3 Results and Discussion

A thorough analysis has been made for multiple slip effects on MHD flow of SA-CNTs non-Newtonian nanofluid inside a vertical channel. The flow of CNTs suspended in Sodium Alginate has been analysed. Yu and Choi model [17] was in consideration to be used for thermal conductivity. However, the experimental research shows that the effect of length and diameter on thermal conductivity cannot be disregard which is not been reflected in Yu and Choi model. Thus, a modified Yu-Choi model established by Jiang et al. [18] is used here to predict thermal conductivity of CNTs. According to experimental results of Jiang et al. [18], the modified Yu and Choi model is enhanced than the previous existing models for forecasting thermal conductivities of CNTs.

The values of different parameters used in modified Yu and Choi model [17] for thermal conductivity are listed in Table 1. Thermophysical properties of SA and CNTs are given in Table 2 [15, 16]. Table 3 shows the impact of diameter d of CNTs on the thermal conductivities of nanofluids. It is observed that thermal conductivity increases with decreasing values of diameter of CNTs. Thermal conductivity of CNTs with diameter 500 pm is greater than those with diameter 50 mm. Smaller the diameter of CNTs, greater will be their specific surface means more Brownian motion which increases their thermal conductivities. The interfacial layer made up of liquid molecules is thicker for CNTs with smaller diameter, which is a factor in increasing thermal conductivities. Jiang et al. [18] observed same behaviour. The effect of length L of CNTs on thermal conductivities is shown in Table 4. It is estimated that smaller the length of CNTs, greater the thermal conductivities of nanofluids. The CNTs with 50 pm has the highest thermal conductivity while lowest for CNT with 0.1 m length.

The variation of temperature profile is shown in Figs. 1, 2 and 3 for both SWCNT and MWCNT. Figure 1(a) and (b) shows that both single and multiple wall CNTs temperature profile increases with increasing values of volume fraction ϕ. Increasing ϕ,

Table 1. Values of different parameters taken from Yu and Choi. [17].

Parameters	Values ($\times 10^{-9}$)
a	25000
b	12.5
c	12.5
t	54
Thermal conductivity of surrounding layer k_s	100

Table 2. Thermophysical properties for base fluids and CNTs [15] and [16].

Physical properties	Base fluid	Nanoparticles	
	Sodium Alginate	SWCNT	MWCNT
$\rho (kg/m^3)$	989	2,600	1,600
$c_p (J/kgK)$	4,175	425	796
$k(W/mK)$	0.6376	6,600	3,000

Table 3. Thermal conductivities of nanofluids for various diameter and Length

Diameter	Thermal conductivity	Length	Thermal conductivity
50 mm	1.072	0.1 mm	0.664
200 μm	1.073	5 μm	0.665
20 μm	1.077	0.5 μm	0.666
5 μm	1.079	5 nm	0.667
50 nm	1.079	50 pm	0.64
500 pm	1.08		

augments its thermal conductivity, hence temperature of the fluid exhibits a significance increase. From Fig. 2(a) and (b), temperature profile decreases with increasing thermal slip parameter β_2 near the wall. Figures 3(a) and (b) show temperature variation with different diameters d of CNTs (SWCNT and MWCNT). Temperature profile increases with increase in diameter of CNTs. But this effect is observed for a greater volume fraction of CNTs (i.e.$\phi = 1.2$) for $\phi < 1$ there was a negligible effect. Figures 4(a) and (b) show that with increasing values of Solutal slip parameter β_3 concentration profile exhibits a notable increase for both type of CNTs.

The factors influencing the velocity are non-Newtonian fluid parameter, volume fraction, radiation parameter, velocity slip parameters. Velocity of SWCNTs is observed to be greater than that of MWCNTs. It is since SWCNTs has greater density and thermal conductivity than MWCNTs that leads to increase in velocity. Figure 5(a) and (b) reveal

Table 4. Thermal conductivities of nanofluids for various volume fractions

Volume fraction ϕ	Thermal conductivity
0	0.638
0.01	0.641
0.02	0.644
0.03	0.648
0.04	0.651
0.05	0.654
0.06	0.658
0.07	0.661

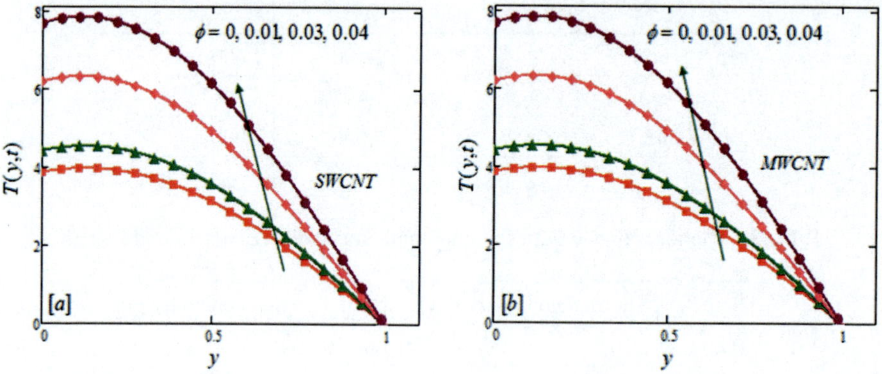

Fig. 1. Temperature profiles variation with ϕ when $N = 1.8$, $\beta = 0.5$, $\beta_2 = 1$

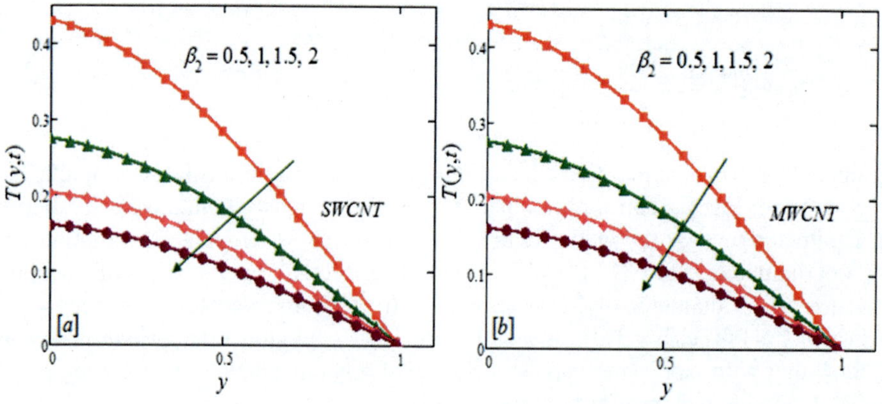

Fig. 2. Temperature profiles variation with β_2 when $N = 1.4$, $\beta = 0.1$, $\phi = 0.04$

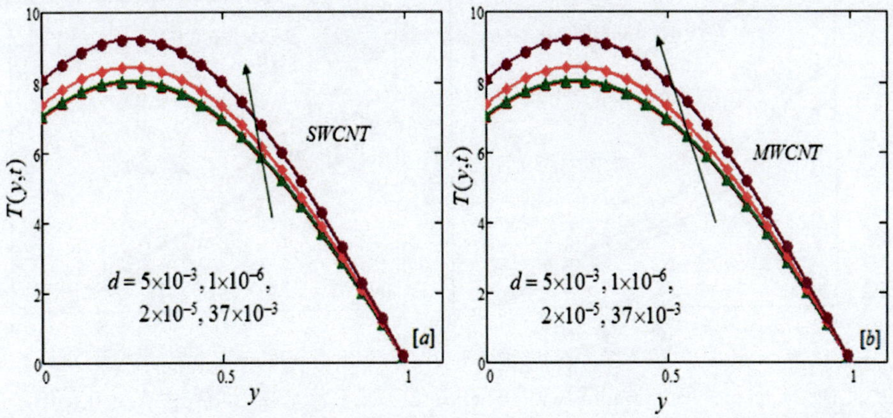

Fig. 3. Temperature profiles variation with d when $N = 2.7$, $\phi = 1.2$, $\beta = 0.5$, $\beta_2 = 2$.

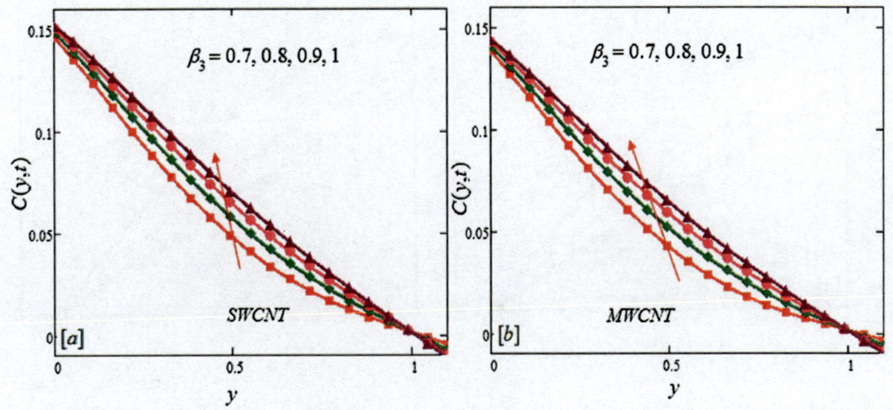

Fig. 4. Concentration profiles variation with β_3 when $\phi = 0.04$, $N = 1.4$, $\beta = 0.5$, $\beta_1 = 0.1$, $\beta_2 = 0.2$, $Ri_T = 3.5$, $\gamma = 1$, $Sr = 1$.

the impact of non-Newtonian parameter β on velocity profile for both types of CNTs. It is detected that velocity of the nanofluids decreases with increase in β which enhances the resistance between the fluid layers, thus lowers the fluid flow as noticed in [20].

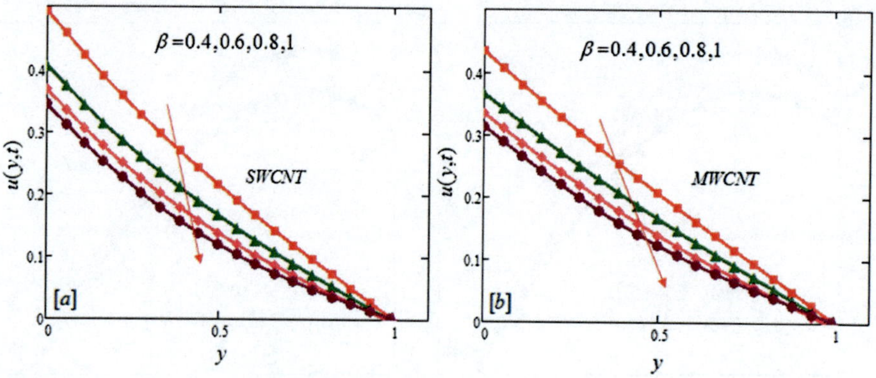

Fig. 5. Velocity profiles variation with β when $N = 1.2, \phi = 0.03, \beta_1 = 0.1, \beta_2 = 0.2, \beta_3 = 0.3, Ri_T = 3.5, Ri_c = 2, Sr = 0.2, \gamma = 0.8$.

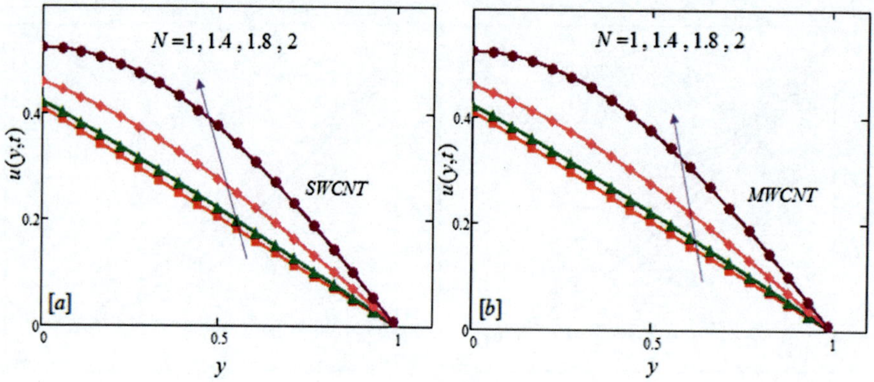

Fig. 6. Velocity profiles variation with N when $\phi = 0.03, \beta = 0.5, \beta_1 = 0.1, \beta_2 = 0.2, \beta_3 = 0.3, Ri_T = 3.5, \gamma = 2, M = 0.015, Ri_c = 2, Sr = 0.2$.

To observe the effect of N on velocity profile Figs. 6(a) and (b) are drawn. It is obvious that fluid flow increases with N due to an elevation in temperature. Figure 7(a) and (b) spotted the impact of velocity slip β_1 on fluid flow. Velocity slip enhances the fluid flow. Figure 8(a) and (b) represent the impact of volume fraction of CNTs on velocity profile. It is spotted that velocity minimizes with maximizing ϕ. SA-CNTs non-Newtonian nanofluids becomes more viscous by increasing ϕ which tends to depress velocity profile.

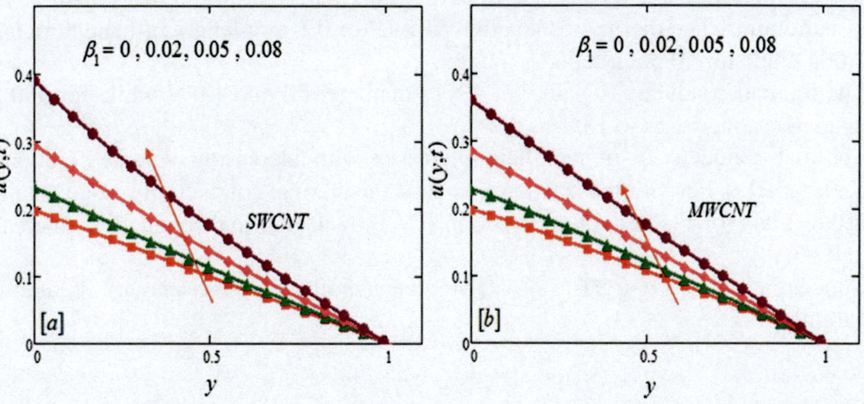

Fig. 7. Velocity profiles variation with β when $N = 1.2, \phi = 0.03$, $\beta = 0.4$, $\beta_2 = 0.2$, $\beta_3 = 0.3$, $Re = 0.8$, $Ri_T = 3.5$, $Ri_c = 2, \gamma = 0.8$, $Sr = 0.2$, $M = 0.015$.

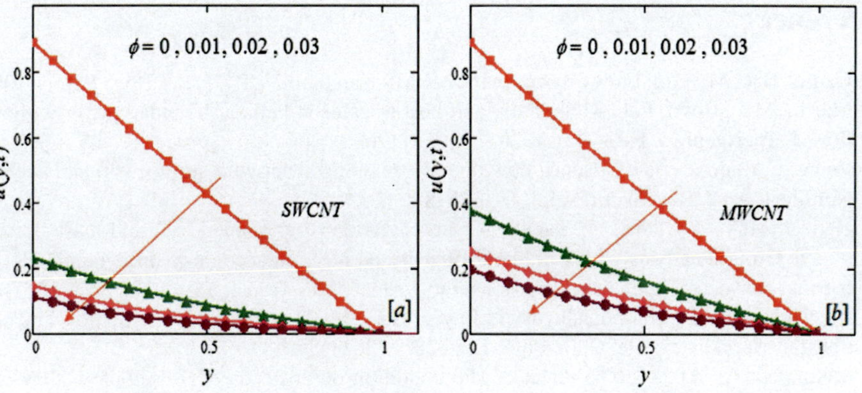

Fig. 8. Velocity profiles variation with ϕ when $N = 0.6$, , $\beta = 0.1$, $\beta_1 = 0.1$, $\beta_2 = 0.2$, $\beta_3 = 0.3$, $Re = 0.3$, $Ri_T = 2.5$, $Ri_c = 2$, $Sr = 0.2$, $\gamma = 1.2$.

4 Conclusion

Effects of multiple slip conditions on SA-CNTs casson nanofluid flow in a vertical channel in the presence of porosity are investigated. The effects of diameter and length of CNTs was observed on the thermal conductivity and temperature of nanofluids using modified Yu and Choi model. We have concluded with the following outcomes:

- Maximizing volume fraction minimizes the velocity of nanofluids because the fluid becomes denser by adding nanoparticles.
- Temperature of SA-CNTs non-Newtonian nanofluids increases with increase in the diameter of CNTs and thermal slip parameter.
- Thermal conductivity maximizes with reducing values of diameter of CNTs.

- It is spotted that smaller the length of CNTs, greater is the thermal conductivities of nanofluids. The thermal conductivity value for 0.1 mm length of nanoparticles is 0.664 while for 50 pm it reduces to 0.64.
- The thermal conductivity value for CNTs diameter 50 mm 1.072 while for 500 pm diameter, it increases to 1.08.
- Thermal conductivity of nanofluid increases with increasing volume fraction of nanoparticles. For Volume fraction $\phi = 0.01$ the thermal conductivity value obtained is 0.641 and for a larger volume fraction $\phi = 0.07$ the thermal conductivity increases to 0.661.
- Concentration of SA-CNTs non-Newtonian nanofluids increases with solutal slip parameter.

Acknowledgments. This research was funded by a grant from Centre of Excellence for Advanced Research in Fluid Flow (CARIFF), Universiti Malaysia Pahang through RDU190376 and PGRS170399.

References

1. Gad-el-Hak, M. (ed.): Flow Physics in the MEMS Handbook. CRC Press, Boca Raton (2002)
2. Martin, M.J., Boyd, I.D.: Momentum and heat transfer in laminar boundary layer with slip flow. J. Thermophys. Heat Transf. **20**, 710–719 (2006)
3. Wang, C.Y.: Analysis of viscous flow due to stretching sheet with surface slip and suction. Nonlinear Anal. Real World Appl. **10**, 375–380 (2009)
4. Hamad, M.A.A., Uddin, M., Ismail, A.: Investigation of combined heat and mass transfer by Lie Group Analysis with variable diffusivity taking into account hydrodynamic slip an thermal convective boundary conditions. Int. J. Heat Mass Transf. **55**, 1355–1362 (2012)
5. Bhattacharyya, K., Mukhopadhyay, S., Layek, G.C.: MHD boundary layer slip flow and heat transfer over a flate plate. Chin. Phys. Lett. **28**, 2 (2011)
6. Mehmood, A., Ali, A.: The effect of slip condition on unsteady MH oscillatory flow of a viscous fluid in a planner channel. Rom. J. Phys. **52**, 85–91 (2011)
7. Qasim, M., Khan, Z.H., Khan, W.A., Shah, I.A.: MHD boundary layer slip flow and heat transfer of ferrofluid along a stretching cylinder with prescribed heat flux. PLoS ONE **9**(1), e83930 (2014). https://doi.org/10.1371/journal.pone.0083930
8. Uddin, M.J., Baig, O.A., Uddin, M.N.: Multiple slip and variable transport property effect on magnetohydrodynamic dissipative thermosolutal convection in a porous medium. Am. Soc. Civ. Eng. **29**, 04016024 (2016). https://doi.org/10.1061/(ASCE)AS.1943-5525.0000614
9. Singh, R.J., Gohil, T.B.: Mixed convection heat transfer in a cavity with rotating cylinder under the influence of magnetic field. In: Saha, P., Subbarao, P.M.V., Sikarwar, B.S. (eds.) Advances in Fluid and Thermal Engineering. LNME, pp. 53–62. Springer, Singapore (2019). https://doi.org/10.1007/978-981-13-6416-7_6
10. Tiwary, B., Kumar, R., Singh, P.K.: Heat transfer enhancement in oblique finned channel. In: Saha, P., Subbarao, P.M.V., Sikarwar, B.S. (eds.) Advances in Fluid and Thermal Engineering. LNME, pp. 157–167. Springer, Singapore (2019). https://doi.org/10.1007/978-981-13-6416-7_15
11. Aman, S., Ilyas, K., Zulkhibri, I., Salleh, M.Z., Alshomrani, A.S., Alghamdi, M.S.: Magnetic field effect on Poiseuille flow and heat transfer of carbon nanotubes along a vertical channel filled with Casson fluid. AIP Adv. **7**, 015036 (2017). https://doi.org/10.1063/1.4975219

12. Ismail, Z., Khan, I., Mohamad, A.Q., Shafie, S.: Second grade fluid for rotating MHD of an unsteady free convection flow in a porous medium. Defect Diffus. Forum **362**, 100–107 (2015)
13. Liakos, I., Rizzello, L., Scurr, D.J., Pompa, P.P., Bayer, I.S., Athanassiou, A.: All-natural composite wound dressing films of essential oils encapsulated in sodium alginate with antimicrobial properties. Int. J. Pharm. **463**, 137–145 (2014)
14. Aman, S., Ilyas, K., Zulkhibri, I., Salleh, M.Z., Al-Mdallal, Q.: Heat transfer enhancement in free convection flow of CNTs Maxwell nanofluids with four different types of molecular liquids. Sci. Rep. **7**, 2445 (2017). https://doi.org/10.1038/s41598-017-01358-3
15. Sheikholeslami, M., Rashidi, M.M.: Ferrofluid heat transfer treatment in the presence of variable magnetic field. Eur. Phys. J. Plus **130**, 115 (2015). https://doi.org/10.1140/epjp/i2015-15115-4
16. Pawar, S.S., Sunnapwar, V.K.: Experimental studies on heat transfer to Newtonian and non-Newtonian fluids in helical coils with laminar and turbulent flow. Exp. Therm. Fluid Sci. **44**, 792–804 (2013)
17. Yu, W., Choi, S.U.S.: The role of interfacial layers in the enhanced thermal conductivity of nanofluids: a renovated Hamilton-crosser model. J. Nanopart. Res. **6**, 355–361 (2004)
18. Jiang, W., Ding, G., Pang, H.: Measurement and model on thermal conductivities of carbon nanotube nano refrigerants. Int. J. Ther. Sci. **48**, 1108–1115 (2009)
19. Xiao, L., Chin, Z., Feng, C., Liu, L., Bai, Z.-Q., Wang, Y., Li, Q., Zhang, Y., Li, Q., Jiang, K., Fan, S.: Flexible, stretchable, transparent carbon nanotube thin film loudspeakers, Nanoletters, **12**(8), 4539–4545 (2008)
20. Mehmood, A.A., Hamid, M.R., Mustafa, G.: Coupled effects of nano-size, stretching and slip boundary conditions on non-linear vibrations of nanotube conveying fluid by the homotopy analysis method. Physica E **52**, 77–85 (2013)
21. Kandasamy, R., Mohamad, R., Ismoen, M.: Impact of chemical reaction on Cu, Al_2O_3, and SWCNTs-nanofluid flow under slip conditions. Eng. Sci. Tech. Int. J. **19**(2), 700–709 (2015). https://doi.org/10.1016/j.jestch2015.11.011
22. Sheikholeslami, M., Sadoughi, M.K.: Simulation of CuO-water nanofluid heat transfer enhancement in presence of melting surface. Int. J. Heat Mass Transf. **116**, 909–919 (2018)
23. Khan, M., Azeem Khan, W., Saleh Alshomrani, A.: Non-linear radiative flow of three-dimensional Burgers nanofluid with new mass flux effect. Int. J. Heat Mass Transf. **101**, 570–576 (2016)
24. Khan, M.R., Ya, H.H., Pao, W., Majid, M.A.A.: Numerical investigation of sand particle erosion in long radius elbow for multiphase flow. In: Awang, M., Emamian, S.S., Yusof, F. (eds.) Advances in Material Sciences and Engineering. LNME, pp. 41–49. Springer, Singapore (2020). https://doi.org/10.1007/978-981-13-8297-0_6
25. Khan, M., Malik, R., Munir, A., Khan, W.A.: Flow and heat transfer to Sisko nanofluid over a nonlinear stretching sheet. PLoS ONE **10**(5), e0125683 (2015). https://doi.org/10.1371/journal.pone.0125683
26. Jiao, J., Cui, Y., Xia, Y.: Improved thermal conductivity of epoxy composites prepared with a mixed filler of multi-walled carbon nanotubes and aluminium nitride particles. High Perform. Polym. **29**(4), 484–492 (2016). https://doi.org/10.1177/09540083l6644036
27. Xue, Q.: Model for thermal conductivity of carbon nanotube-based composites. Phys. B Condens. Matter **368**, 302–307 (2005)
28. Sheikholeslami, M., Rokni, H.B.: Simulation of nanofluid heat transfer in presence of magnetic field: a review. Int. J. Heat Mass Transf. **115**, 1203–1233 (2017)
29. Sheikholeslami, M.: Influence of magnetic field on nanofluid free convection in an open porous cavity by means of Lattice Boltzmann method. J. Mol. Liq. **234**, 364–374 (2017)

30. Sekhar, K.R., Reddy, G.V., Raju, C.S.K., Shehzad, S.A.: Non-uniform heat source/sink and multiple slips on 3D magnetic-Casson fluid in a suspension of copper nanoparticles over a porous slendering sheet. J. Nanofluids **7**(3), 469–477 (2018)
31. Mabood, F., Shateyi, S.: Multiple slip effects on MHD unsteady flow heat and mass transfer impinging on permeable stretching sheet with radiation. Modell. Simul. Eng. **2019**, 1–11 (2019)
32. Raju, C.S.K., Hoque, M.M., Khan, N.A., Islam, M., Kumar, S.: Multiple slip effects on magnetic-Carreau fluid in a suspension of gyrotactic microorganisms over a slendering sheet. Proc. Inst. Mech. Eng. Part E J. Process Mech. Eng. **233**(2), 254–266 (2019)
33. Majeed, A., Zeeshan, A., Xu, H., Kashif, M., Masud, U.: Heat transfer analysis of magneto-Eyring–Powell fluid over a non-linear stretching surface with multiple slip effects: application of Roseland's heat flux. Can. J. Phys. **97**(12), 1253–1261 (2019)
34. Raza, J., Farooq, M., Mebarek-Oudina, F., Mahanthesh, B.: Multiple slip effects on MHD non-Newtonian nanofluid flow over a nonlinear permeable elongated sheet. Multidiscip. Model. Mater. Struct. **15**(5), 913–931 (2019)
35. Titiloye, E.O., Gbadeyan, J.A., Adeosun, A.T.: Heat and mass transfer of MHD dissipative Casson nanofluid flow over a stretching or shrinking sheet with multiple slip boundary conditions. In: Defect and Diffusion Forum, vol. 393, pp. 103–120. Trans Tech Publications Ltd. (2019)

Optimisation of Injection Moulding Process Parameter Using Taguchi and Desirability Function

Vivekanandan Panneerselvam and Faiz Mohd Turan(✉)

Faculty of Manufacturing and Mechatronic Engineering Technology, Universiti Malaysia Pahang, 26600 Pekan, Pahang, Malaysia
faizmt@ump.edu.my

Abstract. The main goal of conducting this research is to primarily investigate the effects of injection moulding parameters on the mechanical properties of plastic part made of recycled plastic. In order to achieve the goal, the significant process parameter is identified with a proper research on previous studies. The mechanical properties of the specimens are measured in term of tensile strength and flexural modulus respectively. Instead of focusing on single quality characteristic, this research also proposed multi response optimization by integrating Taguchi method and desirability functions. The optimal processing parameters and the significance of each processing parameter on the mechanical properties of part produced are determined. From the experimental analysis and results, it is shown that the most significant processing parameter affecting the tensile strength is melt temperature, injection time and holding, time. The most effective process parameter on flexural modulus is melt temperature, holding time and injection pressure. The best set of process parameter are 180 °C of melt temperature, 55 MPa injection pressure, 30 mm/s injection speed, 8 s injection time, 20 MPa holding pressure, 3 s holding time and 25 s cooling time. This parameter will optimise the part quality to 199 kgf/cm^2 of tensile strength and can result in flexural modulus of 10005 kgf/cm^2. From the integration of Taguchi optimization processes and desirability function, the mechanical performances exhibited by the part made of recycled plastic have shown that recycled materials are potentially substituted for virgin material.

1 Introduction

The study carried out by Heidbreder, Bablok, Drews, and Menzel clearly shown the rapid increasing in the consumption of plastics [1]. Around 350 million tons per year has been produced currently which also prove that the production of plastic markedly increases over the last decades. The statistics are very clear to demonstrate that millions of tons of plastic are produced and consumed in the world every year.

It is really sad to see that the increasing of plastic consumption is also lead to the large volume of plastic disposal. Most of these plastics will end up in landfills. The decompensation of these plastic will take a long time to break down like up to hundreds of years due to their non-degradable properties. This can lead to a serious environmental

M. A. Zakaria et al. (Eds.): *Advances in Mechatronics, Manufacturing, and Mechanical Engineering*, LNME, pp. 247–260, 2021.
https://doi.org/10.1007/978-981-15-7309-5_24

pollution around the world. The idea of recycling is really crucial in diminishing the negative effects of disposed plastics on environment. However, todays manufacturing industries are still preferring to use raw material compared to recycled plastics. This situation happens when the mechanical properties of the plastic deteriorate as it is recycled which is a major drawback that limits its usage in the manufacturing production. Takatori stated in his studies that the impurities from crushing and the reheating process is the main reason for this change in mechanical properties of recycled plastics [2]. Even though it is impossible to produce plastic parts that made of recycled plastic that have similar mechanical properties as virgin form, but it is possible to find the optimum point of mechanical properties of these recycled material by controlling all the significant processing parameters during the manufacturing process.

Singh, Pradhan, and Verma found that choice of material, process parameter and mould design are the factors which affect the quality of the product [3]. So, it is crucial in the plastic production to control all the process parameter effectively by utilising appropriate optimisation method. The issues of optimizing the processing parameters have become significant among the manufacturers and researchers, particularly when it involves the recycled material. But the question is how the processing parameters can be optimized and which appropriate methods can be adopted as well. In the actual operations, normally the processing parameters are often selected from references or handbooks, and then adjusted subsequently by a trial and error approach. This is a common practice that has been adopted in industry for economic reasons; mainly the results are quite satisfactory. Even though this approach has it merits, it consumes and wastes lots of materials.

Therefore, as an initial effort towards filling this void and address the stated problems, in the present work, the Taguchi method and desirability function will be integrated in conducting all the experiments to determine the best range of designs for quality and performances. The emphasis will be given on the impact of processing parameters on the mechanical properties of the part produced particularly when it involves recycled plastics. The feasibility of the substitution of recycled plastic for virgin plastic based on mechanical properties point of view will also be investigated.

2 Methodology

This research is divided into three main phase, process parameter and quality characteristic selection, process parameter optimisation and establishing quantitative relationship and feasibility study. Figure 1 below shows the proposed framework for process parameter optimisation. A proper research studies have been done in order to identify the important process parameter. The research on process parameter optimisation in injection moulding started way back in history. Some studied on the optimisation of warpage, shrinkage, sink mark, short-shorts and so on. The others studied on optimisation for different type of materials like polypropylene, HDPE, LDPE, polycarbonate and so on. Review on all this research are done and the important parameter which give a significance effects on the selected quality have been selected for this study based on the previous researches. Table 1 below shows the process parameter that have been selected by previous researchers to study on the parameter optimisation.

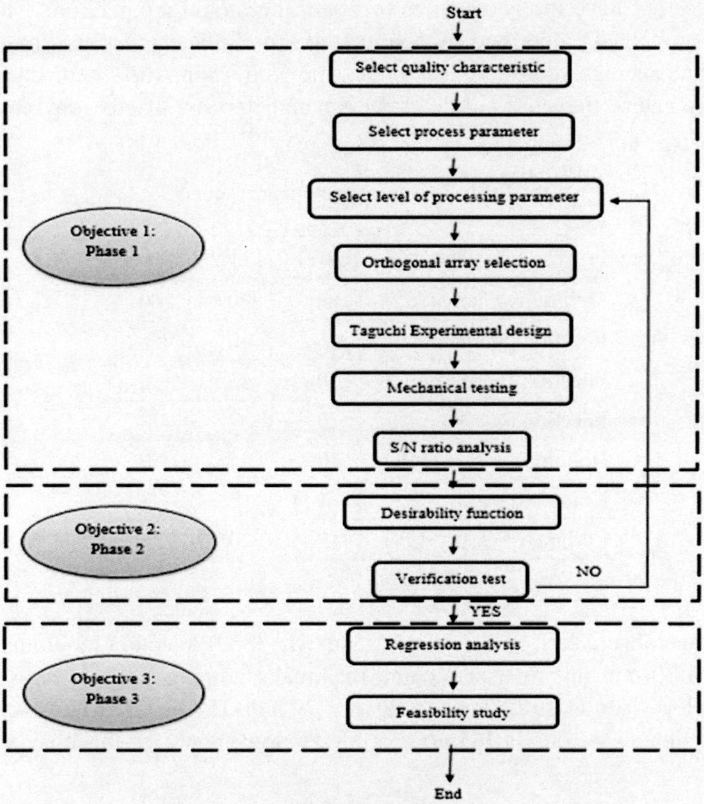

Fig. 1. Flow chart

Table 1. Selected process parameter

Factors	Process parameter
A	Melt temperature
B	Injection pressure
C	Injection speed
D	Injection time
E	Holding pressure
F	Holding time
G	Cooling time
H	Mould temperature

Even though mould temperature appears to have the impact on the parameter in the previous researches, this parameter is not chosen as the processing parameter in this research. This is due to the injection mouldings machines that will be used to produce the part does not have the accessories to verify the mould temperature. Three values for each parameter is identified by conducting a preliminary test and the maximum, minimum and average values are listed out. After confirming the significance of all the process parameters, the values of the process parameters are listed as in Table 2.

Table 2. Process parameter value

Process parameter	Level 1	Level 2	Level 3
Melt temperature (°C)	180	220	260
Injection pressure (MPa)	45	50	55
Injection speed (mm/s)	20	25	30
Injection time (s)	6	7	8
Holding pressure (MPa)	20	35	50
Holding time (s)	1	2	3
Cooling time (s)	15	20	25

Experiments were carried out on a Haitian (MA3800) injection moulding machine, which has a 70 mm diameter screw and maximal clamping force of 380T. Machine dimension (L × W × H) is 7.62 m × 1.96 m × 2.15 m. The injection capacity is 1127 g and its injection speed rate is 385 g/s. Figure 2 below shows the machine used for the testing.

Fig. 2. Injection moulding machine

According to the survey conducted in the plastic manufacturing companies in Malaysia, Wahab, Abidin, and Azhari reported that the consumption rate for PP and PE is the highest among the resin types [4]. The properties of the recycled polypropylene material are summarised in Table 3.

A test mould with one cavity insert for the injection moulding of Plastic pallets with square dimensions of 1700 mm × 1700 mm (length × width) and thickness of 200 mm used in this testing. The diameter of the hot runner had a typical value of 7 mm. The

Table 3. Material properties

Property	Specification
Moisture	18%
Melt flow index	7.44 g/10 min
Density	0.981 kg/m^3
Izod impact strength	57.8938 j/m
Flexural modulus	9475.8 kgf/cm^2
Tensile strength	187.615 kgf/cm^2

mould temperature is maintained at 31°C. The cavity and core side of mould is shown in Fig. 3.

Fig. 3. Mould used for testing

Tensile tests were performed to determine the tensile stress at break, the tensile strain at break and the Young's modulus of the material. The tests were carried out according to ASTM D638 using a Type 1 tensile bar on a tensile test machine with a 5 kN load cell. The flexural properties tested were flexural modulus and flexural modulus, assessed in accordance with ASTM D790 [7, 8].

Bigger the better formulae from Taguchi method were used for both responses to calculate the signal to noise ratio. The equation is shown below [5]:

$$\text{Smaller: MSD} = \frac{1}{n}\sum\nolimits_{i=1}^{n} Y_i^2 \tag{1}$$

$$\text{Nominal: MSD} = \frac{1}{n}\sum\nolimits_{i=1}^{n} \left(Y_i - \bar{Y}\right)^2 \tag{2}$$

$$\text{Bigger: MSD} = \frac{1}{n}\sum\nolimits_{i=1}^{n} \frac{1}{Y_i^2} \tag{3}$$

Where n is the number of trials, y, is the value obtained during each trial and y is the average value. MSD values are then used to calculate, what is called, a signal-to-noise (S/N) ratio. Formulae for S/N ratio are given in Eqs. (4), (5), and (6).

$$\text{Smaller: } \frac{S}{N} = -10\log_{10}(MSD) \tag{4}$$

$$\text{Nominal: } \frac{S}{N} = -10\log_{10}\left[\frac{Y^{-2}}{\sigma^2}\right] \quad (5)$$

$$\text{Bigger: } \frac{S}{N} = -10\log_{10}(MSD) \quad (6)$$

Where a is the standard deviation in the results. MSD and S/N ratios are yardsticks for the analysis of the variability in experimental results obtained under identical processing conditions. A smaller value of MDS and a bigger value of S/N is desirable and represent minimum variability in the measurements [6]. To solve the multi-response optimisation problems, a technique for integrating multiple responses into a dimensionless function, called the overall desirability function (D), is applied. The approach is to convert each response (Y_i) into a dimensionless function, known as the individual desirability function (D_i), that can be between zero and one. If the response Y_i is at its target the most desirable case is obtained ($D_i = 1$), otherwise, $D_i = 0$ (the least desirable case). The desirability function approach assumes that there is a positive number, w, known as the weight factor. To simplify the investigation, the weights for the response variables are considered equal to one. In this study, the individual desirability functions are calculated based on the type of the optimisation functions, i.e. maximization or minimization using Eq. (7)–(9). If the target for the response *yi* is a maximum value, the desirability is based on Eq. (7). If the target is a minimization one, the desirability is based on Eq. (8). Furthermore, if the target is located between the lower and upper limits, the desirability is obtained based on Eq. (9).

$$D_i = \left\{\begin{array}{cc} 0 & Y_i < L_i \\ \frac{Y_i - L_i}{T_i - L_i} & L_i \le Y_i \le T_i \\ 1 & Y_i > T_i \end{array}\right\} \quad (7)$$

$$D_i = \left\{\begin{array}{cc} 1 & Y_i < L_i \\ \frac{T_i - Y_i}{T_i - L_i} & L_i \le Y_i \le T_i \\ 1 & Y_i > T_i \end{array}\right\} \quad (8)$$

$$D_i = \left\{\begin{array}{cc} 1 & \\ \frac{T_i - Y_i}{T_j - L_j} & Y_i < L_i \\ \frac{U_i - Y_i}{U_i - L_i} & L_i \le Y_i \le T_i \\ & Y_i > T_i \\ 0 & \end{array}\right\} \quad (9)$$

Next, the individual desirability functions are integrated as overall (composite or aggregated) desirability (D), which can be between 0 and 1. It is defined as the weighted geometric mean of all the previously defined desirability functions, calculated by Eq. (10), where *wi* is a comparative scale for weighing each of the resulting *di* assigned to the *i*th response, and *n* is the number of responses. The optimal values of the parameters are determined to maximise overall desirability (D), by applying a reduced gradient algorithm with multiple starting points.

$$\mathrm{D} = \left(D_1^{W_1} \times D_2^{W_2} \times D_3^{W_3} \times \ldots \times D_n^{W_n}\right)^{\frac{1}{(w_1+w_2+w_3+\ldots+w_n)}} = \prod_{i=1}^{n} D_i^{w_i \frac{1}{\sum_{i=1}^{n} w^i}} \quad (10)$$

3 Results and Discussion

Process parameter for each level where taken from the master parameter which is already set by machine foreman. From the master parameter the lowest and highest value is determine by running a preliminary testing. Based on the identified process parameter value, a Taguchi experimental table is designed. The Table 4 below shows the Taguchi experimental design.

Table 4. Taguchi experimental design

No.	Melting temperature (°C)	Injection pressure (MPa)	Injection speed (mm/s)	Injection time (s)	Holding pressure (MPa)	Holding time (s)	Cooling time (s)
1	180	45	20	6	20	1	15
2	180	45	20	6	35	2	20
3	180	45	20	6	50	3	25
4	180	50	25	7	20	1	15
5	180	50	25	7	35	2	20
6	180	50	25	7	50	3	25
7	180	55	30	8	20	1	15
8	180	55	30	8	35	2	20
9	180	55	30	8	50	3	25
10	220	45	25	8	20	2	25
11	220	45	25	8	35	3	15
12	220	45	25	8	50	1	20
13	220	50	30	6	20	2	25
14	220	50	30	6	35	3	15
15	220	50	30	6	50	1	20
16	220	55	20	7	20	2	25
17	220	55	20	7	35	3	15
18	220	55	20	7	50	1	20
19	260	45	30	7	20	3	20
20	260	45	30	7	35	1	25
21	260	45	30	7	50	2	15
22	260	50	20	8	20	3	20
23	260	50	20	8	35	1	25
24	260	50	20	8	50	2	15
25	260	55	25	6	20	3	20
26	260	55	25	6	35	1	25
27	260	55	25	6	50	2	15

The test results were evaluated in terms of signal/noise (S/N) ratio. The S/N was calculated by larger is better for compressive strength and smaller is better for part weight. This is to determine the effect of injection parameters on selected quality characteristics. The formula of S/N ratio was shown in equation. The calculated signal to noise ratio for both quality responses is listed as in the Table 5 below.

Table 5. S/N ratio

Experiment	Tensile strength (kgf/cm^2)	Flexural modulus (kgf/cm^2)	Tensile S/N ratio	Flexural S/N ratio
1	194.05	9573.67	45.76	79.62
2	191.10	9870.70	45.63	79.89
3	195.25	9716.47	45.81	79.75
4	193.17	9646.50	45.72	79.69
5	197.26	9875.88	45.90	79.89
6	196.45	9547.03	45.86	79.60
7	196.21	9716.55	45.85	79.75
8	196.43	10090.28	45.86	80.08
9	194.80	10083.18	45.79	80.07
10	188.57	9994.51	45.51	80.00
11	189.92	9587.74	45.57	79.63
12	189.44	9238.54	45.55	79.31
13	186.01	9714.82	45.39	79.75
14	188.41	9636.82	45.50	79.68
15	182.35	9842.94	45.22	79.86
16	190.97	9685.84	45.62	79.72
17	189.75	9782.20	45.56	79.81
18	188.48	9663.60	45.51	79.70
19	171.35	9313.63	44.68	79.38
20	165.01	9273.70	44.35	79.35
21	175.72	9206.43	44.90	79.28
22	172.52	9719.12	44.74	79.75
23	168.48	9194.24	44.53	79.27
24	170.62	9556.44	44.64	79.61
25	167.08	9295.73	44.46	79.37
26	155.33	9166.37	43.82	79.24
27	165.75	9525.45	44.39	79.58

From the signal to noise analysis the most significance parameter that effect the tensile strength and flexural modulus is identified. The delta rank in the response table shows ranks the process parameter from the most influential to the least influential on the selected quality characteristic. For tensile strength, larger the better signal is selected. The most important parameter for maximum tensile strength is Melt temperature followed by injection time then holding time. Injection pressure shows least effects on tensile strength. The main effect plot will show the most effective process value for each selected process parameter. The Fig. 4 shows the most suitable value of each process parameter for tensile strength. The optimal injection moulding conditions for the maximum tensile strength were 180 °C barrel temperature, 8 s injection time, 3 s holding time, 15 s cooling time, 20 mm/s injection speed, 20 MPa holding pressure and 45 MPa injection pressure.

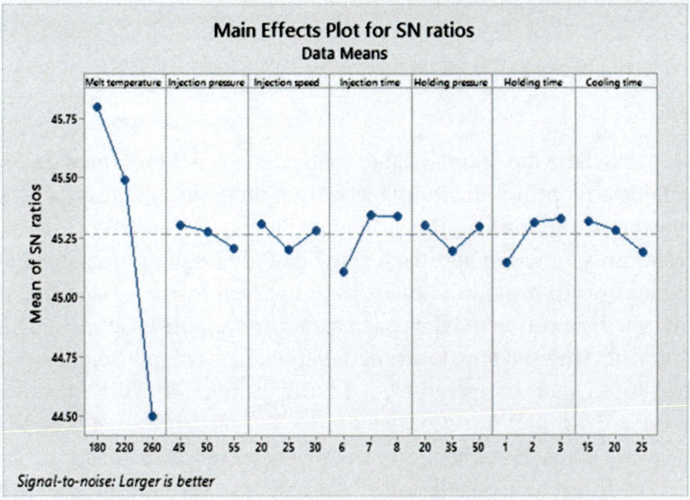

Fig. 4. Main effects plot for SN ratio for tensile strength

However, for flexural modulus, the most effective process parameter was Melt temperature followed by holding time then injection pressure. The least effective process parameter on flexural modulus is holding pressure. The optimal injection moulding conditions for the maximum flexural modulus were 180 °C melt temperature, 2 s holding time, 55 MPa injection pressure, 8 s injection time, 30 mm/s injection speed, 20 s cooling time and 20 MPa holding pressure. The effective process parameter value can be seen in the Fig. 5 below.

Both tensile strength and flexural modulus is converted into dimensionless function using the desirability method. This is to integrate both responses into a dimensionless function called composite desirability. From this composite desirability the most desirable value is considered as the optimal value. The Table 6 below shows the individual desirability for each response and also the composite desirability.

After investigating each response variable as an objective function individually, all response variables are optimised using the desirability function approach, while two response variables are considered as objective functions simultaneously. The results are

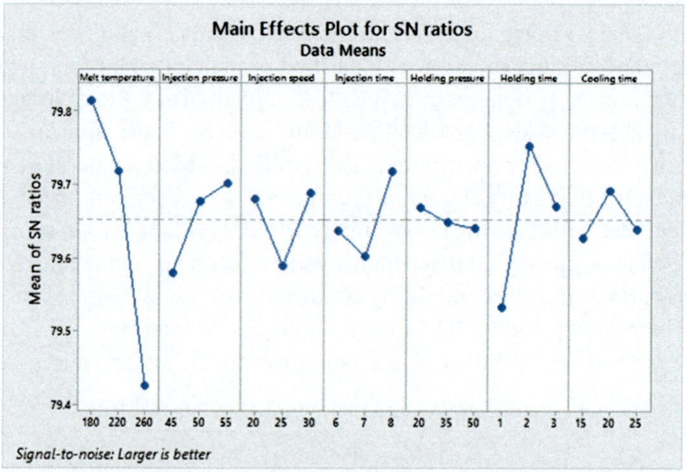

Fig. 5. Main effects plot for SN ratio for flexural modulus

presented in Fig. 6, where the optimal values of the factors are presented. In Fig. 6, all the response variables are optimised simultaneously. Comparing the results obtained from the Taguchi design of experiment, the individual desirability function approach and the composite desirability function approach show that although optimising each response variable individually will provide a better result for each response variable, the optimal parameter values will be different when each response variable is optimised individually. For example, considering the tensile strength variable as a response, the optimal values of factors are 180 °C melt temperature, 3 s holding time, 45 MPa injection pressure, 8 s injection time, 30 mm/s injection speed, 15 s cooling time and 20 Mpa holding pressure. However, when flexural modulus is considered as a response variable, the optimal values are different. Considering all the response variable as objective functions simultaneously in the composite desirability function method, generates one general value for all the parameters of the algorithms, which leads to an optimal value of all the response variables.

The confirmation injection test was set up with the optimal combination using the same material and injection machine. A specimen was moulded and mechanical test was performed in the same way as principal experiment. The average tensile strength and flexural modulus was calculated. The value of average tensile strength and flexural modulus obtained from the confirmation experiment was then compared with the estimated value as shown in Table 7.

All experimental values are within a 5% difference from predicted results. The predicted results had very close values with the experimental results, thus the optimisation approach used in this study is effective and practical.

A comparison between the virgin polypropylene mechanical properties and recycled polypropylene is compared. As can be seen from Fig. 7 and 8, both responses have a slight improvement in strength with optimised process parameter compared to master parameter. Even though it is impossible to produce plastic parts that made of recycled plastic that have similar mechanical properties as virgin form, but it is possible to find

Table 6. Individual desirability and composite desirability

Experiment	Tensile strength (kgf/cm^2)	Flexural modulus (kgf/cm^2)	Tensile individual desirability	Flexural individual desirability	Tensile and flexural composite
1	194.05	9573.67	0.96	0.51	0.70
2	191.10	9870.70	0.97	0.58	0.75
3	195.25	9716.47	0.98	0.66	0.80
4	193.17	9646.50	0.99	0.64	0.80
5	197.26	9875.88	1.00	0.71	0.84
6	196.45	9547.03	1.00	0.79	0.89
7	196.21	9716.55	1.00	0.77	0.88
8	196.43	10090.28	1.00	0.85	0.92
9	194.80	10083.18	1.00	0.92	0.96
10	188.57	9994.51	0.72	0.48	0.59
11	189.92	9587.74	0.83	0.53	0.66
12	189.44	9238.54	0.71	0.36	0.50
13	186.01	9714.82	0.59	0.46	0.52
14	188.41	9636.82	0.69	0.51	0.59
15	182.35	9842.94	0.57	0.34	0.44
16	190.97	9685.84	0.63	0.57	0.60
17	189.75	9782.20	0.74	0.62	0.68
18	188.48	9663.60	0.62	0.45	0.53
19	171.35	9313.63	0.41	0.28	0.34
20	165.01	9273.70	0.30	0.11	0.18
21	175.72	9206.43	0.40	0.16	0.25
22	172.52	9719.12	0.46	0.39	0.43
23	168.48	9194.24	0.34	0.22	0.27
24	170.62	9556.44	0.45	0.27	0.35
25	167.08	9295.73	0.32	0.37	0.35
26	155.33	9166.37	0.21	0.20	0.20
27	165.75	9525.45	0.31	0.25	0.28

the optimum point of mechanical properties of these recycled material by controlling all the significant processing parameters during the manufacturing process. Graph 4.1 and 4.2 below shows the comparison between after and before optimisation and virgin material.

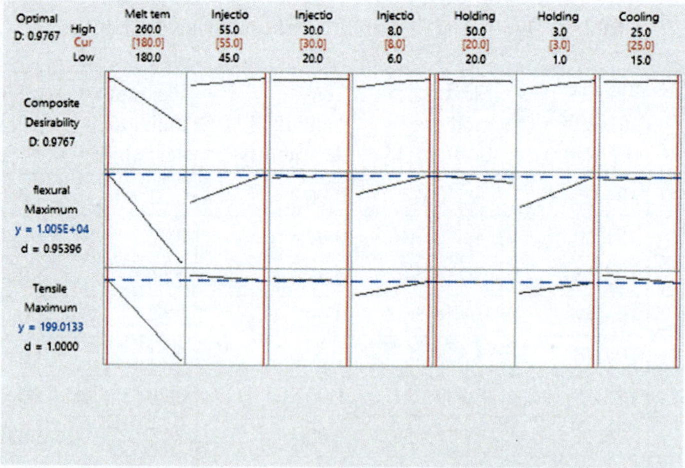

Fig. 6. Optimal process parameter

Table 7. Individual desirability and composite desirability

Responses	Optimisation prediction	Actual result	Residual error
Tensile strength	199 kgf/cm^2	190 kgf/cm^2	4.5%
Flexural modulus	10005 kgf/cm^2	9875 kgf/cm^2	1.3%

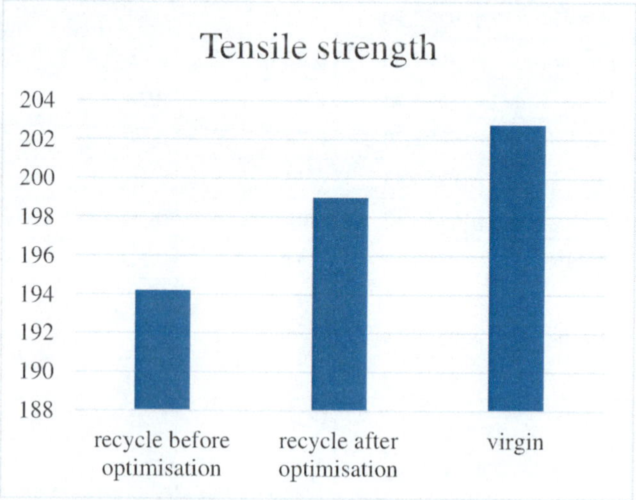

Fig. 7. Comparison of tensile strength

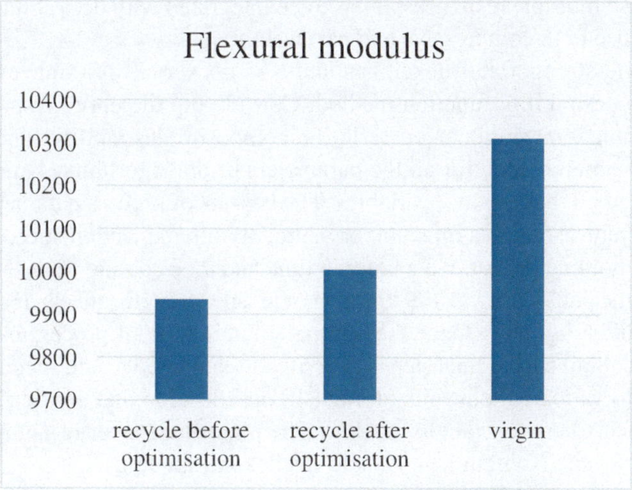

Fig. 8. Comparison of flexural modulus

4 Conclusion

The optimal injection moulding process parameter of recycled polypropylene is identified in this study. First of all, the important quality characteristic for this study is selected, which is tensile strength and flexural modulus. The respective process parameter which directly influences the selected quality characteristic is identified. Several previous researches have been studied to choose these parameters. Seven process parameters have been selected for this study which is Melt temperature, injection pressure, injection speed, injection time, holding pressure, holding time and cooling time. After that, the value of each process parameter is determined by running preliminary testing. The selected process parameters are used to conduct experiment based on Taguchi experimental design. The tensile strength and flexural modulus for each experiment were identified. Those results are used to optimise the process parameter using Taguchi and desirability functions.

For Taguchi optimisation, S/N ratio is calculated for both responses and the optimal process parameter are identified. From the signal to noise analysis the most significance parameter that affects the tensile strength and flexural modulus is identified. The most important parameter for maximum tensile strength is Melt temperature followed by injection time then holding time. Injection pressure shows least effects on tensile strength. The optimal injection moulding conditions for the maximum tensile strength were 180 °C barrel temperature, 8 s injection time, 3 s holding time, 15 s cooling time, 20 mm/s injection speed, 20 MPa holding pressure and 45 MPa injection pressure.

However, for flexural modulus, the most effective process parameter were Melt temperature followed by holding time then injection pressure. The least effective process parameter on flexural modulus is holding pressure. The optimal injection moulding conditions for the maximum flexural modulus were 180 °C melt temperature, 2 s holding time, 55 MPa injection pressure, 8 s injection time, 30 mm/s injection speed, 20 s cooling time and 20 MPa holding pressure. However, Taguchi and signal to noise ratio alone cannot optimise multiple responses. It has to be integrated with desirability functions in order to optimise both compressive and part weight quality.

Both tensile strength and flexural modulus is converted into dimensionless function using the desirability function method. Considering the entire response variables as objective functions simultaneously in the composite desirability function method, generates one general value for all the parameters of the algorithms, which leads to an optimal value of all the response variables. The best set of process parameter are 180 °C of melt temperature, 55 MPa injection pressure, 30 mm/s injection speed, 8 s injection time, 20 MPa holding pressure, 3 s holding time and 25 s cooling time. This parameter will optimise the part quality to 199 kgf/cm^2 of tensile strength and can result in flexural modulus of 10005 kgf/cm^2. Once the optimal combination of process parameters and their level was obtained, the final step is to verify the estimated result. A confirmation test is performed to validate the results of Taguchi optimisation and provide evidence that interaction effects between factors are low. The percentage error of optimised process parameter for tensile strength is 4.5% and for flexural modulus is 1.3%. The response variable of the optimised process parameter had very close values with the experimental results, thus the optimisation approach used in this study is effective and practical.

Acknowledgement. The authors would like to give special thanks to Research & Innovation Department, Universiti Malaysia Pahang, Malaysia for funding this research project (RDU1901158).

References

1. Heidbreder, L.M., Bablok, I., Drews, S., Menzel, C.: Tackling the plastic problem: a review on perceptions, behaviors, and interventions. Sci. Total Envir. **668**, 1077–1093 (2019)
2. Takatori, E.: Material recycling of polymer materials & material properties of the recycled materials. Int. Polym. Sci. Technol. **42**(7), 9–14 (2015)
3. Roy, R.K.: Design of Experiments Using the Taguchi Approach: 16 Steps to Product and Process Improvement. Wiley, Hoboken (2001)
4. Wahab, D.A., Abidin, A., Azhari, C.H.: Recycling trends in the plastics manufacturing and recycling companies in Malaysia. J. Appl. Sci. **7**(7), 1030–1035 (2007)
5. Freddi, A., Salmon, M.: Introduction to the Taguchi method. Design Principles and Methodologies. STME, pp. 159–180. Springer, Cham (2019). https://doi.org/10.1007/978-3-319-95342-7_7
6. Singh, G., Pradhan, M., Verma, A.: Multi Response optimization of injection moulding Process parameters to reduce cycle time and warpage. Mater. Today Proc. **5**(2), 8398–8405 (2018)
7. ASTM C 1958: ASTM Standards (1958)
8. de Sousa Mol, A., Oréfice, R.L.: Preparation of chitin nanofibers (whiskers) and their application as property-recovery agents in re-processed polypropylene. Polym. Bull. **73**(3), 661–675 (2015). https://doi.org/10.1007/s00289-015-1512-3

Sustainable Finished Product Optimization on Quality Response and Attitudinal Parameters

Nur Qurratul Ain Adanan, Faiz Mohd Turan(✉), and Kartina Johan

Faculty of Manufacturing Engineering, Universiti Malaysia Pahang, 26600 Pekan, Pahang, Malaysia
nurqurratulaiyen.eric@gmail.com, faizmt@ump.edu.my

Abstract. In this demanding polymer production industry such as injection molding process where the quality of the product is the most critical concerns, it is necessary to have a systematic production planning and routine analysis. But, contemporary production is no longer possible without an efficient and permanent search for better quality especially on sustainability optimization of the system. So, there is a need for consideration on sustainability capable improving the variables that can affect quality and are a result of the human factors' action must be optimised within the plastic injection moulding (PIM) process. Consequently, this paper focused on the optimisation for quality response and attitudinal attributes that influenced sustainable PIM for both technical and management in order to meet optimum sustainable product quality.

1 Introduction

Over time, plastic injection molding has become one of the main methods for producing plastic parts. Product quality is one of the most important things in plastic injection molding (PIM) and plastic compressive strength is always used as an important criterion to evaluate plastic product quality. Moreover, product quality mainly depends on the choice of materials, mold design, and process variables. This research is aim at finding the optimal set of parameters to maximize plastic compressive strength. The variables are Melt temperature, Injection speed and Holding pressure. The research goal is to maximize plastic compressive strength using design of experiment (DOE), desirability functions and regression analysis to find the optimal parameter settings. The biggest advantage of plastic injection molding is that the manufacturing process could be suitable for any plastic products with complicated shapes. People make all kinds of products with different geometry shape by designing different mold. Moreover, PIM has higher efficiency than other type of molding once people have a designed mold. And PIM can also manufacture more than one product in one mold by co-injection. But the shortage is also clear, one mold can just fit one product and some complicated mold may cost a lot.

2 Research Background

PIM product quality depends on the choice of materials, mold design, and process parameters. Different materials have different property, so choosing the suitable kind

M. A. Zakaria et al. (Eds.): *Advances in Mechatronics, Manufacturing, and Mechanical Engineering*, LNME, pp. 261–268, 2021.
https://doi.org/10.1007/978-981-15-7309-5_25

of materials will do well to product quality. But people should do a lot experiment on different type of materials to pick a most suitable one, moreover, the material choose is the most basic step to make sure the product quality, people still cannot make the best product quality only by material choose. Plastic product is shaped by mold, so that a well-designed mold is very important for product quality. Also, some other details in mold like cooling channel and vent will have effect on product quality. But a well-designed mold may cost a lot time and money, even for some complex product. Compare to the previous two means, choosing the optimal set of process parameters to improve the product quality can be an efficient and universally applicable way. So this study aims to improve product quality by finding the optimal set of parameters.

Compressive strength and weight is very important factor to evaluate plastic product quality. Better plastic product usually has higher plastic strength. The plastic compressive strength is highly related to the process operation parameters in PIM Chen, Nguyen, Chiu, Chen, and Tai [1] proposed a method to optimize process parameters using Taguchi method, RSM, and hybrid FA-PSO. Melt temperature, Injection velocity, packing pressure, packing time and cooling time are the process parameters in their study. Their study used mechanical device to measure the warpage and shrinkage of the products. Oliaei [2] optimized warpage and shrinkage in PIM using Taguchi, ANOVA and artificial neural network methods. Chiang and Chang [3] using the response surface methodology (RSM) to analyze the shrinkage and warpage in PIM, the product they focused is thin shell feature. Gao and Wang [4] reduced plastic warpage in PIM by surrogate-based process optimization. In their study, a cellular phone cover is investigated, where mold temperature, melt temperature, injection time, packing time and packing pressure are selected to be the design variables. But if total packing time is too short, the performance of parts may be affected.

Sudsawat and Sriseubsai [5] optimized warpage and shrinkage in PIM by using response surface methodology (RSM) with genetic algorithm and firefly algorithm techniques, their study claims that firefly algorithm created better optimal solution than genetic algorithm. Kitayama, Miyakawa, Takano, and Aiba [6] introduced multi-objective optimization to optimize plastic volume shrinkage and cycle time for PIM by using conformal cooling channel. However, the optimized cooling channel may not suitable for all product with different shape. Xu and Yang [7] used soft computing and grey correlation analysis to optimize the process parameters in PIM, multiple objectives were involved in their optimization. Kitayama and Natsume, [8] also introduced multi-objective optimization to optimize plastic volume shrinkage and clamping force for PIM via sequential approximate optimization. In multi-objective optimization, the optimized process parameters combination is one optimal solution chose among the pareto-frontier, the selection of optimal solution is based on people's preference.

So, as of the objectives for this research are (i) to analyzed the effect of finished product for molding operation process parameters for quality response. (ii) to develop mathematical model on quality response and attitudinal parameters for sustainable finished product of molding operation process. (iii) to develop multi-objective optimization of integration model using response surface methodology (RSM). The experiment and test run is done at plastic pallets molding company. The polymer material used during the experiments was recycled polypropylene. The properties of this PP material are summarized in table below. The grade of the PP was MH-418 (Petrochemicals Inc., Turkey) with a melt index of 4.5 g/10 min (at 230 °C). A rectangular-shaped specimen was injection molded with a 2400-ton injection molding machine (JU2400, Haitian), which performs the injection process by adjusting the experimental parameters via a control program. Materials were dried for 4 h at 60 °C prior to injection.

3 Results and Discussion

This research consists of three main stages of process parameter optimization. Based on the Fig. 1, the stages are Phase 1, Phase 2, and Phase 3. Before optimizing the process parameter, the variables that need to be considered must be known in order to have a successful parameter. The first phase is about determining all the component values required in design regarding the first objectives. The 2nd phase is to develop the mathematical model of the identified parameters on Phase 1 and the last phase is to validate the optimized parameters.

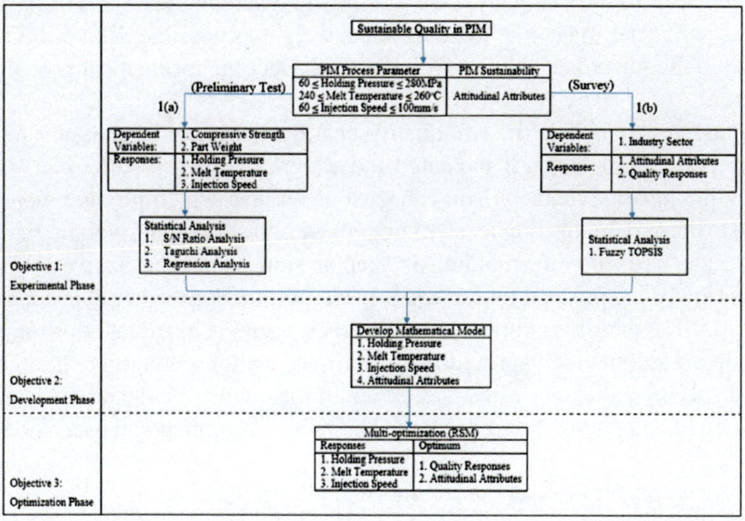

Fig. 1. Research's framework

The first phase is categorized as three stage which are parameter selection, data collection and statistical analysis. First, the objective is to investigate the effect of PIM process parameters for quality response and attitudinal attributes for sustainability. So, the first step in phase one is to select the quality of the product which we want to study and optimized, there are so many quality problems in plastics injections moldings and we tend to choose the compressive strength of the product or well known as the ultimate load the product can withstand. Then we proceed to study on the parameters which can affect this quality. Several previous research works had been gone thru to determine the influential parameters. Once selected, some preliminary test had to be done to identify the maximum, minimum and average value of the parameter to construct the Taguchi experiment table.

The concept of product quality is ambiguous and there are several product quality definitions that can be classified into the following three categories: (a) dimensional properties (for e.g., weight, length, and thickness), (b) surface properties represented by the appearance of surface defects (for e.g., sink marks and jetting), and (c) mechanical or optical properties (for e.g., tensile and impact strength). Yang and Gao [9] claimed that the performance of a manufacturing process and its quality control are monitored through product weight because quality is inversely proportional to variability and this is reflected in the product weight variation while product weight is closely related to other quality properties. For example, Chu, Wel and Shih [10] point out a strong linear correlation exists between product length and weight, and Min [11] indicated a strong correlation between product shrinkage and weight. Therefore, product weight is a good indicator of process stability.

Product weight plays an important role in quality characteristic. Every product is set to specific weight before it is decided into production. However, there will be a variation in product weight due to some errors in production parameter selection. Higher product weight will cause loss to the manufacturing company, even extra 100 g per product will cause huge losses to the company. Due to this, the machine operator will try to adjust the process parameter to get the lowest product weight as possible. However, it can't be minimized to the lower because it will indirectly affect the mechanical property of the product.

Compressive strength is the rite quality characteristic can be measure to identify the plastic pallets capability. It indicated the weight or force a pallet can withstand. Measuring this quality characteristic can give us an idea which product weight is the optimized to the maximum compressive strength with lower product weight. Each plastic pallet is weighted rite after the molding and kept in store for more than five hours for it to cool down slowly and proceed for next quality testing. Compression test is done for each molded parts. The part is compressed until a crack sound is heard and the compression force is stopped automatically providing the ultimate compression force. It indicates the forces that specific pallet can withstand. After all the quality characteristic is measured like part weight and compressive strength. The data is compiled and used for S/N ratio calculation.

There are a number of machine settings that allows the control of all steps of slurry or melt preparation, injection into a mold cavity and subsequent solidification. Proper selections of all the process parameter put direct impact on the quality and productivity of the plastic product so by considering all these factors some important process parameters like melt temperature, holding pressure and injection speed are selected and for conducting the experiments some set of definite values of all the process parameters are taken in Table 1. Three values for each parameter is identified by con-ducting a preliminary test and the maximum, minimum and average values are listed out. After confirming the significance of all the process parameters, the values of the process parameters are listed as a table below.

Then, two analysis had been done, S/N analysis and Regression analysis. From the first analysis, the most influential process parameter is identified and the value of the

Table 1. Preliminary test

Process parameter	Level 1	Level 2	Level 3
Melt temperature	240	250	260
Injection speed	60	80	100
Holding pressure	60	70	280

factors which gives the best compressive strength and part weight is also identified. The regression analysis is done and the quantitative relationship between the process and quality is identified. Both of the analysis is done using the help of Minitab 18 software. The selected process parameter and their calculated parameter values are tabulated using L9 orthogonal array design system. There are 9 set of parameters which is used to run the experiment. Each experiment has 3 molded products which is used for compression testing which shown in Table 2. The compression testing is done for each set of parameters after 5 h. This 5 h waiting is to the plastic pallet to cool down and undergo the shrinkage.

Table 2. Taguchi L9 orthogonal design

No	Melting temperature	Injection speed	Holding pressure
1	240	60	60
2	240	80	70
3	240	100	80
4	250	60	70
5	250	80	80
6	250	100	60
7	260	60	80
8	260	80	60
9	260	100	70

As the signal to ratio for each test is calculated, the better chosen criteria are the larger compressive strength but smaller part weight value. The calculated signal to noise ratio for both quality responses. From the signal to noise analysis the most significance parameter that effect the compressive strength is identified. From the Table 3, it can be seen that, the melt temperature is the most important parameter which effect the responses higher than the other two parameters. Then holding pressure and injection speed effect least on the responses. From the s/n ratio analysis it also can be seen which parameter are the best and optimized. The graph show that the melt temperature of 260 °C, injection speed of 80 mm/s and holding pressure of 11 MPa is the best set of parameter which optimized the response characteristic.

Table 3. S/N ratio analysis for

Level	Melt temperature	Injection speed	Holding pressure
1	74.04	74.10	74.83
2	73.22	74.46	74.21
3	75.35	74.05	73.57
Delta	2.12	0.41	1.27
Rank	1	3	2

From the signal to noise analysis the most significance parameter that effect the part weight is identified. From the Table 4, it can be seen that, the melt temperature is the most important parameter which effect the responses higher than the other two parameters. Then holding pressure and injection speed effect least on the responses. From the s/n ratio analysis it also can be seen which parameter are the best and optimized. The graph show that the melt temperature of 260 °C, injection speed of 80 mm/s and holding pressure of 22 MPa is the best set of parameter which optimized the response characteristic.

Table 4. S/N ratio analysis for

Level	Melt temperature	Injection speed	Holding pressure
1	−21.33	−21.36	−21.39
2	−21.36	−21.39	−21.39
3	−21.44	−21.38	−21.36
Delta	0.11	0.03	0.03
Rank	1	3	2

As in a case study on the key process parameter of injection molding which directly influences on mechanical property of product is identified which is melt temperature, injection speed and holding pressure. Taguchi method was used to determine the optimum combinations of the process conditions for the best quality of PP material. The result showed that the holding pressure was found to be the least effective factor. The relationship between process parameters and the ensuing product quality is expressed in a mathematical equation using regression equation with an R2 of 84%. All the factors of process explain 84% of the differences in the quality responses.

4 Conclusion

As in case study result focusing on sustainable finished product, which considering optimization for both factor which are quality and sustainability attitudinal in manufacturing processes, there is possible for modern production to developed an efficient and system. Simultaneously, as highlighted by Judi, Jenal, and Genasan [12] the consideration on sustainability able to improve the critical quality and productivity problem in manufacturing process. In order to achieve this, variables that can affect quality and are a result of the human factors' action [12] will be optimized within the plastic injection molding (PIM) process. So, the survival step need to be taken in order to provide sustainable high quality product where the company have to ensure they included the human factors "management" and "employees" as mentioned by Kanwarpreet and Ahuja [13] and optimize the process parameter statically. Accordingly, the result still uncompleted to shows optimization part where both of the quality responses act independently and not correlated to each other. So, before the multi response optimization part is done in future, both quality responses and sustainable attributes of Fuzzy TOPSIS will be integrated by formulated mathematical model. And lastly will be optimized using Response Surface Methodology (RSM) in order to meet optimum sustainable product quality.

Acknowledgement. The authors would like to give special thanks to Research & Innovation Department, Universiti Malaysia Pahang and Ministry of Higher Education Malaysia (Fundamental Research Grant Scheme, FGRS - RDU150120) for funding this research project.

References

1. Chen, W.-C., Nguyen, M.-H., Chiu, W.-H., Chen, T.-N., Tai, P.-H.: Optimization of the plastic injection molding process using the Taguchi method, RSM, and hybrid GA-PSO. Int. J. Adv. Manufact. Technol. **83**(9-12), 1873–1886 (2015). https://doi.org/10.1007/s00170-015-7683-0
2. Shai, O.: Transforming engineering problems through graph representations. Adv. Eng. Inform. **17**(2), 77–93 (2003)
3. Oliaei, E., et al.: Warpage and shrinkage optimization of injection-molded plastic spoon parts for biodegradable polymers using Taguchi, ANOVA and artificial neural network methods. J. Mater. Sci. Technol. **32**(8), 710–720 (2016)
4. Gao, Y., Wang, X.: Surrogate-based process optimization for reducing warpage in injection molding. J. Mater. Process. Technol. **209**(3), 1302–1309 (2009)
5. Sudsawat, S., Sriseubsai, W.: Optimized plastic injection molding process and minimized the warpage and volume shrinkage by response surface methodology with genetic algorithm and firefly algorithm techniques. Indian J. Eng. Mater. Sci. **24**(3), 228–238 (2017)
6. Kitayama, S., Miyakawa, H., Takano, M., Aiba, S.: Multi-objective optimization of injection molding process parameters for short cycle time and warpage reduction using conformal cooling channel. Int. J. Adv. Manuf. Technol. **88**(5-8), 1735–1744 (2016). https://doi.org/10.1007/s00170-016-8904-x
7. Xu, G., Yang, Z.: Multiobjective optimization of process parameters for plastic injection molding via soft computing and grey correlation analysis. Int. J. Adv. Manuf. Technol. **78**(1-4), 525–536 (2014). https://doi.org/10.1007/s00170-014-6643-4
8. Kitayama, S., Natsume, S.: Multi-objective optimization of volume shrinkage and clamping force for plastic injection molding via sequential approximate optimization. Simul. Model. Pract. Theor. **48**, 35–44 (2014)
9. Yang, Y., Gao, F.: Injection molding product weight: online prediction and control based on a nonlinear principal component regression model. Polym. Eng. Sci. **46**(4), 540–548 (2006)

10. Chu, C.-P., Wei, J.-H., Shih, M.-C.: Adaptive model following control of the mold filling process in an injection molding machine. Polym. Eng. Sci. **31**(15), 1123–1129 (1991)
11. Min, B.H.: A study on quality monitoring of injection-molded parts. J. Mater. Process. Technol. **136**(1–3), 1–6 (2003)
12. Judi, H.M., Jenal, R., Genasan, D.: Quality control implementation in manufacturing companies: motivating factors and challenges. Appl. Exp. Qual. Control, 495–508 (2011)
13. Singh, K., Ahuja, I.S.: An evaluation of transfusion of TQM-TPM implementation initiative in an Indian manufacturing industry. J. Qual. Maint. Eng. **21**(2), 134–153 (2015)

An Information Gain and Hierarchical Agglomerative Clustering Analysis in Identifying Key Performance Parameters in Elite Beach Soccer

Rabiu Muazu Musa[1(✉)], Anwar P. P. Abdul Majeed[2], Azlina Musa[1], Mohamad Razali Abdullah[3], Norlaila Azura Kosni[4], and Mohd Azraai Mohd Razman[2]

[1] Centre for Fundamental and Liberal Education, Universiti Malaysia Terengganu, 21030 Kuala Nerus, Terengganu, Malaysia
rabiu.muazu@umt.edu.my

[2] Innovative Manufacturing, Mechatronics and Sports Laboratory, Universiti Malaysia Pahang, 26600 Pekan, Pahang Darul Makmur, Malaysia

[3] East Coast Environmental Research Institute, Universiti Sultan Zainal Abidin, 21300 Kuala Nerus, Terengganu, Malaysia

[4] Faculty of Sport Science and Recreation, Universiti Teknologi MARA, Bandar Tun Abdul Razak Jengka, Malaysia

Abstract. The current investigation is aimed at identifying the relevant essential performance parameters (PP) that could discriminate winning and losing performance through the application of multivariate and machine learning analysis. A set of 20 different PP was collected from the Asian beach soccer tournament that constituted information such as, tactical and technical strategies, winning and losing performances. An information gain (IG) analysis is applied to extract the features that could best describe winning and losing performance. Hierarchical agglomerative cluster analysis (HACA) is used to create two different clusters based on the initial 20 PP, and the extracted features from IG whilst a canonical discrimination function analysis was used to ascertain the level of separation ability between the two aforesaid clusters. The IG identified a set of 11 PP that could best describe the winning and losing performances and the HACA formed two distinctive clusters. It is shown that the clusters formed using the 11 PP identified was able to offer an excellent separation of the two teams as opposed to the use of the initial 20 PP. The Canonical Correlation function provided a discrimination power of 0.80 for the 11 PP in comparison to the 0.70 obtained when using the 20 PP. The techniques employed in the present study serve useful in identifying key PP that could best describe winning and losing performances in the elite Asian beach soccer tournament which could assist the coaches in modifying playing strategies to ensure victory.

Keywords: Beach soccer · Information gain · Hierarchical agglomerative clustering · Performance parameters · Discrimination function

M. A. Zakaria et al. (Eds.): *Advances in Mechatronics, Manufacturing, and Mechanical Engineering*, LNME, pp. 269–275, 2021.
https://doi.org/10.1007/978-981-15-7309-5_26

1 Introduction

The simplicity of beach soccer has contributed to making the game more appealing and largely aid its spread all over the world. The nature of the game is attributed to the relatively smaller pitch, with small goals and a sandy floor. Typically, the pitch measures not more than 28 to 30 yards in width, by a minimum of 38 yards in length [1]. The sandy surface of the playing pitch couple with the aesthetics display of skills by the players contributed to the rapid popularity of the game by attracting many spectators. The soft sandy playing pitch that accounts for an erratic bounce of the ball warrants more aerial passes between players as compared to more ground passes normally observed in the game played on a normal grass pitch [2]. It is against this background that the players are required to play with their bare feet for the purpose of ensuring easy movement as well as ball manoeuvring on the sand.

The game is epitomized with high momentum which necessitates the players to possess adequate skills and stamina to run, dribble as well as score on the sand. The constant ballistic movement and manoeuvring of the ball in the air make the game essentially unique that requires the players to adopt more aerial play for scoring and passing [3]. Thus, for a team to deliver a successful performance in this sport, some relevant performance parameters (PP) might be necessary for defining a winning strategy.

Based on the nature of this game, it could easily presume that several skills are mandatory in order for a team to emerge winner in a competition, especially at the elite level. Hence, the current investigation is aimed at identifying the relevant essential performance parameters that could discriminate winning and losing performance through the application of multivariate and machine learning analysis.

2 Methodology

2.1 Participants

The participants of the beach soccer sport in this study consist of all the teams that were involved in the AFC beach soccer tournament 2017. A sum of 12 teams took part in the tournament which comprised of China PR, Bahrain, Afghanistan, Malaysia, Oman, Lebanon, Thailand, Japan, United Arab Emirates (UAE), Iraq, Iran as well as Qatar. The overall performances of the teams involved were notated and analyzed throughout the competition period.

2.2 Performance Parameters (PP)

A total number of 20 technical, as well as tactical performance parameters, were considered in the present investigation. The performance parameters which included shot at front third, chances created, goals scored at the first period, goals scored at the second period, goals scored at the third period, goals scored at extra time, shot at back third, pass at back third, interception, turn over, complete save by the keeper, uncomplete save by the keeper, shot inside a box as well as shot outside the box were adopted from the previous study which reported the parameters to be relevant to the game of beach soccer [1]. A details explanation on the reliability as well as the developmental process of the parameters could be obtained from the following study [4].

2.3 Information Gain Analysis for Features Extraction

Information Gain (IG) is regarded as an entropy-based feature evaluation technique that is largely employed in the domain of machine learning. The IG method is essentially applied to extract information provided by one or more features with respect to a given categorical dependent variable [5]. It is worth noting that IG is applied in the present study to ascertain the features that could be utilized in providing information in a view to measure the importance of such variable for the classification or discrimination tasks. In the present investigation, the IG is used to extract information that reveals the importance of the features, i.e. the performance parameters in describing the winning and losing performances.

2.4 Hierarchical Agglomerative Clustering Analysis (HACA)

Hierarchical agglomerative cluster analysis (HACA) is considered as an exploratory as well as an unsupervised method in which a hierarchy of clusters is developed with regards to a particular observation and consequently a number of similar observations are formed into a distinct observation [6]. It is worth to mention that the learning process in this algorithm is determined by the merges as well as the splits of a dataset which often took place in a greedy manner such that similar observations are segregated and demonstrated in a dendrogram [7]. It should be noted that in HACA, the number of clusters is displayed by the dendrogram based on the proximity ascertained by a given or pre-determined clusters. The Cosine's distance was utilized in this study, and the validation technique of the clustering was carried out by means of class centroids [8]. On the other hand, a canonical discriminant function analysis was employed to determine the discriminating ability of the clusters developed. In the present study, the data analysis is carried out by means of Orange statistical software version 3.2 for Windows.

3 Results and Discussion

Table 1 reveals the features gain information scores based on the performance parameters evaluated in the study. From the table, related information on the ranking of the parameters such as information gain, information ratio as well as chi-square values is shown. It could be observed that a total of 11 performance parameters namely; uncomplete save by the keeper, chances created, goals scored first, second and third period, shot front third, pass front third, pass mid-third, shot inbox fouls committed as well as tackling are ranked as the most contributed performance parameters that could describe winning and losing performance of a team in a beach soccer competition.

Ranking methods through the application of information gain technique has been reportedly successful in practical usage owing to its simplicity and extraction of useful information from a set of features [9]. In this approach, a suitable ranking criterion is applied to grade the features whilst a threshold is used to eliminate the features that fell below the predefined threshold [10]. It is worth noting that the main aim of the ranking of the features is to identify the relevance of the features in relation to the class attributes. It essentially asserts that if a feature is to be relevant it can be independent of

Table 1. Features gain information score on the performance parameters examined.

Performance parameters	Infor. gain	Infor. ratio	χ^2
Incomplete Save by Keeper	**0.544**	0.277	17.754
Chances Created	**0.381**	0.191	9.615
Goal Scored 1st Period	**0.298**	0.152	14.254
Goal Scored 2nd Period	**0.285**	0.146	14.000
Shot Front Third	**0.282**	0.142	13.636
Pass Front Third	**0.224**	0.112	9.941
Goal Scored 3rd Period	**0.197**	0.101	6.667
Pass Mid Third	**0.119**	0.060	5.388
Shot inbox	**0.110**	0.055	2.449
Fouls Committed	**0.080**	0.041	3.947
Tackling	**0.078**	0.039	2.600
Goal Scored Extra Time	0.070	0.078	0.333
Passing Error by Keeper	0.068	0.035	3.082
Complete Save by Keeper	0.065	0.033	0.211
Passing Back Third	0.058	0.029	2.970
Shot Back Third	0.054	0.027	0.014
Shot Out Box	0.054	0.027	0.229
Turnover	0.044	0.022	1.351
Interception	0.024	0.012	0.015
Shot Mid Third	0.013	0.007	0.500

the input data but cannot be independent of the class labels, in other words, the feature that has no influence on the class labels can be removed [5]. The purpose of applying the feature ranking via the information gain technique in this study is to identify one or more features that could better explain or otherwise have an influence of the class attributes, i.e. winning and losing performance.

Figure 1 displays the initial clustering of the data in the study based on all the 20 performance parameters. It could be observed from the figure that the clustering analysis is unable to provide better segregation of the classes investigated i.e. winner or loser. Moreover, a canonical correlation function of 0.70 was found to be associated with the discrimination power of the aforesaid classes, which is rather trivial.

Figure 2 highlights the grouping of the team performances based on the 11 ranked features identified via the IG analysis. It could be seen from the figure that the separation of the classes (winner or loser) is rather precise as compared to that provided in Fig. 1 with all the 20 performance parameters. Similarly, a canonical correlation function of 0.80 is observed, which reflect the ability of the clustering technique to separate well

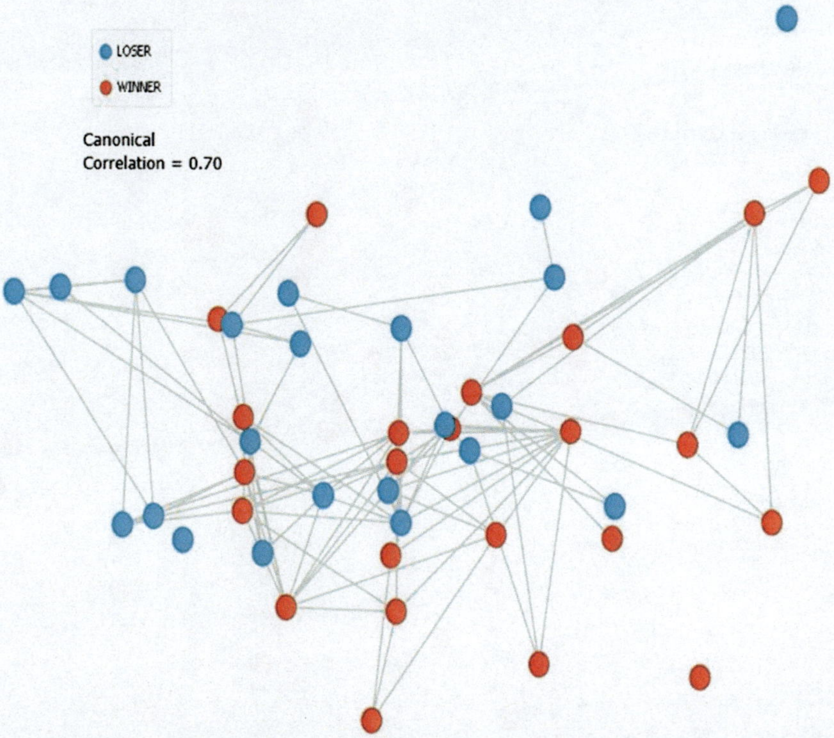

Fig. 1. Clustering based on the 20 sets of performance parameters before feature extractions

the two performance classes of the teams with respect to the 11 performance parameters. The improvement in the ability of the HACA to group the classes effectively could be attributed to the efficacy of the performance parameters identified through the employment of the IG technique in the current investigation.

The performance parameters identified in the current investigation namely; uncomplete save by the keeper, chances created, goals scored in the first, second and third period shot front third, pass front third, pass mid-third, shot inbox fouls committed as well as tackling have proven to be non-trivial in the determination of the match outcome in the elite Asian beach soccer competition. It has been reported previously that to ensure good performance in this game; the players are required to equip themselves with right techniques of passing the ball mainly via mid-air kicks as ground passes on the sand, could accentuate dragging the ball [1, 11]. Moreover, the passing ability is considered as one of the most important skills the players hitherto need to master. Beach soccer players must be very accurate when scoring since the goalposts are a relatively small and also the risk of the ball slowing down on the sandy ground while attempting to score by means of a ground kick often demands special manoeuvring of the ball.

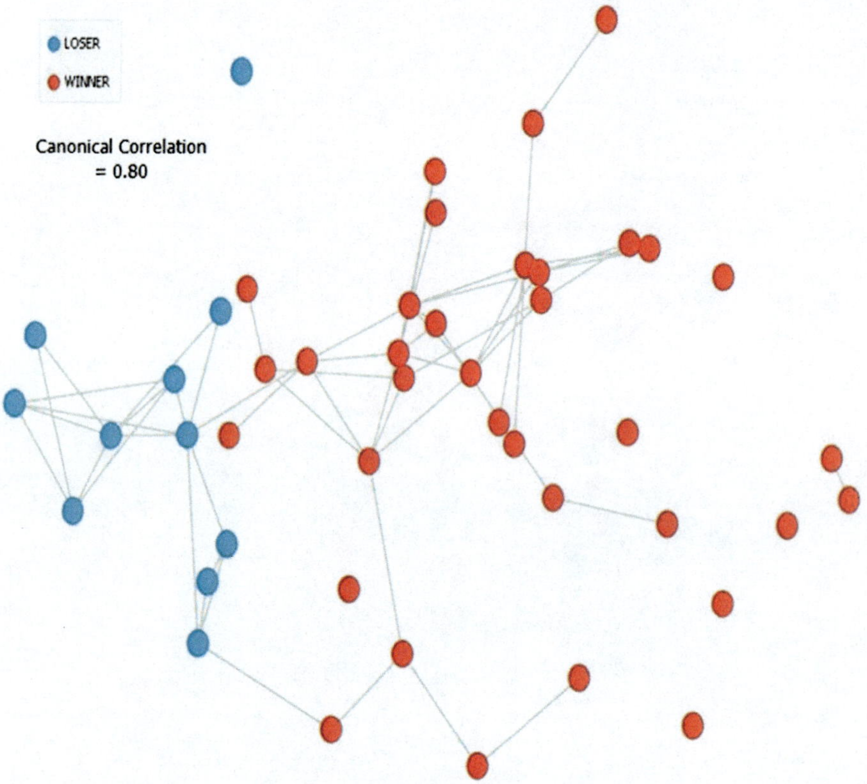

Fig. 2. Clustering based on the 11 sets of performance parameters after features extractions

4 Conclusion

The game of beach soccer is one of the fastest-growing sports as well as a very fascinating game largely because it takes more than just dribbling skills and speed to succeed in the game. The present study has successfully identified the essential performance parameters that could differentiate between winning and losing performance in beach soccer through the employment of machine learning and multivariate analysis. It has been demonstrated from the study that performance parameters that comprise of uncomplete save by the keeper, chances created, goals scored first, second and third-period shot front third, pass front third, pass mid-third, shot inbox fouls committed as well as tackling are non-trivial in determining the outcome of a match in this game. The Information gain technique couple with the multivariate analysis of clustering analysis as well as discriminant function serve useful in providing information on the relevant performance parameters that could enhance team performance in beach soccer. The findings from the present investigation could be essential to the coaches, team managers as well as performance analysts in mapping out technical and tactically playing strategy that might ensure wining in this game

Acknowledgement. The authors would like to acknowledge the coaches and managers for their cooperation towards the accomplishment of this study.

References

1. Muazu Musa, R., Abdul Majeed, A.P.P., Abdullah, M.R., Ab Nasir, A.F., Arif Hassan, M.H., Mohd Razman, M.A.: Technical and tactical performance indicators discriminating winning and losing team in elite Asian beach soccer tournament. PLoS One **14**: e0219138 (2019). https://doi.org/10.1371/journal.pone.0219138
2. Scarfone, R., Minganti, C., Ammendolia, A., Capranica, L., Tessitore, A.: Match analysis and heart rate of beach soccer players during amateur competition. Grad. J. Sport. Exerc. Phys. Educ. Res. **3**, 1–12 (2015)
3. Leite, W., Barreira, D.: Are the teams sports soccer, futsal and beach soccer similar ? **4**, 75–84 (2014). https://doi.org/10.5923/s.sports.201401.11
4. Abdullah, M.R., Musa, R.M., Maliki, A.B.H.M., Kosni, N.A., Suppiah, P.K.: Development of tablet application based notational analysis system and the establishment of its reliability in soccer. J. Phys. Educ. Sport. **16**, 951–956 (2016). https://doi.org/10.7752/jpes.2016.03150
5. Alhaj, T.A., Siraj, M.M., Zainal, A., Elshoush, H.T., Elhaj, F.: Feature selection using information gain for improved structural-based alert correlation. PLoS ONE **11**, e0166017 (2016)
6. Maimon, O., Rokach, L.: Data Mining and Knowledge Discovery Handbook (2005). https://doi.org/10.1007/b107408
7. Musa, R.M., Abdullah, M.R., Maliki, A.B.H.M., Kosni, N.A., Mat-Rasid, S.M., Adnan, A., Juahir, H.: Supervised pattern recognition of archers' relative psychological coping skills as a component for a better archery performance. J. Fundam. Appl. Sci. **10**, 467–484 (2018)
8. Muazu Musa, R., Abdul Majeed, A.P.P., Taha, Z., Abdullah, M.R., Husin Musawi Maliki, A.B., Azura Kosni, N.: The application of artificial neural network and k-nearest neighbour classification models in the scouting of high-performance archers from a selected fitness and motor skill performance parameters. Sci. Sport (2019). https://doi.org/10.1016/j.scispo.2019.02.006
9. Chandrashekar, G., Sahin, F.: A survey on feature selection methods. Comput. Electr. Eng. **40**, 16–28 (2014)
10. Lei, S.: A feature selection method based on information gain and genetic algorithm. In: 2012 International Conference on Computer Science and Electronics Engineering, pp. 355–358. IEEE (2012)
11. Scarfone, S., Tessitore, A., Minganti, C., Ferragana, A., Ammedolia, A.: Match demands of beach soccer between amateur and professional competition. In: International Conference of Sport Science, pp. 8–10 (2010)

Performance Indicators Defining Goal Scoring Opportunities in Elite Asian Beach Soccer: An Artificial Neural Network Approach

Rabiu Muazu Musa[1(✉)], Anwar P. P. Abdul Majeed[2], Muhammad Zuhaili Suhaimi[1], Mohamad Razali Abdullah[3], Mohd Azraai Mohd Razman[2], and Siti Musliha Mat-Rasid[4]

[1] Centre for Fundamental and Liberal Education, Universiti Malaysia Terengganu, 21030 Kuala Nerus, Terengganu, Malaysia
rabiu.muazu@umt.edu.my

[2] Innovative Manufacturing, Mechatronics and Sports Laboratory, Universiti Malaysia Pahang, 26600 Pekan, Pahang Darul Makmur, Malaysia

[3] East Coast Environmental Research Institute, Universiti Sultan Zainal Abidin, 21030 Kuala Nerus, Terengganu, Malaysia

[4] Faculty of Sports Science and Coaching, Universiti Pendidikan Sultan Idris, Perak, Malaysia

Abstract. The game of beach soccer at the elite level is characterised with a high tempo that leads to the scoring of many goals in a match. A draw cannot decide the outcome of a match, and thus, a winner is decided by a team that scores more goals than the opponent hence, investigating performance indicators (PIs) that could lead to a higher goal-scoring opportunity in this game is non-trivial in ensuring the success of a team during a match. In the present study, an analysis of performance indicators was carried out through which a total number of 16 relevant performance indicators were considered, and multiple linear regression (MLR) was used to extract the significant PI that could lead to chances of scoring goals. An Artificial Neural Network (ANN) was used to predict the opportunities for scoring goals based on the extracted PIs. The MLR was able to extract shot at back third, pass at front third, short inside box, turnover, tackling, as well as fouls committed as the most significant PI whilst a robust predictive model was obtained via the ANN with an R^2 of 0.91 and a very low mean absolute error of 1.76 amongst the other model predictive evaluation parameters. It is postulated from the present study that the identified PIs are essential in increasing the chances of scoring goals in the elite Asian beach soccer competition.

Keywords: Beach soccer · Performance indicators · Artificial Neural Network · Goal scoring opportunities

1 Introduction

In an association football such as soccer, the tactical, technical as well as the overall strategies are reported to be an essential determiner that often influences the outcome of

M. A. Zakaria et al. (Eds.): *Advances in Mechatronics, Manufacturing, and Mechanical Engineering*, LNME, pp. 276–282, 2021.
https://doi.org/10.1007/978-981-15-7309-5_27

the game and subsequently the final results [1]. The term strategy could be seen as the holistic plan that is created and utilized for the purpose of achieving a specific objective which is normally attained through the integration of certain tactics [2, 3]. For instance, a team could implement a holistic approach that involves the integration of both attacking and defensive strategy that is aimed at fostering their chances of succeeding during a competition. In a nutshell, a strategy of play may be seen as the overall conduct of a team to accomplish the set target of attacking and defending objectives in a game. It is worth noting that performance indicators are a collection of action parameters that attempt to evaluate the characteristics of performance which can be correlated with attacking as well as a defensive strategy in a soccer game that is often evaluated through performance analysis [4, 5].

Evidence has demonstrated that performance analysis is often utilized as a diagnostic tool for the assessment of team or players' performances [6–8]. The employment of performance analysis to quantify as well as discriminate successful and unsuccessful performances have been extensively studied in various sports. A few of such studies are the identification of defensive strategies in the 2010 FIFA world cup, the establishment of performance parameters in club level Gaelic football, development of notational analysis system in elite soccer and its corresponding reliability respectively [3, 9, 10]. A different study also investigated the essential performance indicators in elite male soccer with respect to a different position of players [7]. A similar study looked into the performance indicators that could distinguish between the winning and losing teams in the sport of beach volleyball [11].

The game of beach soccer at an elite level is characterized by a high tempo that leads to the scoring of many goals in a match. Unlike the game of football, It is important to note that a draw cannot decide the outcome of a match in this game and thus, a team is announced winner only by scoring more goals than the opponent hence, investigating performance indicators that could lead to a higher goal-scoring opportunity in the game is non-trivial in ensuring the success of a team during a match to which the present study is embarked upon.

2 Methodology

2.1 Participants

The participants of the beach soccer sport in this study consist of all the teams that were involved in the AFC beach soccer tournament 2017. A sum of 12 teams took part in the tournament which comprised of China PR, Bahrain, Afghanistan, Malaysia, Oman, Lebanon, Thailand, Japan, UAE, Iraq, Iran as well as Qatar. The overall performances of the teams involved were notated and analyzed throughout the competition period. A total number of 23 matches from the 12 teams were analyzed. It is worth to mention that prior to the commencement of data collection, all the coaches, managers as well as the organizing committee were informed about the aim of the study and verbal consent was obtained via Terengganu State 2017 Beach Soccer Technical Committee (AFC2017-TSG).

2.2 Performance Indicators (PIs)

A total number of 16 technical, as well as tactical performance indicators, were established. The performance indicators were utilized to notate the performances of the teams. It is essential to note that four professional coaches who possessed an average of 10 years of coaching experience validated the performance indicators developed. A StatWatch application; an android based application for notational analysis was used as a tool for the analysis of the teams' performances in accordance with the procedures previously documented by the precedent investigators [3]. For further information on the selection process as well as the operational definition of each performance parameters, the readers are encouraged to refer to the previous study [5].

2.3 Classification

In the present study, a correlation-based multiple linear regression analysis is carried out to extract the performance indicators that are important in describing goals scoring. In this analysis, the total number of goals scored was used as the dependent variables whilst the 16 selected performance indicators were utilised as the independent variables. It should be noted that performance indicators that significantly correlated with the goals scored are exclusively considered for further analysis via the ANN.

2.4 Development of ANN Model

A regression-based model of the multi-layer perceptron (MLP) was developed to predict the goals scored with respect to the identified performance indicators through MLR analysis described previously. The data was divided into a ratio of 70:15:15 for the training, test as well as validation respectively. A single hidden layer feed-forward network trained with Relu activation function was utilized. The SGD solver was utilized, and the regularization method, α was set to 0.6. The number of suitable hidden neuron was determined heuristically, and 25 was found to be effective for the model trained. In the current investigation, an Orange analysis software version 3.2.0 was used as a platform for the data analysis. All the performance metrics measures for the model's evaluation are explained in the subsequent section.

2.5 Model Evaluations

The models developed in the present study was evaluated by means of mean absolute error (MAE), mean square error, root mean square error (RMSE), as well as R^2. The MAE is the average of the absolute values deviated between the predicted and the actual values. It should be noted that as the deviation is absolute, the negative, as well as the positive offset, is often a tradeoff. It is worth to mention that the MAE is not sensitive to the existence of anomalies; however, it serves useful in the projection of actual error of a prediction. The generalised equation for the calculation of MAE is given in Eq. (1), where n is the number of samples, y_i and $\hat{y}_i$ is the actual goals scored and that predicted by the model for the i-th samples respectively.

$$MAE = \frac{1}{n}\sum_{i=1}^{n}\left|\left(y_i - \hat{y}_i\right)\right| \quad (1)$$

The RMSE is a measure that is often used to assess and compare the predictive error of a model. It is the standard deviation of the residuals, which explained how concentered the data is accumulated around the line of best fit. The lower the RMSE value would indicate the better the forecasting ability of a model. Nonetheless, it should be noted that the presence of a few large errors could lead to a larger value of RMSE. The generalized equation for RMSE is given in Eq. (3)

$$RMSE = \sqrt{\frac{1}{n}\sum\nolimits_{i=1}^{n}\left(y_i - \hat{y}_i\right)^2} \tag{2}$$

The R^2 is an indicator that is commonly employed to ascertain how well one or more features describe a target value. In the present investigation, the R^2 is used to evaluate the overall effect one or more performance indicators towards the prediction of goals scored. It is worth noting that the value of R^2 ranges from 0 to 1. The closer the value is to 1, the stronger the degree of the interpretation and vice versa. The R^2 is computed in the present study using the formula shown in Eq. (3), where n refers to the number of samples, $\hat{y}_i$ represents the actual goals scored of the i-th sample, yi serves as the predicted value whilst $\bar{y}$ represents the mean of all the predicted values.

$$R^2\left(y, \hat{y}\right) = 1 - \frac{\sum_{i=1}^{n}\left(yi - \hat{y}_i\right)^2}{\sum_{i=1}^{n}(yi - \bar{y})^2} \tag{3}$$

3 Results and Discussion

Table 1 projects the important performance indicators that significantly correlated with the goals scoring. It could be observed from the table that a number of six performance indicators are found to be significantly correlated with the goal scored. These performance indicators viz. shot back third, pass front third, shot inbox, turnover, tackling as well as fouls committed are considered to be most essential in creating chances of scoring goals in the elite beach soccer tournament.

Figure 1 displays the MSE, MAE, RMSE, as well as the R^2 values for evaluating the prediction efficacy of the developed model. It is demonstrated in Fig. 1 that the MAE, RMSE, and MSE values of the ANN model are significantly lower. The R^2 value of 0.91 is observed from the ANN model in the prediction of the goals scored based on the extracted performance indicators. It is non-trivial to note that in the general evaluation guideline, an R^2 of 0.91 suggests an excellent prediction accuracy.

The ability of the ANN model to provide a reasonably good prediction of the goals scored could be primarily attributed to the importance of the extracted performance indicators as well as the efficacy of the model in understanding the non-linearity relationship of the data involved in the present study. It has been documented in the literature that machine learning models such as ANN are essential in mitigating the problem of prediction with a nonlinear dataset [12–14]. As demonstrated in Fig. 1, the ANN model was able to reasonably pick up the trends closely with the actual goals scored across all the predicted samples.

Table 1. The essential performance indicators extracted via multiple linear regression-based correlations.

Performance indicators	Goals scored
Shot Back Third	**0.309**
Shot Mid Third	−0.024
Shot Front Third	−0.234
Pass Back Third	0.026
Pass Mid Third	0.004
Pass Front Third	**0.380**
Shot in Box	**0.506**
Shot Out Box	−0.097
Chances Created	−0.058
Interception	−0.135
Turnover	**0.914**
Tackling	**0.389**
Fouls Committed	**0.549**
Complete Save by Keeper	0.045
Incomplete Save by Keeper	0.158
Passing Error by Keeper	−0.052
Goals Scored	**1**

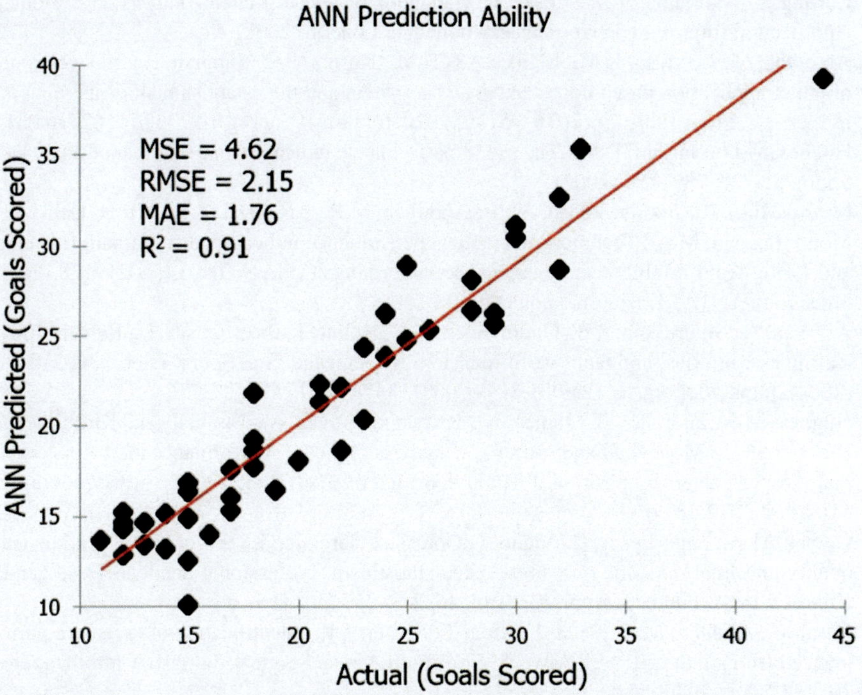

Fig. 1. Comparison of the ANN predictive efficacy with actual samples after parameters extraction

4 Conclusion

The findings from the current investigation demonstrated that performance indicators that comprise of shot back third, pass front third, shot inbox, turnover, tackling as well as fouls committed are the most important predictors of scoring goals in the elite Asian beach soccer tournament. The ANN-based regression model employed in the present investigation is able to provide an excellent prediction efficacy with respect to all the evaluation metrics measured. The model developed in the present study provides information on the essential performance indicators that could increase the chances of scoring goals in the elite Asian beach soccer tournament. The findings from the present investigation serve useful to the coaches, team managers, as well as, performance analysts in identifying the essential performance indicators that could ensure winning in this game.

Acknowledgement. The authors would like to acknowledge the coaches and managers for their cooperation towards the accomplishment of this study.

References

1. Yiannakos, A., Armatas, V.: Evaluation of the goal scoring patterns in European Championship in Portugal 2004. Int. J. Perform. Anal. Sport. **6**, 178–188 (2006)

2. Carling, C., Williams, A.M., Reilly, T.: Handbook of Soccer Match Analysis: A Systematic Approach to Improving Performance. Routledge, London (2007)
3. Abdullah, M.R., Musa, R.M., Maliki, A.B.H.M., Kosni, N.A., Suppiah, P.K.: Development of tablet application based notational analysis system and the establishment of its reliability in soccer. J. Phys. Educ. Sport **16**, 951–956 (2016). https://doi.org/10.7752/jpes.2016.03150
4. Hughes, M.D., Bartlett, R.M.: The use of performance indicators in performance analysis. J. Sports Sci. **20**, 739–754 (2002)
5. Muazu Musa, R., Abdul Majeed, A.P.P., Abdullah, M.R., Ab. Nasir, A.F., Arif Hassan, M.H., Mohd Razman, M.A.: Technical and tactical performance indicators discriminating winning and losing team in elite Asian beach soccer tournament. PLoS One **14**, e0219138 (2019). https://doi.org/10.1371/journal.pone.0219138
6. Arnason, A., Sigurdsson, S.B., Gudmundsson, A., Holme, I., Engebretsen, L., Bahr, R.: Physical fitness, injuries, and team performance in soccer. Med. Sci. Sports Exerc. **36**, 278–285 (2004). https://doi.org/10.1249/01.MSS.0000113478.92945.CA
7. Hughes, M., Caudrelier, T., James, N., Redwood-Brown, A., Donnelly, I., Kirkbride, A., Duschesne, C.: Moneyball and soccer - an analysis of the key performance indicators of elite male soccer players by position. J. Hum. Sport Exerc. **7**, 402–412 (2012). https://doi.org/10.4100/jhse.2012.72.06
8. Gomez, M.A., Lago-Peñas, C., Viaño, J., González-Garcia, I.: Effects of game location, team quality and final outcome on game-related statistics in professional handball close games. Kinesiol. Int. J. Fundam. Appl. Kinesiol. **46**, 249–257 (2014)
9. Casal, C., Andujar, M., Losada, J., Ardá, T., Maneiro, R.: Identification of defensive performance factors in the 2010 FIFA World Cup South Africa. Sports **4**, 54 (2016). https://doi.org/10.3390/sports4040054
10. McGuigan, K., Hughes, M., Martin, D.: Performance indicators in club level Gaelic football. Int. J. Perform. Anal. Sport **18**, 780–795 (2018). https://doi.org/10.1080/24748668.2018.1517291
11. Medeiros, A.I.A., Marcelino, R., Mesquita, I.M., Palao, J.M.: Performance differences between winning and losing under-19, under-21 and senior teams in men's beach volleyball. Int. J. Perform. Anal. Sport **17**, 96–108 (2017). https://doi.org/10.1080/24748668.2017.1304029
12. Taha, Z., Musa, R.M., Abdul Majeed, A.P.P., Alim, M.M., Abdullah, M.R.: The identification of high potential archers based on fitness and motor ability variables: a support vector machine approach. Hum. Mov. Sci. **57**: 184–193 (2018). https://doi.org/10.1016/j.humov.2017.12.008
13. Chang, S.W., Abdullah, M.R., Abdul Majeed, A.P.P., Ab. Nasir, A.F., Taha, Z., Muazu Musa, R.: A machine learning approach of predicting high potential archers by means of physical fitness indicators. PLoS One **14**, e0209638 (2019). https://doi.org/10.1371/journal.pone.0209638
14. Muazu Musa, R., Abdul Majeed, A.P.P., Taha, Z., Abdullah, M.R., Husin Musawi Maliki, A.B., Azura Kosni, N.: The application of Artificial Neural Network and k-Nearest Neighbour classification models in the scouting of high-performance archers from a selected fitness and motor skill performance parameters. Sci. Sport (2019). https://doi.org/10.1016/j.scispo.2019.02.006

The Classification of Wink-Based EEG Signals: The Identification of Significant Time-Domain Features

Jothi Letchumy Mahendra Kumar[1], Mamunur Rashid[2], Rabiu Muazu Musa[3], Mohd Azraai Mohd Razman[1], Norizam Sulaiman[2], Rozita Jailani[4], and Anwar P. P. Abdul Majeed[1](✉)

[1] Innovative Manufacturing, Mechatronics and Sports Laboratory, Faculty of Manufacturing and Mechatronics Engineering Technology, Universiti Malaysia Pahang (UMP), 26600 Pekan, Pahang Darul Makmur, Malaysia
amajeed@ump.edu.my

[2] Faculty of Electrical and Electronics Engineering Technology, Universiti Malaysia Pahang (UMP), 26600 Pekan, Pahang Darul Makmur, Malaysia

[3] Centre for Fundamental and Liberal Education, Universiti Malaysia Terengganu (UMT), 21030 Kuala Nerus, Terengganu Darul Iman, Malaysia

[4] Faculty of Electrical Engineering, Universiti Teknologi MARA (UiTM), 40450 Shah Alam, Selangor Darul Ehsan, Malaysia

Abstract. Brain-Computer Interface (BCI) has become popular with physically challenged individuals, particularly in enhancing their activities of daily living. Electroencephalogram (EEG) signals are used to control BCI-based devices. Nonetheless, it is worth noting that the use of a multitude of features may impede the real-time execution of BCI devices. The present study aims at identifying significant time-domain based features that could provide a reasonable classification of the right or left wink based on EEG signals evoked by the aforesaid facial expressions. The Emotiv Insight mobile EEG system was used to capture the EEG signals acquired from the winking of the left and right eye of five healthy subjects between the age of 23 and 27 years old. Nine statistical time-domain based features were extracted, namely maximum (Max), minimum (Min), mean, median, standard deviation (SD), variance, skewness, kurtosis, and root mean square (RMS) on five channels. An ensemble learning method, i.e. Extremely Randomised Trees, was used to identify the significant features. The feature selection effect towards wink classification was evaluated via the *k*-Nearest Neighbours (*k*-NN) classifier. The training to test ratio of the extracted signals was set to 70:30. It was shown from the study, that five features were found to be significant, viz. Max_AF4, SD_AF4, skewness_AF3, kurtosis_AF4 and kurtosis_AF3, respectively. The training classification accuracy (CA) by considering all features and selected features was ascertained to be both 100%, respectively, whilst, the test CA was also found to be identical for both models with no misclassification transpired. Therefore, it could be established from the study that a comparable classification efficacy is attainable through the identification of significant features. The findings are non-trivial, particularly with respect to the implementation of the developed classifier in real-time.

M. A. Zakaria et al. (Eds.): *Advances in Mechatronics, Manufacturing, and Mechanical Engineering*, LNME, pp. 283–291, 2021.
https://doi.org/10.1007/978-981-15-7309-5_28

Keywords: EEG · BCI · Machine learning · Classification · Feature selection

1 Introduction

It has been reported that cerebrovascular disease, particularly stroke, is the third leading cause of disability globally [1]. In addition, neuromuscular disorders such as Amyotrophic lateral sclerosis (ALS) has also been reported to be on the rise [2]. More often than not, patients that are affected with such diseases are left with the inability to control their motor functions, that in turn, limits their potential to take part in activities of daily living (ADL) [3]. Nonetheless, with the advancement of technology, particularly in the area of Brain-Computer Interfaces (BCI), allows such patients to regain their ADL through the control of peripheral devices, for instance, wheelchair and exoskeleton amongst others [4–6]. It is worth noting that the control of BCI exploits the classification of Electroencephalogram (EEG) and its associative signals [7].

Although there is aplenty literature that reports the classification of EEG based signals, nevertheless, it is worth to note that limited literature is available with regards to the classification of wink-based EEG signals. Rudas and Laki [8] reported the classification of wink-based EEG signals amongst other 11 activities from five subjects, with each activity are performed ten times via Emotiv Epoch+. The authors transformed the data to the frequency domain via Fast Fourier Transform (FFT) and employed a dimensional reduction technique, i.e., Principal Component Analysis for feature extraction. Different features based on the removal of idle window apart from windowing techniques were investigated. A number of classifiers were evaluated based on the features namely Random Forest, Gradient Boosting Trees, Support Vector Machine as well as XG Boost. It was shown from the study that different classifiers perform well on different conditions.

Rashid et al. [9] investigated the classification of left, right and no wink EEG signals of five subjects captured by Emotiv Insight on different classifiers namely, Support Vector Machine, Linear Discriminant Analysis (LDA) and *k*-Nearest Neighbors (*k*-NN). Two features were investigated in the study namely the between the maximum and minimum of the raw signals (time-domain) of the electrodes investigated as well as the mean absolute value (MAV) of the frequency-domain data transformed via FFT. The features were evaluated independently on the classifiers. It was shown from the study that for the frequency-domain based MAV feature, the LDA classifier performed better against other classifiers with a test classification accuracy (CA) of 83.8%. Conversely, by considering the time-domain based range feature, the SVM and *k*-NN model attained a test CA of 96.7%.

It could be seen from the previous studies, particularly with regards to wink-based EEG signals, that the employment of a systematic means of identifying significant features is somewhat limited. Therefore, this study aims at identifying significant statistical EEG-based time-domain signals extracted from the left and right wink via an ensemble method. Two *k*-Nearest Neighbors (*k*-NN) models were developed based on the different set of features (all and significant) to evaluate its effect towards the classification of the winks.

2 Methodology

2.1 Data Acquisition Device and Configuration

In this study, a five-channel Wireless Emotiv Insight mobile headset (as shown in Fig. 1) was used to collect the EEG signals attained from the left and right winking expression. The sensors are made up of Hydrophilic semi-dry polymer. The bandwidth allowed for the device is between 0.5 to 43 Hz. Each channel has a resolution of 0.51 μV. The signals were collected at a sampling rate of 128 samples per second from each channel. The EEG recording setup was utilised based on the International 10–20 EEG system. The channels are designed to read the electrical potential differences produced by electrodes positioned at AF3, AF4, T7, T8, and Pz to collect the wink signals. The reference electrode is placed at the left mastoid.

Fig. 1. Five-channel emotiv insight device.

2.2 Subjects

Five healthy subjects (three male and two female) aged between 23 and 27 years old were chosen to participate in this experiment. The subjects had no known medical ailments and had normal vision. The subjects do not have any prior history of neurological or psychiatric disorder disease. None of the subjects has any form of experience related to the experiment that will be carried out. Informed consent was taken from the subjects prior to the experiment. The study has been sanctioned by an institutional research ethics committee (FF-2013-327).

2.3 Experimental Protocol

The experiment was carried out in a designated room located at the Faculty of Electrical and Electronics Engineering Technology, Universiti Malaysia Pahang to avoid any form of disturbance to the subjects from the environment. The subjects were seated on an ergonomic chair and were instructed to sit in a relaxed position and stay relaxed throughout the experiment without carrying out any other apparent or abrupt physical movement.

The subjects were then given instructions through a cue displayed on the LCD, which was approximately one meter away from the subject. A paradigm which consists of 5 trials of left and right wink was used throughout the experiment, i.e., for the first trial, only left wink is considered. The subjects were instructed to follow the cue displayed and act accordingly. The first slide (display a black 'dot', referred as 'Blank' in Fig. 2) shows that the subjects should be in the rest position (the subjects are asked to concentrate on the 'dot' to minimise EEG artifacts [10]) for 5 s followed by winking either from the left or right eye for 5 s (as depicted in Fig. 2). The sequence is then repeated as per the display, and the overall duration of the experiment is 60 s per subject.

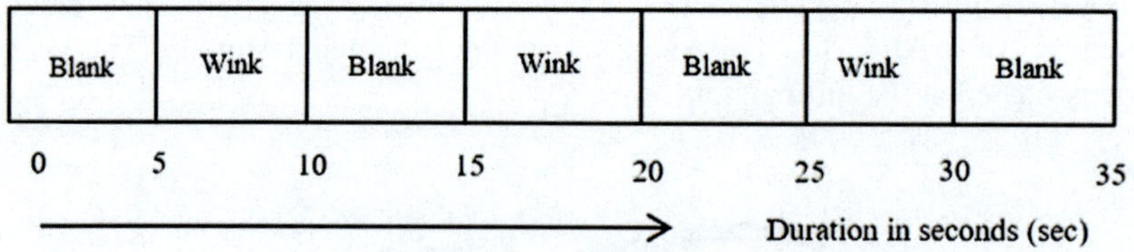

Fig. 2. The experiment paradigm for EEG signal acquisition.

2.4 Signal Preprocessing, Feature Extraction and Selection

The raw EEG signals were filtered using a digital notch filter at 50 Hz prior to a fifth-order digital Sinc filter during the preprocessing stage. A sample of the per-processed signal of a subject performing a left wink as per the experimental paradigm is depicted in Fig. 3. The signals were segmented based on the windows that demonstrate the winking action (no wink signals are omitted), and hence, fifty samples were separated. Once the time-domain signals have been preprocessed and separated, nine typical statistical-based features [11–13], i.e., maximum (Max), minimum (Min), mean, median, standard deviation (SD), variance, skewness, kurtosis, and root mean square (RMS) were extracted from all the five channels, making it a total of 45 features.

In the present study, an ensemble learning method, namely an Extremely Randomised Tree classifier (also known as Extra Trees Classifier) was employed to identify significant features. The quality of the split is based on information gain or the decrease in entropy (a means of measuring impurity) after a dataset is split on an attribute [14, 15]. Essentially, it computes the statistical dependency between the input and target variables. The scikit-learn Python library was used to evaluate the feature importance of the aforesaid extracted features, where the criterion is set to 'entropy' whilst other parameters as default.

2.5 Classification

The *k*-nearest neighbours (*k*-NN) classifier was used to appraise the effect of the evaluated features towards its efficacy in classifying the two-class designated winks. This type of machine learning model has been regarded as one of the simplest model owing to the limited number of hyperparameters that are required to be tuned [16–18]. Two *k*-NN

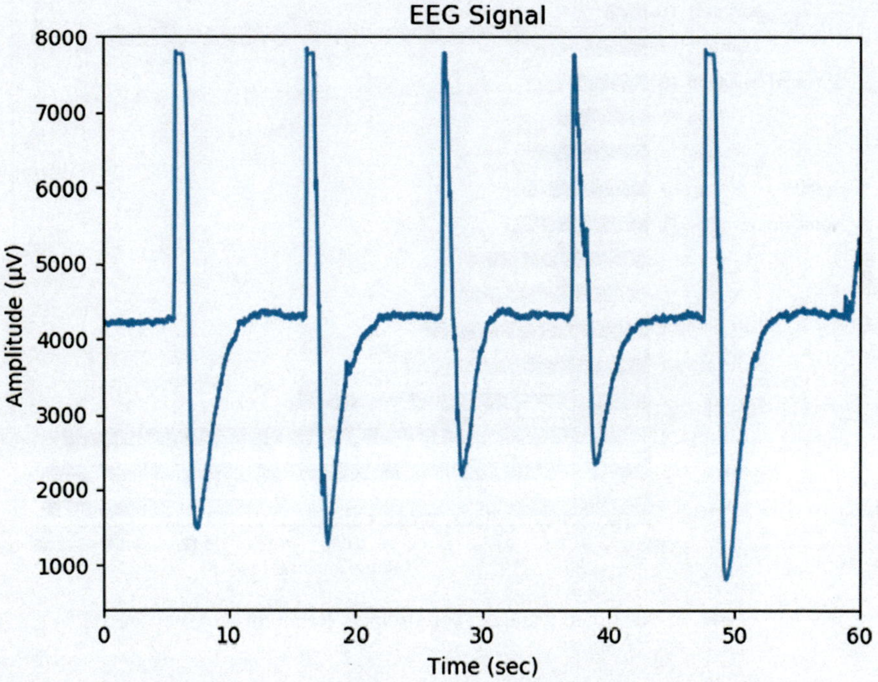

Fig. 3. Preprocessing left wink EEG signal.

models were developed, with a *k* of five and by utilising the Minkowski distance metric. The scikit-learn Pythom library version 0.22.2 was used to develop the models. The ratio for training and testing was set to 70:30 ratio from a total of 50 samples, whilst the performance of the models is assessed via the classification accuracy (CA) as well as the confusion matrix.

3 Result and Discussion

It could be observed through the feature importance investigation depicted in Fig. 4 via the ensemble technique, particularly Extremely Randomised Tree that Max_AF4, SD_AF4, skewness_AF3, kurtosis_AF4 and kurtosis_AF3, were found to be significant. It is also evident that channel AF4 is somewhat dominant in comparison to other channels. Both the identified features set as well as the original features (all) were then used to classify the left and right winks. It is important to note at this juncture that hyperparameters of the *k*-NN models were configured to be identical to provide an unbiased evaluation of the classification efficacy of the models based on the feature set. It evident from the bar chart depicted in Fig. 5 that with regards to the training accuracy both models were found to be able to attain a 100% CA. Conversely, on the test dataset, it appears that both sets of features also yields no misclassification. There is no apparent sign of overfitting that transpired on both models developed. Moreover, from the confusion matrix (illustrated in Fig. 6 and Fig. 7, respectively) on the test dataset (eight left and seven right winks) for both features investigated, no misclassified transpired.

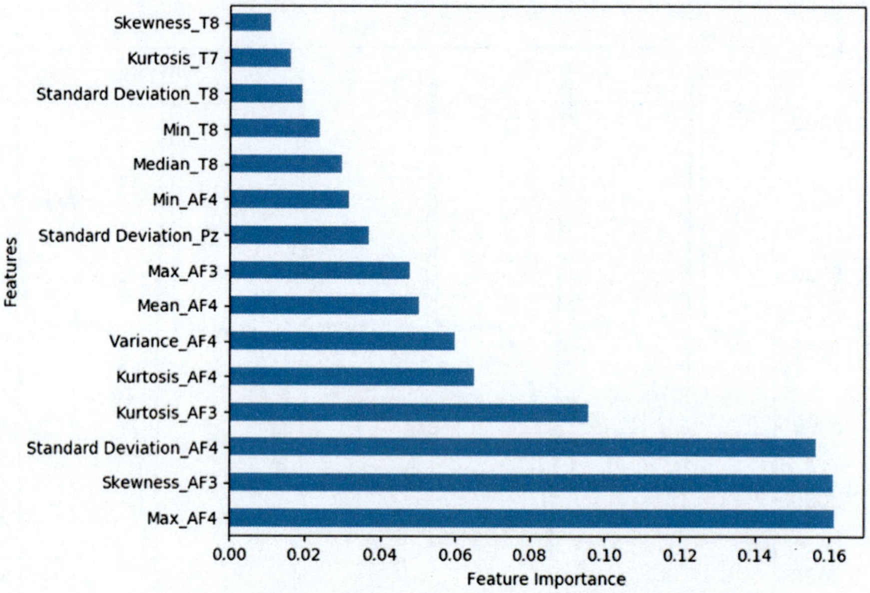

Fig. 4. Identified features via Extremely Randomised Trees

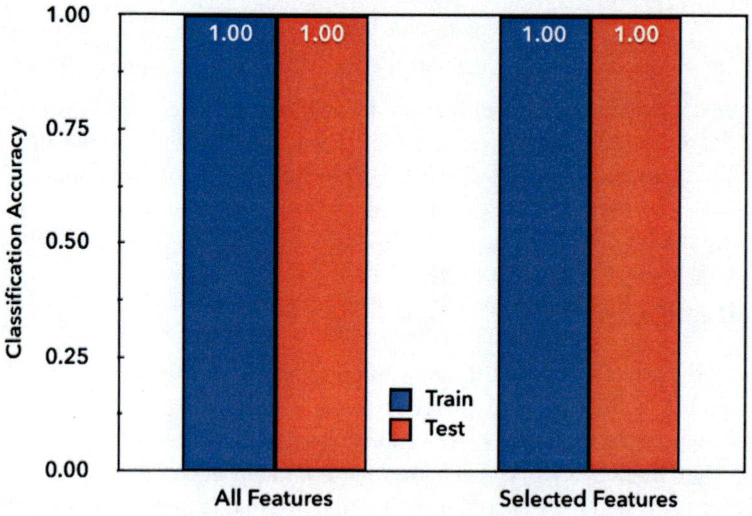

Fig. 5. Classification accuracy based on the different feature set evaluated

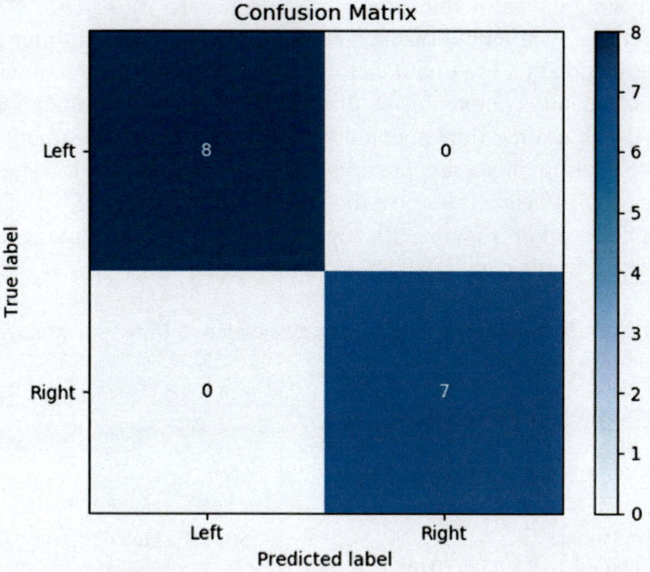

Fig. 6. Confusion matrix of the test dataset by using all features

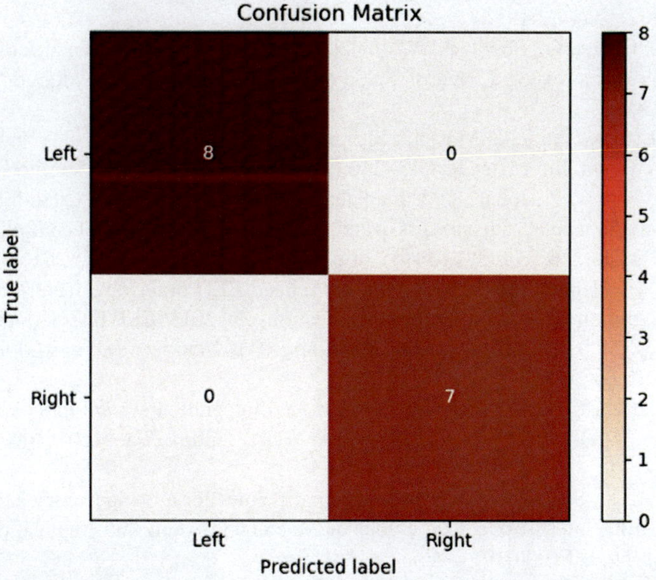

Fig. 7. Confusion matrix of the test dataset by using selected features

4 Conclusion

In this study, a number of statistical-based time-domain features were extracted from the wink-based EEG signals. Subsequently, the essential features are identified via an ensemble learning algorithm, i.e., the Extremely Randomised Trees. It was shown that the top five dominant or significant features out of 45 features are, Max_AF4, SD_AF4, skewness_AF3, kurtosis_AF4 and kurtosis_AF3, respectively. It was further demonstrated that with the selection of a few significant features, the classification of winking is possible and in fact, equally comparable by utilising all features. This finding is non-trivial, especially in BCI real-time implementation as the computational time is significantly reduced by considering only a few features. It is worth noting that the study is rather in its preliminary phase; hence, future studies shall investigate the effect of other feature selection techniques apart from the efficacy of different types of classifiers as well as its hyperparameters towards the classification of the winks.

Acknowledgement. The authors would like to acknowledge Universiti Malaysia Pahang for funding this study via RDU180321.

References

1. Johnson, W., Onuma, O., Owolabi, M., Sachdev, S.: Stroke: a global response is needed. Bull. World Health Organ. **94**, 634 (2016)
2. Rose, L., McKim, D., Leasa, D., Nonoyama, M., Tandon, A., Bai, Y.Q., Amin, R., Katz, S., Goldstein, R., Gershon, A.: Trends in incidence, prevalence, and mortality of neuromuscular disease in Ontario, Canada: a population-based retrospective cohort study (2003–2014). PLoS One. **14**, e0210574 (2019)
3. Mathis, S., Goizet, C., Soulages, A., Vallat, J.M., Le Masson, G.: Genetics of amyotrophic lateral sclerosis: a review. J. Neurol. Sci. **399**, 217–226 (2019). https://doi.org/10.1016/j.jns.2019.02.030
4. Abdulkader, S.N., Atia, A., Mostafa, M.S.M.: Brain computer interfacing: applications and challenges. Egypt. Inform. J. **16**, 213–230 (2015). https://doi.org/10.1016/j.eij.2015.06.002
5. Lange, G., Low, C.Y., Johar, K., Hanapiah, F.A., Kamaruzaman, F.: Classification of electroencephalogram data from hand grasp and release movements for BCI controlled prosthesis. Procedia Technol. **26**, 374–381 (2016). https://doi.org/10.1016/j.protcy.2016.08.048
6. Mattar, E.A., Al-Junaid, H.J.: Manipulation related EEG brainwave feature extraction and events recognition for robotics learning applications. In: 2018 IEEE International Conference on Robotics and Biomimetics, ROBIO 2018, pp. 2051–2056 (2019). https://doi.org/10.1109/ROBIO.2018.8665318
7. Lin, J.S., Hsieh, C.H.: A wireless BCI-controlled integration system in smart living space for patients. Wirel. Pers. Commun. **88**, 395–412 (2016). https://doi.org/10.1007/s11277-015-3129-0
8. Rudas, Á., Laki, S.: On activity identification pipelines for a low-accuracy EEG device. In: 2019 18th IEEE International Conference on Machine Learning and Applications (ICMLA), pp. 1454–1459. IEEE (2019)
9. Rashid, M., Sulaiman, N., Mustafa, M., Bari, B.S., Sadeque, M.G., Hasan, M.J.: Wink based facial expression classification using machine learning approach. SN Appl. Sci. **2**, 183 (2020)
10. Maxwell, R.J.: Pattern recognition analysis of 2–2 (2006). https://doi.org/10.1049/ic:19970472
11. Khairuddin, I.M., Na'im Sidek, S., Majeed, A.P.P.A., Puzi, A.A.: Classifying motion intention from EMG signal: a k-NN approach. In: 2019 7th International Conference on Mechatronics Engineering (ICOM), pp. 1–4. IEEE (2019)

12. Mohd Razman, M.A., Majeed, A.P.P.A., Musa, R.M., Taha, Z., Susto, G.-A., Mukai, Y.: Time-series identification on fish feeding behaviour. In: Machine Learning in Aquaculture, pp. 37–47. Springer (2020)
13. Abdullah, M.A., Ibrahim, M.A.R., Shapiee, M.N.A.B., Mohd Razman, M.A., Musa, R.M., Majeed, A.P.P.A.: The classification of skateboarding trick manoeuvres through the integration of IMU and machine learning. In: Symposium on Intelligent Manufacturing and Mechatronics, pp. 67–74. Springer (2020)
14. Sharaff, A., Gupta, H.: Extra-tree classifier with metaheuristics approach for email classification. In: Advances in Computer Communication and Computational Sciences, pp. 189–197. Springer (2019)
15. Geurts, P., Ernst, D., Wehenkel, L.: Extremely randomized trees. Mach. Learn. **63**, 3–42 (2006)
16. Razman, M.A.M., Susto, G.A., Cenedese, A., Abdul Majeed, A.P.P., Musa, R.M., Abdul Ghani, A.S., Adnan, F.A., Ismail, K.M., Taha, Z., Mukai, Y.: Hunger classification of Lates calcarifer by means of an automated feeder and image processing. Comput. Electron. Agric. **163** (2019). https://doi.org/10.1016/j.compag.2019.104883
17. Musa, R.M., Majeed, A.P.P.A., Taha, Z., Abdullah, M.R., Maliki, A.B.H.M., Kosni, N.A.: The application of Artificial Neural Network and k-Nearest Neighbour classification models in the scouting of high-performance archers from a selected fitness and motor skill performance parameters. Sci. Sports **34**, e241–e249 (2019)
18. Muazu Musa, R., Abdul Majeed, A.P.P., Taha, Z., Chang, S.W., Ab. Nasir, A.F., Abdullah, M.R.: A machine learning approach of predicting high potential archers by means of physical fitness indicators. PLoS One **14**, e0209638 (2019). https://doi.org/10.1371/journal.pone.0209638

Surface Resistivity and Ultrasonic Pulse Velocity Evaluation of Reinforced OPC Concrete and Reinforced Geopolymer Concrete in Marine Environment

M. B. H. Ab Manaf[1(✉)], Z. Yahya[2], R. Abd Razak[2], A. M. Mustafa Al Bakri[4], N. F. Ariffin[3], M. M. Ahmad[2], and Y. C. Chong[2]

[1] School of Environmental Engineering, Universiti Malaysia Perlis, 02600 Arau, Perlis, Malaysia
badrulmanaf@unimap.edu.my
[2] Department of Civil Engineering Technology, Faculty of Engineering Technology, Universiti Malaysia Perlis, 02100 Perlis, Padang Besar, Malaysia
[3] Faculty of Civil Engineering and Earth Resources, Universiti Malaysia Pahang, 26300 Gambang, Pahang, Malaysia
[4] Green Design and Manufacture Research Group, Center of Excellence Geopolymer and Green Technology (CEGeoGTech), Universiti Malaysia Perlis, 01000 Kangar, Perlis, Malaysia

Abstract. The concrete structures that are built along the seaside often suffer from reduced service life due to inadequate durability against deterioration. This research reports the findings of concrete resistivity and quality using two Non-Destructive Testing (NDT) measures applied to Reinforced Geopolymer and Ordinary Portland Cement (OPC) concrete in the marine environment. In addition, the relationship between Reinforced Geopolymer and Reinforced OPC concrete was statistically discussed in-terms of strength and direction. The testing was carried out using a Proceeq Resipod Wenner 4-probe to measure Surface Resistivity (SR) and Ultrasonic Pulse Velocity (UPV), respectively. The testings were carried out on beam shaped samples of OPC and Geopolymer concrete that were immersed in seawater over a period of 90 days with similar curing condition. It was found from the present investigation that the maximum SR and maximum UPV values acquired for both the Reinforced OPC and Reinforced Geopolymer concrete are 2.73 kΩcm and 2.07 kΩcm, as well as 4.18 km/s and 4.05 km/s, respectively. It is apparent from the study that both concrete is comparable in terms of quality and surface resistivity.

Keywords: Reinforced concrete · Geopolymer concrete · Surface resistivity · Ultrasonic pulse velocity · Non-destructive test · Marine environment · Correlation · Relationship

1 Introduction

Concrete structures such as wharves and jetties are susceptible to concrete degradation and deterioration despite high durability and compressive strength owing to its exposure

M. A. Zakaria et al. (Eds.): *Advances in Mechatronics, Manufacturing, and Mechanical Engineering*, LNME, pp. 292–298, 2021.
https://doi.org/10.1007/978-981-15-7309-5_29

to the corrosive environment. Consequently, the inherent traits of the environment (high chloride concentration and humidity) will reduce the operational service life of the concrete [1, 2]. These structures require regular maintenance to ensure the building remains functional. Conventional durability testing for concrete structure utilizes data obtained from destructive testing that cause damages to structures. It is worth noting that the application of Non-Destructive Testing (NDT) has received considerable attention in the field of civil engineering, mainly as the name implies, it is not destructive [2]. Two common non-destructive testing methods employed in evaluating concrete strength and durability are Surface Resistivity (SR) testing and ultrasonic pulse velocity (UPV).

UPV testing methods are widely used for inspection and monitoring of concrete structure [2]. Three conditions of concrete that can be assessed through this method are depth of surface cracks, discontinuity in cross-section (i.e., cracks, cover concrete delamination and qualitative assessment of the strength of concrete) as well as the gradation in different locations of structural members [3].

Electrical resistivity is one of the non-invasive and non-destructive methods that can be used to evaluate the microstructure of concrete. Diffusion coefficients of chloride ions and water permeability can be predicted by utilizing volume fraction of pores and conductivity of pore solution. The data acquired from monitoring electrical resistivity can be used to determine the properties of concrete in different conditions as a simple form of quality control [4]. The Wenner 4-probe is one of the devices that could be used to acquire the electrical resistivity data of a given concrete [5].

It is noteworthy to mention that curing conditions of concrete is relatively essential to evaluate the rate of degradation or deterioration of a material, for instance, the temperature of the surroundings, relative humidity and the duration of concrete immersion. Rabie et al. [7] reported that an increase in both temperature and moisture would provide better ion mobility in pore solution, which, in turn, contributes to a lower resistivity value. Conversely, the effect of water immersion towards type I cement-based concrete was investigated by Osterminski [8]. Different immersion level of tap water, namely 10, 15, 20, and 50 mm was assessed. It was shown from the study that the SR would show an increasing trend for all concrete cover until it is exposed where the SR will then show a decreasing trend.

The aim of the present investigation is to determine concrete quality between Reinforced Ordinary Portland Cement (OPC) and Reinforced Geopolymer concrete in the marine environment using UPV testing. In addition, the corrosion rate between Reinforced OPC and Reinforced Geopolymer concrete was also investigated by means of the SR testing under the same environmental conditions. Moreover, the relationship between Surface Resistivity and UPV Test of both Reinforced OPC and Reinforced Geopolymer concrete is also examined.

2 Methodology

2.1 Raw Material

Two types of concrete were used in this research, namely, OPC and Geopolymer concrete. Both concrete used coarse and fine aggregates where the maximum size for coarse aggregates is 10 mm to prevent block up of aggregates larger than 10 mm between formwork and reinforcement bars. There are two types of fly ash that was utilized in the study, i.e., Class C and Class F [9]. Low calcium fly ash (ASTM Class F) was chosen as it has been reported that the performance of low calcium fly ash was better than that of high calcium with regards to the polymerization process [10]. Fly ash used for this research acquired from the Cement Industry of Malaysia Berhad, Chuping, Perak. Unlike OPC that uses water for binding effect, Geopolymer concrete depends on an alkaline activator, which consists of sodium silicate and sodium hydroxide solutions for the binding. The molarity of sodium hydroxide is fixed at 12 M and 99.0% purity for this research. In addition, high tensile reinforcement bar with a diameter of 16 mm and mild steel reinforcement bar with a diameter of 6 mm were cut into a length of 500 mm.

2.2 Mix Design and Process

With regards to the OPC concrete, the mixing process involved the mixing of coarse aggregates, fine aggregates, water and cement, whilst for Geopolymer concrete, it involves the mixing of coarse aggregates, fine aggregates, alkaline activator and fly ash. The alkaline activator will be mixed to a homogenous state before it is poured into the mixer. The mix design of the OPC concrete with the proportions of each raw materials are shown in Table 1. Geopolymer concrete has a different mix design with the OPC concrete where fly ash will replace cement, and alkaline activator replaces water. However, both concrete was aimed to have a similar compressive strength which is 35 MPa. Reinforcement bars will be tied with a 25 mm spacer to ensure adequate concrete cover upon concreting and inserted into wooden formwork. Fresh concrete was poured into two wooden beam formwork with a dimension of 600 mm × 150 mm × 150 mm and nine steel moulds with a size of 100 mm × 100 mm × 100 mm. Both OPC and geopolymer concrete will be cured for 28 days after demould to provide enough time for the concrete to build-up strength.

Table 1. OPC concrete mix composition

Ingredients	Kg/m^3
Ordinary Portland Cement	457
Water	233
Fine aggregates: Sand	760
Course aggregates	930

In order to simulate the effect of the marine environment, seawater was acquired from a local beach at Kuala Perlis. After curing for 28 days, the developed concretes are tested and immersed into seawater for 90 days. The data will be recorded for 30, 45, 60, 75 and 90 days. A control sample where tap water was used as the immersion solution was designed to compare the data acquired between normal and the marine environment. Both solutions were stored in different containers and placed outside of the laboratory with sunlight exposure.

2.3 Testing and Analyzing

Profometer testing was done to determine the location of reinforcement bars lay beneath the concrete surface by providing relevant data such as concrete cover, location of reinforcement bars and diameter of reinforcement bars. After locating reinforcement bars, a yellow mark is drawn to ease SR and UPV testings.

SR test was done to determine the corrosion resistance of concrete as per AASHTO TP 95-11 [11] where Proceeq Resipod Wenner 4-probe with electrodes spacing of 38 mm was used. The surface of concrete was splashed with water to deliver current into the concrete while the SR reading will be measured by acquiring the potential difference between the inner electrodes.

Next, the UPV testing was carried out as per BS 1881: Part 203: 1986 to determine the internal quality of concrete [12]. Different conditions of the concrete could be determined through the test, viz. surface cracks, discontinuity in cross-section and gradation at different locations. The direct transmission was assessed for the given period of 1, 30, 45, 60, 75, and 90 days immersion in seawater.

The correlation between surface resistivity and ultrasonic pulse velocity value was determined through a commercial statistical software package, i.e., IBM SPSS. A linear regression model was developed to evaluate the correlation between UPV and SR for both concrete.

3 Result and Analysis

Based on Fig. 1 and Fig. 2, the maximum result attained for OPC concrete was slightly higher than Geopolymer concrete for both tests. The maximum SR results obtained for Reinforced OPC and Geopolymer concrete is 2.73 kΩcm and 2.07 kΩcm, respectively. Conversely, the maximum UPV recorder for Reinforced OPC and Geopolymer concrete is 4.18 km/s and 4.05 km/s, respectively. It is evident that the difference between the two concrete does not vary much and are comparable in terms of quality and surface resistivity. In addition, both concrete demonstrated low surface resistance, and this is possibly due to moisture content in the concrete as it was immersed in the water. The findings are in agreement with the ones reported in [7, 8].

Furthermore, based on Fig. 1 and Table 2, the Pearson correlation, R between the SR and UPV of OPC concrete in seawater was 0.818 or $R^2 = 0.669$, which shows a strong positive relationship between each other. As for Geopolymer concrete, as shown in Fig. 2 and Table 2, the R between SR and UPV was 0.970 or $R^2 = 0.941$ which also shows a strong positive relationship that demonstrates that the SR increases as UPV increases. The adjusted R-squared values, as well as the correlation equations, are also shown in Table 2, suggesting the strong association between the evaluated parameters for both types of concrete.

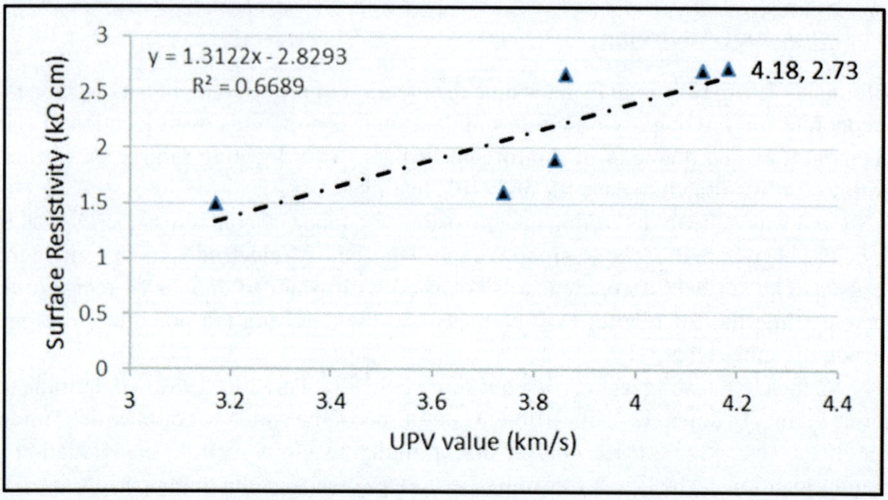

Fig. 1. Correlation between surface resistivity value and UPV value of OPC concrete in seawater.

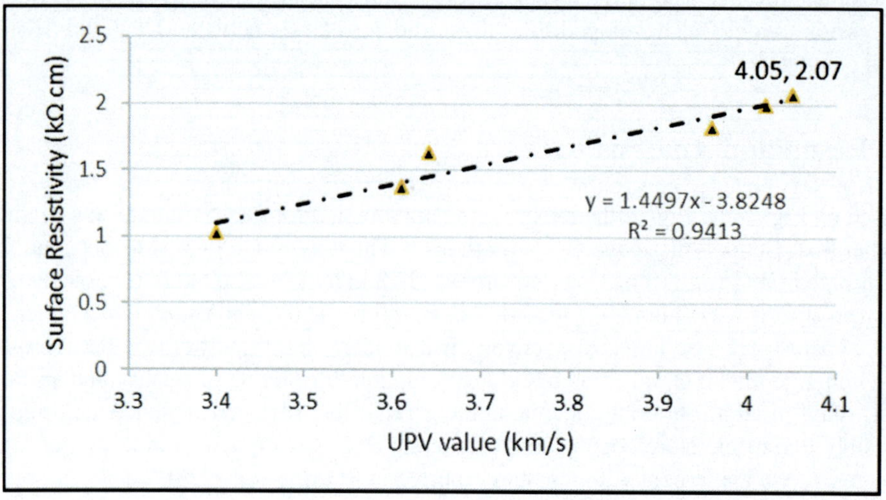

Fig. 2. Correlation between surface resistivity value and UPV value of Geopolymer concrete in seawater.

Table 2. Model summary of surface resistivity against UPV for OPC concrete and Geopolymer in seawater

Model	Concrete type	R	Adjusted R square	Representing equation
1	UPV of OPC concrete	.818	.586	y = 1.312x – 2.829
2	UPV of Geopolymer concrete	.970	.928	y = 1.450x – 3.825

4 Conclusion

It could be concluded from the present investigation that a close association between SR and UPV was established for both Reinforced Geopolymer and Reinforced OPC concretes. The strength quality of both concrete is not compromised as the UPV results were greater than that of 4 km/s, suggesting the dense internal structure that both concrete possesses. In addition, both concrete demonstrated low SR owing to the immersion of the concrete in seawater. It could also be observed that both concrete appears to be not affected by seawater as shown by the increasing trend of both the UPV and SR. It is worth noting that the study is in its preliminary stage, hence a longer test duration is required to evaluate the actual performance of the concretes against marine environment.

Acknowledgements. The authors would like to acknowledge the Department of Civil Engineering Technology, Faculty of Engineering Technology, UniMAP for the lab facilities and Research Management & Innovation Centre, UniMAP for the funding provided for the present study.

References

1. Mehta, P.K.: Durabiliy of concrete: fifty years of progress? In: Proceedings of the 2nd International Conference on Durabilitay of Concrete, pp. 1–31 (1991)
2. Mackechnie, J.: Predictions of reinforced concrete durability in the marine environment. Department of Civil Engineering, University of Cape Town, Cape Town (2001)
3. Lorenzi, A.T.: Ultrasonic pulse velocity analysis in concrete specimens. UFRGS Porto Alegre, Rio Grande do Sul, Brazil, October 2007
4. Abdelouaheb, G., Abdelhalim, B.: Investigation of concrete segregation by ultrasonic pulse velocity. J. Archit. Eng. Technol. **5**, 169 (2016). https://doi.org/10.4172/2168-9717.1000169
5. Ghosh, P., Tran, Q.: Correlation between bulk and surface resistivity of concrete. Int. J. Concr. Struct. Mater. **9**(1), 119–132 (2014). https://doi.org/10.1007/s40069-014-0094-z
6. Yoon, I.S., Nam, J.W.: Influence of chloride content of on electrical resistivity in concrete. J. Korea Inst. Struct. Maintenance Inspection **6**, 90–96 (2014). https://doi.org/10.11112/jksmi.2014.18.6.090
7. Rabie, S., Nassif, H., Na, C., Salvador, M.: Evaluation of surface resistivity indication of ability of concrete to resist chloride ion penetration. Rutgers Infrastructure Monitoring and Evaluation (RIME) Group Rutgers, New Jersey (2015)
8. Osterminski, K.: Long term behaviour of the resistivity of concrete. HERON **57**(3), 211–230 (2012)

9. AASHTO TP 95: Standard method of test for surface resistivity of concrete's ability to resist chloride ion penetration. American Association of State Highway and Transportation Officials, Washington, DC (2011)
10. BS 1881: Testing Concrete - Part 203: recommendations for the measurement of velocity ultrasonic pulses in concrete. British Standards Institution, London (1986)
11. Gamage, N., Liyanage, K., Fragomeni, S., Setunge, S.: Overview of different types of Fly Ash and their use as a building and construction material. In: International Conference of Structural Engineering, Construction and Management, pp. 1–8. Kandy, Sri Lanka (2011)
12. Abdul Razak, R., Abdullah, M.M.A.B., Hussin, K., Ismail, K.N., Hardjito, D., Yahya, Z.: Optimization of NaOH molarity, LUSI Mud/Alkaline activator, and Na2SiO3/NaOH ratio to produce lightweight aggregate-based geopolymer. Int. J. Mol. Sci. **16**(5), 11629–11647 (2015)

Explosion of Undried and Dried Rice Flour with Ignition Time of 20 ms

W. Z. Wan Sulaiman[1(✉)], M. F. Mohd Idris[1], J. Gimbun[2,3], and S. Z. Sulaiman[4]

[1] Faculty of Industrial Science and Technology, Universiti Malaysia Pahang, 26300 Gambang, Pahang, Malaysia
zaiton@ump.edu.my

[2] Engineering College, Universiti Malaysia Pahang, 26300 Gambang, Pahang, Malaysia

[3] Centre of Excellence for Advanced Research in Fluid Flow (CARIFF), Universiti Malaysia Pahang, 26300 Gambang, Pahang, Malaysia

[4] Faculty of Chemical and Processes Engineering Technology, Universiti Malaysia Pahang, 26300 Gambang, Pahang, Malaysia

Abstract. In this work, the explosion characteristics of rice flour towards difference concentration at ignition time of 20 ms were analyzed. A series of experiments were performed in a 20 L spherical chamber to obtain a maximum overpressure (P_{max}), rate of the pressure rise (dP/dT), and deflagration index (K_{st}) of undried and dried commercial rice flour. The dust sample and air were ignited by two chemical ignitors. Kistler piezoelectric pressure sensor was used to determine the propagation of pressure wave during the explosion. The moisture content of the samples was measured via proximate analysis. The P_{max} was obtained at the highest pressure over the range of concentrations. P_{max} for undried rice flour was 10.9 bar at concentration of 1000 kg/m^3. Pmax for dried rice flour was 14.4 bar at concentration of 1000 kg/m^3. The highest dP/dT rise was 103 bar/s for undried flour achieved at concentration of 750 kg/m^3 and 202 bar/s for dried flour achieved at concentration of 1000 kg/m^3. K_{st} for undried and dried rice flour are 27.96 bar.m/s and 54.83 bar.m/s respectively. It was found that the explosion severity increased as the dust flour concentration increases.

Keywords: Rice flour · Explosion · Maximum overpressure · Rate of pressure rise · Deflagration index

1 Introduction

Flour particle may cause an explosion when the dust explosion pentagon criteria is fulfilled. Dust explosion involving flour is a major hazard in industry related to the grain and food processing [1]. For instance, in 1785 the first recorded explosion involving flour occurred at Giacomelli's Bakery Warehouse, an Italian flour mill when the incident triggered from a contact of flour with a mounted lamp, which injured two workers [2]. Rice flour is the milled form of rice scientifically named *Oryza sativa*, which come from a crop that has been grown globally for long time as it has been a staple food for about half of the world's population [3]. Rice flour has been recognized to be an

M. A. Zakaria et al. (Eds.): *Advances in Mechatronics, Manufacturing, and Mechanical Engineering*, LNME, pp. 299–305, 2021.
https://doi.org/10.1007/978-981-15-7309-5_30

alternative for foods contain gluten such as cereals, cakes, cookies or any beverages made from wheat, barley and rye [4]. A lot of Malaysian food uses rice flour as main food ingredients, such as desserts, cakes, cookies, and savoury cooking. Many studies for various types of dust associated with food were carried out experimentally and numerically in an explosion chamber or silo to study the severity and explosibility of food dust explosion [5–12]. Apart from that, over 4300 types of tested organic and inorganic dusts, with basic characteristics and explosibility can be obtained from GESTIS-DUST-EX, a German Association Database, BIA [13]. However, the data obtained from the database only contain very basic physical properties without further information on other characteristics of dust [14]. Thus, this work aims to provide more information on rice flour dust explosion. This paper presents analysis of the effect of a different dust concentration for undried and dried rice flour at ignition time of 20 ms towards the severity and sensitivity of dust explosion.

Characteristics of the dust and experiment setting significantly affect the explosion behaviour. The analysis of dust explosion with different characteristics and experiment setting is crucial in order to apply an effective protection and safety systems available to prevent and mitigate the dust explosion in industries. The experiments were done in a 20 L spherical explosion chamber at ignition time of 20 ms. Maximum explosion overpressure (P_{max}) and rate of pressure rise (dP/dT) of undried and dried commercial rice flour were derived in the experiments. The maximum explosion overpressure, P_{max} was obtained from the highest corrected value of explosion overpressure over a wide range of fuel concentration. The rate of the pressure rise (dP/dT) was defined from the maximum slope of the tangent through the point of inflexion in the rising portion of the pressure versus time curve [15]. The results of dP/dT will be significant when multiplied by the cube root of the chamber to obtain deflagration index (Kst). It is also known as volume-normalized maximum rate of pressure rise [16]. The results of explosion severity may be used to design the basis for explosion protection and mitigation such as explosion relief venting and explosion suppression but it depends entirely on the validity of the cube root law [17, 18].

2 Methodology

2.1 Preparation and Characterization of Samples and Materials

The rice flour sample used in this work is a commercial rice flour produced by Eng Heap Seng Rice & Flour Mill (M) Sdn. Bhd which is located in Penang, Malaysia. The sample preparation was performed according to the operating instructions of the 20 L explosion apparatus [15]. Upon testing, the dried samples would be dried at 75 °C for two hours in an oven at ambient pressure to get rid of the moisture [15]. The particle size distribution (PSD) was measured by using a Malvern Mastersizer fitted with an automated dry powder dispersion unit (Scirocco 2000). The particle size distribution was characterized by the volume weighted mean. The moisture content of the samples was carried out according to British Standard 1016 Part 6; Analysis and testing of coal and coke: Proximate analysis of coal [19]. The moisture content was calculated from the weight difference of the samples weighted before and after drying in an oven for one hour at 105 ± 5 °C [19].

2.2 Development of Dust Explosion

The maximum explosion overpressure (P_{max}) and rate of pressure rise (dP/dT) reported here were obtained in the 20 L spherical chamber as shown in Fig. 1. The chamber was made of stainless steel and was rated to withstand up to 30 bar (static pressure). The explosion experiments were performed by using two chemical ignitors. The ignitors were cut and the wire was exposed by trimming the insulation layer by using scissors before it was connected to the ignition leads. The ignition delay time t_v was set at 20 ms. The pressure inside the spherical chamber was measured by two piezo-electric pressure sensors. The pressure sensors were installed on the wall of the inside chamber. In the experiments, dusts were loaded directly to the storage container and would be dispersed with the rebound nozzle connected to an outlet valve located at the bottom of the chamber by using compressed air pressurized at 20 bar (gauge). A water jacket surrounds the spherical bomb for the control of the internal wall temperature. The dust concentration loading was started at 10 g before gradually increased until constant pressure achieved. The chamber was interfaced with a computer, which controls the dispersion/firing sequence and data collection by using control system named KSEP. As part of the experimental programme, two repeat tests would be performed on each test and this would be enough for good reproducibility.

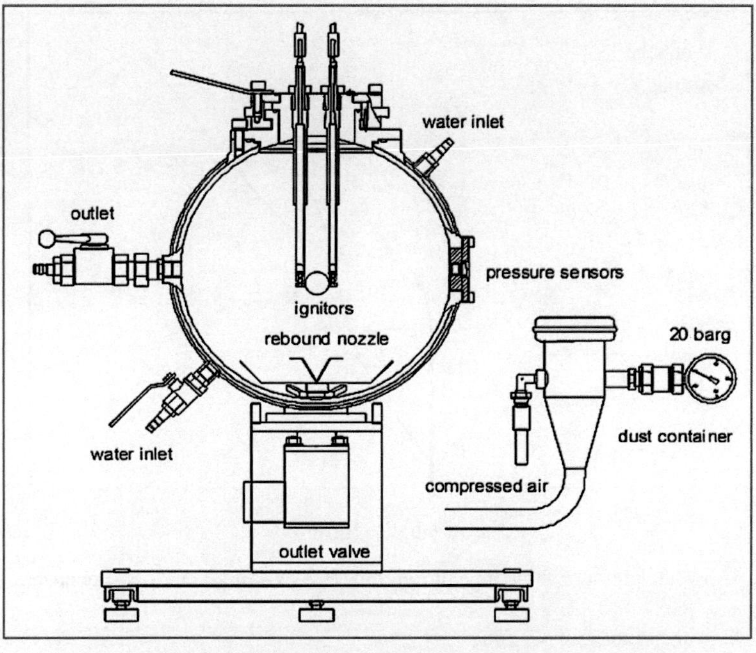

Fig. 1. Schematic diagram of Siwek 20 L spherical chamber [15]

3 Result and Discussion

Figure 2 shows the result of absolute pressure at the time of ignition (P_m) as a function of dust concentration for undried and dried rice flour at ignition time of 20 ms. The graph shows that absolute pressures of both undried and dried rice flour increase as the concentrations increase with a little bit of fluctuation. The moisture content is 7.79% for undried rice flour and 2.47% for dried rice flour. P_{max} for both undried rice flour and dried one were obtained at the highest level of concentration of 1000 kg/m^3. P_{max} for undried rice flour is 10.9 bar, lower than P_{max} of dried rice flour, 14.4 bar. It shows that the severity of the samples, represented by P_{max} are quite high when the samples were exploded at ignition energy of 20 ms. The trend of the graph of dried rice flour in Fig. 2 shows that at concentration of 750 kg/m^3, the pressure of undried rice flour is higher that pressure of dried rice flour. Other than that, the pressures of dried rice flour are higher than undried rice flour. This has been influenced by the presence of moisture which is higher in undried rice flour than dried one. Yuan et al. [20] found that a higher moisture content may cause agglomeration of finer particles, which can increase the uncertainty of explosion tests. This explain the fluctuation of the graph in Fig. 2 for both undried and dried samples.

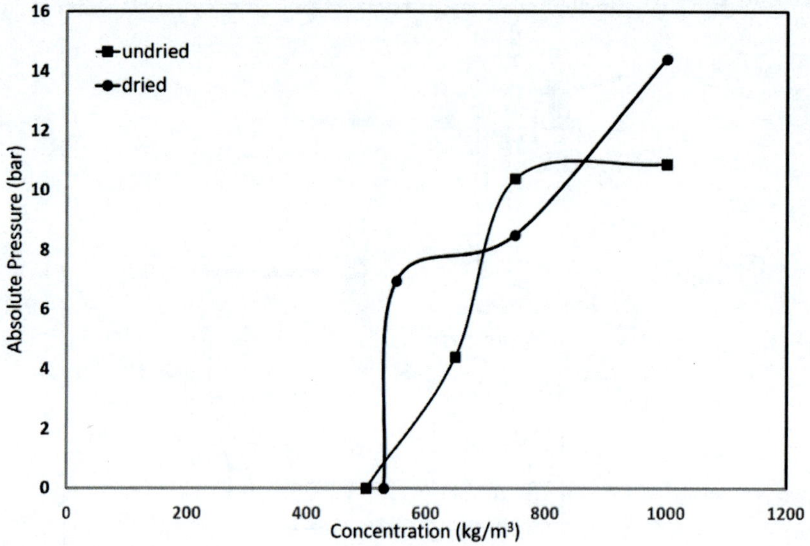

Fig. 2. Absolute pressure at the time of ignition (P_m) as a function of dust concentration

Figure 3 depicted that the rate of the pressure rise (dP/dT) of the undried rice flour presents a trend of first increase and then started at concentration of 750 kg/m^3, it decreases with the increase of dust concentration. The rate of pressure rise for dried rice flour increases with the increase of dust concentration until reaching its maximum at 1000 kg/m^3. The maximum rate of pressure rise for undried and dried rice flour are

103 bar/s and 202 bar/s respectively. The Figure shows that the rate of pressures of dried rice flour for all concentrations are much higher than rate of pressures of undried rice flour. The value of dP/dT is very crucial to get the value of dust constant or cube root law, K_{st}. The value of K_{st} is one of the parameters to measure the severity and regarded as fundamental parameter to calculate the vent sizing [21].

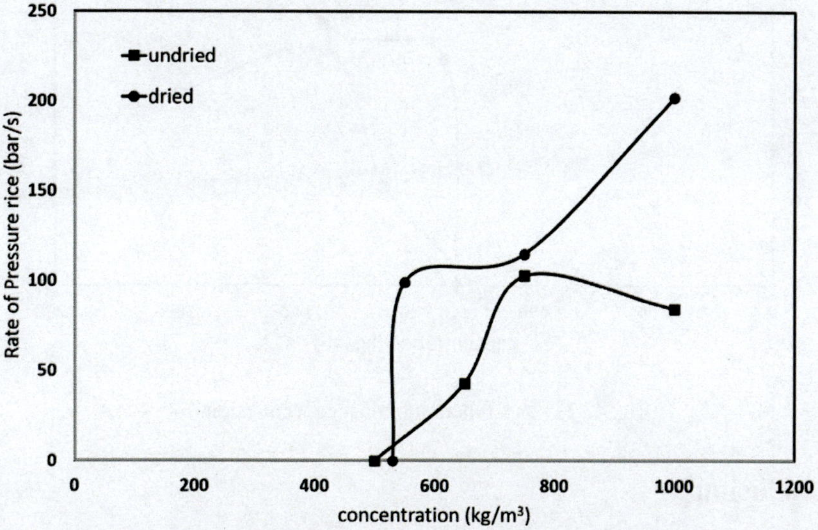

Fig. 3. Rate of pressure rise (dP/dT) as a function of dust concentration

The deflagration index (K_{st}) were obtained from multiplication of the dP/dT value with the cube root of the chamber volume. K_{st} is also known as volume-normalized maximum rate of pressure rise [16]. Figure 4 shows the K_{st} value for undried and dried rice flour at ignition time of 20 ms. The values of K_{st} for undried and dried rice flour are 27.96 bar.m/s and 54.83 bar.m/s, respectively. The values of K_{st} for the undried and dried samples are found to be in the group of ST1 base on the severity ranked by Kst, 0 < Kst < 200 which is considered weak [22].

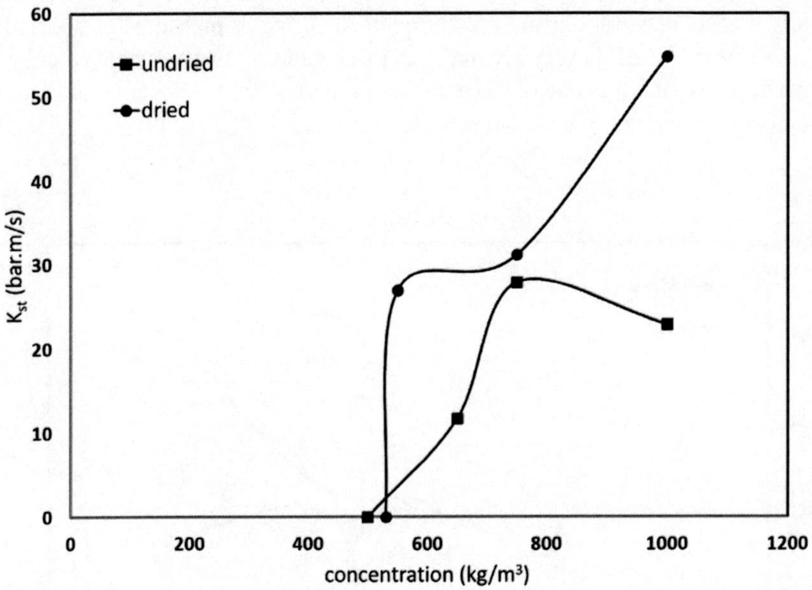

Fig. 4. K_{st} as a function of dust concentration

4 Conclusion

The severity of dust explosion for rice flour samples were evaluated from a series of absolute pressures measured in 20 L spherical chamber. P_{max} for undried rice flour is 10.9 bar while P_{max} for dried rice flour is much higher than the undried rice flour with 3.5 bar difference. The trend of graph clearly shows that the severity for dried rice flour is much higher than undried rice flour. The results of maximum dP/dT also shows that maximum dP/dT of dried rice flour is higher than undried rice flour at 202 bar/s and 103 bar/s respectively. K_{st} for undried and dried rice flour are 27.96 bar.m/s and 54.83 bar.m/s respectively. The trend of the whole results of dP/dT and K_{st} for dried rice flour generally shows that dried rice flour is severe than undried rice flour. This study concluded that, the severity of the samples increases as the concentration increases and the presence of moisture has pronounced influence towards severity as it may decrease the severity of the rice flour.

Acknowledgement. The first author thanks Ministry of Education and Universiti Malaysia Pahang for the scholarship. The authors are grateful for the funding from grant RDU1703112 from Universiti Malaysia Pahang.

References

1. Santon, R.C., (ed.) Mist fires and explosions - an incident survey. In: Institution of Chemical Engineers Symposium Series (2009)

2. Jain, P., Pasman, H.J., Waldram, S.P., Rogers, W.J., Mannan, M.S.: Did we learn about risk control since Seveso? Yes, we surely did, but is it enough? An historical brief and problem analysis. J. Loss Prev. Process Ind. **49**, 5 (2017)
3. Wu, T., Wang, L., Li, Y., Qian, H., Liu, L., Tong, L., et al.: Effect of milling methods on the properties of rice flour and gluten-free rice bread. LWT **108**, 137–144 (2019)
4. Dennis, M., Lee, A.R., McCarthy, T.: Nutritional considerations of the gluten-free diet. Gastroenterol. Clin. North Am. **48**(1), 53 (2019)
5. Chen, T., Zhang, Q., Wang, J., Liu, L., Zhang, S.: Flame propagation and dust transient movement in a dust cloud explosion process. J. Loss Prev. Process Ind. **49**, 572–581 (2017)
6. Tascón, A., Aguado, P.J.: CFD simulations to study parameters affecting dust explosion venting in silos. Powder Technol. **272**(Supplement C), 132 (2015)
7. Tascón, A., Aguado, P.J.: Simulations of vented dust explosions in a 5m3 vessel. Powder Technol. **321**, 409–418 (2017)
8. Klippel, A., Schmidt, M., Muecke, O., Krause, U.: Dust concentration measurements during filling of a silo and CFD modeling of filling processes regarding exceeding the lower explosion limit. J. Loss Prev. Process Ind. **29**(Supplement C), 122 (2014)
9. Daniel, V., Andrés, P., Mariangel, A., Nicolás, R., Felipe, M., Carlos, M., et al.: CFD as an approach to understand flammable dust 20 L standard test: effect of the ignition time on the fluid flow. AIChE J. **64**(1), 42 (2018)
10. Bind, V.K., Roy, S., Rajagopal, C.: A reaction engineering approach to modeling dust explosions. Chem. Eng. J. **207–208**, 625 (2012)
11. Kuracina, R., Szabova, Z., Pangracova, D., Balog, K.: Determination of explosion characteristics of wheat flour. J. Slovak Univ. Technol. **25**(40), 9 (2017)
12. Zhang, H., Chen, X., Zhang, Y., Niu, Y., Yuan, B., Dai, H., et al.: Effects of particle size on flame structures through corn starch dust explosions. J. Loss Prev. Process Ind. (2017)
13. BIA-Report 13/97. HVBG, Sankt Augustin (1997)
14. Tascón, A.: Influence of particle size distribution skewness on dust explosibility. Powder Technol. **338**, 438–445 (2018)
15. Cesana, C., Siwek, R.: Operating Instructions 20 L Apparatus, 6th edn. Kuhner AG, Birsfelden (2000)
16. Amyotte, P.R., Eckhoff, R.K.: Dust explosion causation, prevention and mitigation: an overview. J. Chem. Health Saf. **17**(1), 15 (2010)
17. Reyes, O.J., Patel, S.J., Mannan, M.S.: Quantitative structure property relationship studies for predicting dust explosibility characteristics (Kst, Pmax) of organic chemical dusts. Ind. Eng. Chem. Res. **50**(4), 2373 (2011)
18. Eckhoff, R.K.: Dust Explosions in the Process Industries. Gulf Professional Publishing, Burlington (2003)
19. British Standard I. BS 1016-104. British Standards Institution, London (1999)
20. Yuan, J., Wei, W., Huang, W., Du, B., Liu, L., Zhu, J.: Experimental investigations on the roles of moisture in coal dust explosion. J. Taiwan Inst. Chem. Eng. **45**(5), 2325 (2014)
21. Fumagalli, A., Derudi, M., Rota, R., Copelli, S.: Estimation of the deflagration index KSt for dust explosions: a review. J. Loss Prev. Process Ind. **44**, 311 (2016)
22. Hazard Communication Guidance for Combustible Dusts (2009)

Author Index

M. A. Zakaria et al. (Eds.): *Advances in Mechatronics, Manufacturing, and Mechanical Engineering*, LNME, pp. 307–308, 2021.
https://doi.org/10.1007/978-981-15-7309-5